Springer Proceedings in Earth and Environmental Sciences

Series Editor

Natalia S. Bezaeva, The Moscow Area, Russia

The series Springer Proceedings in Earth and Environmental Sciences publishes proceedings from scholarly meetings and workshops on all topics related to Environmental and Earth Sciences and related sciences. This series constitutes a comprehensive up-to-date source of reference on a field or subfield of relevance in Earth and Environmental Sciences. In addition to an overall evaluation of the interest, scientific quality, and timeliness of each proposal at the hands of the publisher, individual contributions are all refereed to the high quality standards of leading journals in the field. Thus, this series provides the research community with well-edited, authoritative reports on developments in the most exciting areas of environmental sciences, earth sciences and related fields.

More information about this series at http://www.springer.com/series/16067

Nikolay Vladykin
Editor

Alkaline Rocks, Kimberlites and Carbonatites: Geochemistry and Genesis

Proceedings of the XV International Seminar "Deep-Seated magmatism, Its Sources and Plumes", 1–7 September 2019, Russia, Saki.

Springer

Editor
Nikolay Vladykin
Vinogradov Institute
of Geochemistry SB RAS
Irkutsk, Russia

ISSN 2524-342X ISSN 2524-3438 (electronic)
Springer Proceedings in Earth and Environmental Sciences
ISBN 978-3-030-69669-6 ISBN 978-3-030-69670-2 (eBook)
https://doi.org/10.1007/978-3-030-69670-2

This Springer imprint is published by the registered company Springer Nature Switzerland AG
The registered company address is: Gewerbestrasse 11, 6330 Cham, Switzerland

Editor's Preface

Recently, due to the intensive development of the plume tectonics hypothesis throughout the world, including Russia, there has been an increased interest in the field of petrology and geochemistry of high-alkalinity intracontinental magmatism. These include rocks of alkaline complexes, kimberlites, carbonatites and alkaline basalts.

The fact is well accepted that alkaline rocks represent unique formations on the Earth. They have been long attractive for research because large Nb, Ta, Zr, Y, TR, Cu and P deposits; gemstones of charoite; Cr-diopside and dianite are associated with them. For instance, in Australia, diamonds are recovered in lamproites. The complicated processes of their formation provoked scientific disputes still going on. The newly developed analytical methods and techniques provided abundant information on the composition of alkaline rocks. The data on geochemistry of isotopes confirm the evidence on the mantle sources of the substance of alkaline rocks. The new concepts of plume tectonics are applied by scientists when studying alkaline rocks as the deep-seated geodynamics of the Earth is interpreted based on these data.

The problems of petrology of deep-seated magmatism (alkaline rocks, kimberlites and their mantle xenoliths) as well as the association of their sources with plume processes were discussed at 15 International Seminars held by the Institute of Geochemistry SB RAS in Moscow, Siberia, the Urals and the Far East, as well as in Italy and Turkey within the time span from 2001 to 2019. The 15th Seminar was held in Saki, Russia, and the reports from this seminar are published in the present issue.

The issues discuss fundamental problems of deep-seated magmatism. It is shown that the source of the largest apatite and rare metal deposits is the enriched mantle reservoirs. The articles discuss geochemical features, sources and ore content of the Zhidoi and Zashikhinsky deposits (East Sayan), signs of plume magmatic activity in the middle part of the Russian Platform (Ustyansk diamond-bearing region). The present issues also consider the features in composition of kimberlites from deep levels of Yubileinaya pipe and mineralogical characteristics of the mantle substrate of the Nakyn kimberlite field (Yakutia), the mechanisms of concentration of rare elements in komendite melts of the Early Mesozoic Sant (Central Mongolia) bimodal association and mineralogical-geochemical features of the Tsagan-Ul pantellerite belt (North Mongolia). The crystal-chemical features of rare and complex silicates from charoite rocks of the deep-seated Murun massif and the composition of pyroxenes of the Inagli massif are considered. The data on thermobarogeochemistry of mantle rocks and the composition and parameters of the source magma for ultramafic Uitkomst (South Africa) complex and komendites of Mongolia are given.

The book is of interest to petrologists, geochemists and specialists in deep alkaline and kimberlite magmatism, as well as University students and teachers.

Irkutsk, Russia Prof. Nikolay Vladykin

Contents

Enriched Mantle Reservoirs as a Source of the Largest Apatite and Rare-Metal Deposits

L. N. Kogarko

Abstract

Alkaline magmatism has occurred since 2.7–3Ga and its abundance has continuously increased throughout the Earth's history. Alkaline rocks appeared on the Earth with changes in the geodynamic regime of our planet, i.e., when plume tectonics was supplemented by plate tectonics. Global scale development of plate tectonics at the Archean–Proterozoic boundary initiated subduction of already significantly oxidized oceanic crust enriched in volatiles and largescale mantle metasomatism caused the formation of enriched reservoirs as sources of alkaline and carbonatite magmatism. Study of metasomatized mantle material showed the occurrence of traces of primary carbonatite melts, which are strongly enriched in rare elements, according to ionmicroprobe analyses. The results obtained allowed us to propose a new two-stage genetic model for Ca-rich carbonatites including (1) metasomatic wehrlitization and carbonatization of mantle material and (2) partial melting of wehrlitized mantle with formation of carbonaterich melts or three immiscible liquids (at high alkali contents), i.e., silicate, carbonatitic, and sulfide (at high sulfur activity).

Keywords

Oceanic crust · Plate tectonic · Mantle material · Enrich mantle · Alkaline magmatism

L. N. Kogarko (✉)
Vernadsky Institute of Geochemistry and Analytical Chemistry, ul. Kosygina 19,
Moscow 119991, Russia
e-mail: kogarko@geokhi.ru

N. Vladykin (ed.), *Alkaline Rocks, Kimberlites and Carbonatites: Geochemistry and Genesis*, Springer Proceedings in Earth and Environmental Sciences,
https://doi.org/10.1007/978-3-030-69670-2_1

1 Introduction

The appearance of alkaline magmatism in the Earth's history is related to the initiation of plate tectonics and, as a consequence, of large-scale mantle metasomatism, which led to the formation of enriched mantle reservoirs. Ion-microprobe studies of products of carbonate metasomatism in mantle material from some localities demonstrated that the metasomatizing mantle fluids had very high rare-element concentrations, which are comparable to those in the Lovozero giant deposit and some carbonatites. The results allowed us to propose a new two-stage genetic model for to current data of seismic tomography and heat flow, the Ca-rich carbonatites and related deposits. According metasomatizing fluids enriched in rare elements were probablyderived from plumes ascending from anomalous mantle existing in the Earth's interior.

2 Enriched Reservoirs in the Earth's History

Alkaline igneous rock associations are the most productive for various valuable elements. Modern data on geochemistry, isotopy, and seismic tomography show that alkaline magmas were generated from enriched mantle reservoirs. Giant complex deposits of rare metals (Nb, Ta, Zr, REE, Sr, Ba), radioactive elements, phosphorus, and aluminum are concentrated in alkaline rocks. Almost 100% of world reserves of magmatic Sr and >85% of Nb, Zr, and Ce are related to alkaline igneous rocks and carbonatites. Demand for these elements continuously increases, particularly in industrial countries; thus, alkaline rocks can be regarded as materials of the future.

The alkaline magmatism is most typical of stable regions (platforms) and is controlled by rift structures and zones with abruptly decreasing thickness of the continental lithosphere. Note, however, that alkaline rocks originate in almost all geodynamic settings, including oceanic ones. Evolution of the magmatic regimes of the Earth is primarily determined by global differentiation of the mantle, as the thickest layer in the Earth's interior.

The post accretionary history of the Earth comprised multiple episodes of mantle melting resulting in the formation of continental and oceanic crust. During this process, the mantle gradually lost CaO, Al_2O_3, TiO_2, rare lithophile elements and transformed into strongly depleted substrate, which could produce only highly magnesian magmas depleted in rare elements. However, such evolution is not typical in the Earth's history. On the contrary, many data show enrichment of mantle magmas (basalts and picrites) in Ti, LREE, and incompatible elements [1].

Isotopic studies show the existence of mantle reservoirs significantly enriched in rare elements, which could not be possible in the model of continuous global melting of mantle in the post accretionary Earth evolution. The ideas developed in the last decades suggested active mantle–crust interaction [2] with largescale exchange of materials and formation of the mantle reservoirs enriched in rare elements. The appearance of alkaline magmatism with extreme concentrations of rare elements is undoubtedly related to the generation of such zones in the mantle.

The analysis of literature and our data [3] shows that alkaline magmatism appeared in the Earth's history at about 2.5–2.7 Ga (see Fig. 1), and alkaline rocks older than 2.7 Ga are extremely rare. The figure shows continuously increasing activity (abundance and volume) of alkaline magmatism with the Earth's evolution. For example, the total volume of young platophonolites in Kenya exceeds 50000 km^3 [4], which is much larger than the volume of alkaline rocks of all other geologic epochs. The increasing scale of alkaline magmatism with time was also noted by Lazarenkov [5]. A similar tendency is characteristic of the evolution of oceanic magmatism. The database for igneous rocks of oceanic islands and seamounts of the Atlantic Ocean, primarily of alkaline composition, helped us to reveal that the intensity of intraplate magmatism significantly grows with time. The appearance of alkaline rocks at the Archean–Proterozoic boundary coincides with several global geologic events, and this boundary became an important benchmark in the Earth's history.

Most researchers believe that the geodynamic regime of our planet changed at this boundary, and plume tectonics was supplemented by plate tectonics [6].

The oxygen atmosphere of the Earth originated particularly at this boundary [7], generally due to hydrogen dissipation to cosmic space and the activity of living organisms. At the earlier stages, the Earth's atmosphere consisted of CH_4, H_2, and CO_2 [8].

At the Archean–Proterozoic boundary, the continental crust extensively grew and stable, cratonic regimes were developed.

Fig. 1 Distribution of intensity of continental alkaline magmatism with time

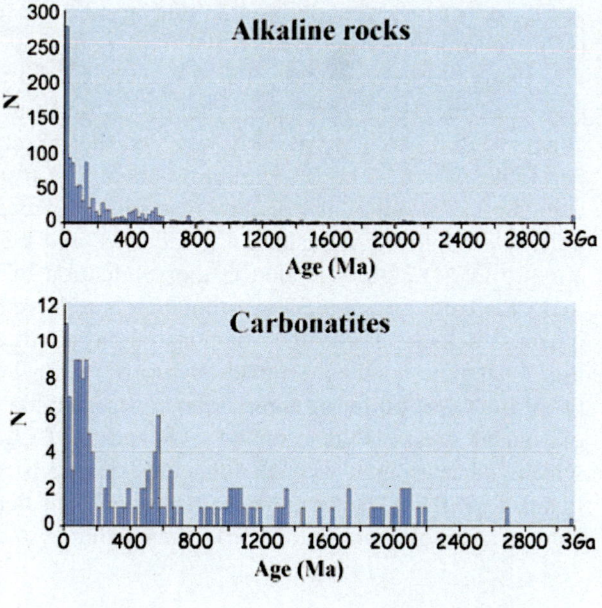

The oxygen bearing atmosphere caused the formation of oxidized marine sediments (jaspilites, carbonates) and the oxidation of paleosoils [7]. Note that the Archean giant sedimentary pitchblende and pyrite deposits of the Witwatersrand type have no analogues in the later epochs, because the minerals composing their ores are unstable in an oxygen atmosphere [7, 9]. Global development of plate tectonics at the Archean–Proterozoic boundary initiated the subduction of already significantly oxidized oceanic crust with elevated contents of volatiles (generally H_2O and CO_2), which, in turn, resulted in the involvement of these components in mantle cycles. Active mantle–crust interaction increased the concentrations of volatile components in the mantle, which had lost these components during high-temperature accretion at the earlier stages of Earth evolution. Intensive studies of the mantle fluid regime revealed mantle heterogeneity with respect to oxidation–reduction potential. According to recent data [11], oxygen fugacity in the mantle substrate varies within five–six logarithmic units around the quartz–fayalite–magnetite (QFM) buffer system. However, despite the significant redox heterogeneity of the lithospheric and astenospheric mantle of the Earth, there are some regularities. Most researchers accept the idea that the oldest Archean lithospheric mantle (a diamond source) had low $f\ O_2$ values. Diamonds are much older than host kimberlites [12–14], >3 Ga in some cases. They contain inclusions of Ni-free metallic iron [15], moissanite, magnesiowuestite [16], and methane [17], which indicates strongly reduced conditions in the old Archean mantle. These data are consistent with data on the V contents in Archean tholeiites [18] and the significant growth of the Ti/V ratio during differentiation caused by fractionation of pyroxenes concentrating reduced trivalent V. Our recent studies revealed the occurrence of Cu-bearing Fe–Ni alloys in the mantle, i.e., the phases composing the Earth's core [19]. Their stability requires very low oxygen fugacities, by three logarithmic units lower than the wuestite–iron buffer [20]. Such conditions probably existed in early geological history at about 4.6 Ga during the core formation. Arculus and Delano [10] experimentally determined the oxidation–reduction regime for mantle xenoliths using electrochemical cells and demonstrated that low $f\ O_2$ values for some rocks and megacrysts in basalts are also close to the wuestite–iron buffer. On the contrary, Bullhaus [21] recently revealed the more oxidized state of most mantle peridotites (from –2 to 0 logarithmic units with respect to QFM), particularly the mantle rocks strongly modified by metasomatic processes (2 and higher logarithmic units relative to QFM). Island ark andesites and basalts (+3 logarithmic units relative to QFM) closely related to the subduction of oceanic crust and the alkaline rock of oceanic island (+2 logarithmic units relative to QFM) are the most oxidized igneous rocks [21]. Our data demonstrate a significant degree of oxidation of the highly alkaline lavas of Trindad Island [22] with the oxygen fugacity exceeding $f\ O_2$ of the QFM buffer by three orders of magnitude. Study of the oxidation degree of volcanic gases [8] determined $f\ O_2$ and CO/CO_2 at 0.01 and 0.03 respectively, which indicates high oxygen fugacities during magma formation in the modern mantle. All these data emphasize the significant heterogeneity of the mantle with respect to redox conditions, verify the existence of relicts of older reduced mantle domains within more oxidized mantle material, and may indicate global processes of mantle oxidation in the Earth's history.

Mantle oxidation might be generally caused by iron disproportionation into the Earth's core [23]. According to Galimov [23], about 95% of the iron core was formed over about 100 m.y. of the earliest evolution of the Earth. Further growth of the core by consumption of FeO from the mantle resulted in oxygen generation over the next 100–500 m.y. of the Earth's history in the amount sufficient for mantle oxidation to QFM buffer values. Some researchers believe and our calculations verify their ideas that subduction accompanied by influx of oxidized H_2O−fluid into mantle cycles could also increase oxygen fugacity in mantle fluids. This process is distinctly pronounced in zones of global mantle metasomatism. A significant increase of volatile influx into the mantle due to the activation of subduction at the beginning of the Proterozoic stimulated large-scale mantle metasomatism, because the oxidized fluid phase ($H_2O + CO_2$) is capable of transporting great amounts of silicate material [24]. According to [25] and our calculations, annual addition of H_2O and CO_2 into mantle due to the subduction of oceanic crust with hydrous minerals and carbonates is 10^{14} g mol/y and 6×10^{12} g mol/y, respectively. Note that carbonatites containing 20% C compose no more than 6–27% in alkaline igneous associations. Relation of carbonatites with the oceanic crust recycling is verified by the wide scatter of δC_{13} values for carbonatite complexes worldwide [26]. Thus, changes in the geodynamic regime of the Earth and the strong activation of the mantle–crust interaction led to the global metasomatic migration of material and the formation of zones enriched in rare elements in the mantle substrate, a source of alkaline magmas. Coincidence of these events could not be accidental and probably indicates a basic relation of these processes. Alkaline igneous rocks appeared at the end of Archean as a result of the above events and were not abundant during the whole Proterozoic, at least until the end of the Vendian (see Fig. 1). Some small peak occurred only at the end of the Middle Proterozoic, between 1300 and 1000 Ma. Instead, the activity of alkaline magmatism continuously increased from 600 Ma, with peaks at about 550, 400, and 250 Ma, and particularly strongly from 200–180 Ma, reaching the maximum at about 40 Ma in the Eocene. A similar increase in the abundance of alkaline rocks is also typical of individual continents, i.e., North and South America, Eastern Europe, Siberia, and Africa. The role of kimberlites, lamproites, and carbonatites also strongly increased since the end of the Ordovician with a maximum in the Late Cretaceous and Early Paleogene. It should be mentioned, however, that the primary distribution could be disturbed because of the significant consumption of old alkaline rocks of oceanic islands and seamounts in subduction zones. It primarily concerns the Mesozoic igneous rocks of the eastern Pacific Ocean. Thus, continental alkaline massifs are probably best preserved. If the subducted rocks could be taken into account, the temporal distribution of alkaline rocks would be less contrasting, but woold not principally change. Hence, this distinctly heterogeneous distribution with a strong increase of alkaline rock abundance in the Panerozoic requires some specific interpretation. Most alkaline rock massifs were emplaced in intraplate environments due to mantle plume activity. Only some of them are located at spreading axes above hot spots, e.g., the Azores archipelago in the Atlantic Ocean. According to seismic tomography data, there is a close relationship between descending slabs of the subducted oceanic lithosphere and nearby ascending mantle plume jets at the deep mantle level near the mantle–core boundary. As a result, even the volcanics emplaced within the lithospheric plates show isotopic and geochemical signatures of recycled crustal material, which was revealed, for example, for Hawaiian volcanoes [27]. One more suggestion should also be considered. It is possible that periods of the whole-mantle convection and two-layer convection (separately for the lower and upper mantles) alternated in the Earth's history. The two-layer convection dominated during epochs of assembly of supercontinents (Pangeas), while the whole-mantle convection operated during the supercontinental epochs and breakup initiation. The alkaline rocks appeared with the formation of the first (sufficiently well documented) Pangea-0 supercontinent. According to Ringwood's ideas [28], the slabs subducted during periods of two-layer convection could be retained at the upper and lower mantle boundary, where they could be entrapped by plumes. There is a dependence between mantle viscosity and convection structure, and probability of the whole-mantle convection increases with increasing viscosity caused by gradual cooling of the Earth. The whole-mantle convection is probably operating now, while the two-layer convection was typical of the Late Archean, and plumes were initiated at the base of the transitional zone between the lower and upper mantles at depths of about 660–670 km. Significant changes in deep geodynamics are suggested at the end of the Neoproterozoic. They corresponded to the initiation of the Rhodinia breakup (early Late Riphean) and may have been expressed in dominant whole-mantle convection. The latter may have further developed later, at 400–250 and, particularly, 200–180 Ma, during the epoch of the initiation of Wegener's Pangea breakup. According to seismic tomography data, many slabs of the subducted oceanic lithosphere (eastern Pacific periphery, Mongolia–Okhotsk and South Anui paleooceans [29]) began to reach at that time the D'' layer at the mantle base and to stimulate plume ascent from this depth. Thus, the fluid-rich material from the mantle base (1000–1200 km), rather than from the lower/upper mantle boundary (670 km), could ascend to the surface.

The demonstrated increase in the activity and abundance of alkaline magmatism in the Earth's history may have been caused by the gradual involvement of progressively deeper mantle horizons into the whole-mantle convection and advection (plumes), progressively increasing material exchange between the Earth's surface and crust and its deeper interior due to recycling, and progressively increasing filtration of metasomatizing fluid flows through the mantle.

3 Global Mantle Metasomatism and Genesis of Enriched Mantle Reservoirs

Large-scale interstitial film melting by the mechanism suggested by Ringwood [28] may have occurred as a geochemical consequence of the global degassing of the subducted material and the release of oxidized fluids (H_2O and CO_2). We imply that the melts enriched in H_2O and CO_2, generated at low-melting degree and capable of transporting great amounts of incompatible rare elements and alkalis, presented the main metasomatizing agent in the mantle. Experimental data show [30] that aqueous fluid can contain at high pressure a few tens of percent of alkali rich silicates.

It was demonstrated earlier (e.g., [31]) that metasomatic processes play a highly important role in the generation of alkaline magmas. Strong enrichment of alkaline magmas in rare lithophile elements and depletion with respect to radiogenic isotopes can be accounted for only by the metasomatic influx of rare lithophile elements into zones of alkaline magma generation. This idea can be verified, for example, by our isotopic data [32] on the worlds largest alkaline complexes on the Kola Peninsula (Khibina and Lovozero massifs) accompanied by giant rare metal deposits. These data show strong depletion of the initial mantle source of alkaline magmas with respect to rare elements. In the εNd–εSr diagram, the rocks and ores of the Khibina and Lovozero massifs fall within the field of depleted compositions [31, 32]. This paradox can only be explained by the very rapid addition of metasomatizing material strongly enriched in incompatible rare elements to the melted mantle domain when there was not enough time for the accumulation of radiogenic isotopes (^{87}Sr and ^{143}Nd), and the Sr and Nd isotopic ratios still corresponded to the initial depleted compositions. Recent detailed studies of mantle material from various regions [33, 34] and our data [3, 37] demonstrated the significant role of metasomatism in the geochemical history of the mantle. Crystallization of amphiboles, phlogopite, apatite, primary carbonates, minerals of the lindsleyite– mathiasite series, and other minerals concentrating rare elements is related to metasomatic reactions in mantle substrate, most likely, to low-degree melts. As was shown experimentally by Fransis [35], partial melting of strongly amphibolized metasomatized mantle can generate alkaline melts. Highalkali glasses in metasomatized mantle xenoliths were also described by Edgar [36]. When studying strongly metasomatized and carbonatized nodules from volcanics of the Montana Clara Islands (Canary Islands) and Fernando de Noronha (Brazil), we found glasses (in association with primary carbonates) corresponding in composition to phonolites and trachites oversaturated in alkalis. Thus, alkaline magmas could have been generated during the partial melting of metasomatized mantle.

Recently, we actively studied the processes of mantle carbonate metasomatism in various regions. In the oceanic mantle of some Atlantic islands [3, 37], we revealed primary carbonates developed metasomatically by the following reactions:

$$2MgSiO_3 + CaMg(CO_3)_2 = 2Mg_2SiO_4 + CaMgSi_2O_6 + 2CO_2,$$
$$3CaMg(CO_3)_2 + CaMgSi_2O_6 = 4CaCO_3 + 2Mg_2SiO_4 + CO_2.$$

Metasomatic clinopyroxenes and olivines are products of these reactions. Using a CAMECA ion microprobe, we determined concentrations of metasomatic pyroxenes in oceanic and continental carbonatized mantle (Canary Islands, Fernando de Noronha, and East Antarctica). With the partition coefficients of rare elements in pyroxene–carbonate melt equilibria [38, 39], we calculated the concentrations of rare and rare earth elements in the metasomatizing mantle carbonate melts. Note that the average composition of the Lovozero massif accompanied by giant rare metal deposits fall within the interval of concentrations calculated for metasomatizing mantle fluids (see Fig. 2). These data are consistent with the compositions of carbonate melts in the metasomatized mantle of some other regions [39]. All melts are enriched to a various degree in LREE and incompatible elements. This REE distribution (with LREE enrichment) indicates that metasomatic meltfluids were generated in crystal melt equilibria, because the LREE partition coefficients in most mantle rockforming minerals, particularly in garnets, are notably lower than the HREE partition coefficients. Most calculated compositions of metasomatic meltfluids show minima of HFSE (Ti, Zr, Nb) (see Fig. 2). This is consistent with experimental data [38] on the low solubility of these elements in

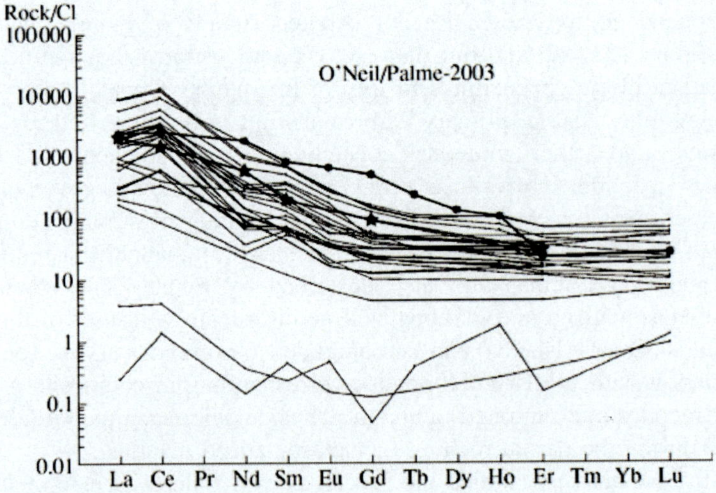

Fig. 2 Spidergram for the mantle metasomatizing melt fluids (calculated by ion microprobe analyses of metasomatic clinopyroxenes of mantle rocks from Fernando de Noronha Island, East Antarctica, and Canary archipelago). Stars—average composition of rocks of the Lovozero alkaline massif, asterisks—average composition of carbonatites

hightemperature carbonatite liquids. However, HFSE maxima (Ti, Zr) were revealed in some cases (see Fig. 2), which could indicate alkaline carbonate–silicate compositions of metasomatizing agent, because HFSE solubility strongly increases in highly alkaline silicate melts [39]. Despite some differences in the compositions of metasomatizing melts, all of them are enriched in incompatible rare elements and LREE. Universal enrichment of mantle fluids in rare elements is a basic geochemical feature of the Earth's mantle.

The ascending mantle plumes induce low-degree melting of the surrounding mantle (particularly at the base of the plumes), which produces carbonate melts enriched in rare elements and volatile components. These melts have very low viscosities, rapidly penetrate to upper horizons, and metasomatically replace mantle peridotites (stage 1). Further ascent of the diaper causes the partial melting of metasomatized mantle and the generation of alkaline and carbonatite magmas (stage 2). The proposed model of the formation of enriched reservoirs is generally consistent with the geochemistry of metasomatizing meltfluids, but requires very low melting degrees of mantle material. If the mantle source of giant rare-metal deposits of the Kola Peninsula was strongly depleted in rare elements [32], the mass of the depleted material producing metasomatizing fluids should be huge. It is more logical to suggest that the plume itself, significantly enriched in rare elements as compared even to primitive mantle, was the source of rare elements for the giant alkaline intrusions. Let us consider these problems in the next section.

4 Depth of Formation of Enriched Mantle Reservoirs

According to modern concepts, two global reservoirs, upper and lower, can be distinguished in the Earth's mantle. However, recent data of seismic tomography [40] indicate the occurrence of new structural heterogeneity in the lower mantle below about 1000 km until D" layer at 2300 km. P wave velocities significantly decrease in this interval [40]. These data suggest the existence of three dynamic regimes in the Earth's mantle. The above heterogeneity continues over about two thirds of the transitional lower mantle–core layer (see Fig. 3) and comprises 20–30% of the whole Earth's mantle. The heterogeneity is related to elevated viscosity and density of the material at the base of the lower mantle. The size and shape of this region (megaplume) may indicate the chemical heterogeneity of the lower mantle. If this seismic heterogeneity had thermal origin, it could not exist for a long time because of tremendous heat flux from the core. Density in the anomalous zone probably increases due to the decomposition of silicates and the formation of high-density oxides (magnesiowuestite, stishovite) at pressures >650 kbar. Crystallization of magnesiowuestite enriched in Fe is another mechanism of the formation of heavy mantle material. The suggestion of the existence of an anomalous layer in the lower mantle is consistent with modeling of the thermal regime of the Earth. The heat flux of the whole Earth is estimated at 44 TW; the heat flux of the crust is 6 TM due to high concentrations of radioactive elements [40], and the

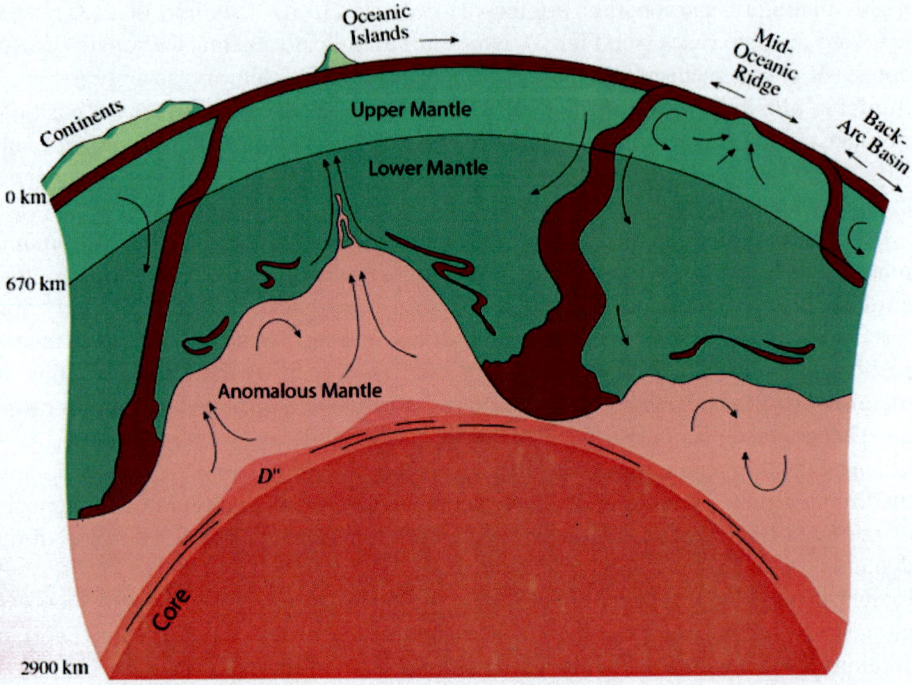

Fig. 3 Structural modified scheme of the Earth's upper and lower mantle (after [41])

remaining 38 TW should be supplied by the mantle and core. Even if the whole Earth was chondritic in composition, its heat flux would be only 31 TW [41]. It is established now that 25–90% of mantle substrate (MORB sources) is depleted in rare and radioactive elements, and, thus, heat flux of the Earth should be lower than 31 TW [41], but it is actually much higher (44 TW). Hence it follows that a layer producing huge heat amount corresponding to U concentration of 25.6 ppm should exist in the mantle [41].

The origin of this zone enriched in rare and radioactive elements in the mantle is not clear. A proposed model suggests the longterm development of a reservoir isolated since the early stages of the Earth's evolution and originating either during the evolution of the global magmatic ocean or by the recycling of Archean crust enriched in mafic components and incompatible rare elements. Elevated contents of radioactive elements should cause active heat generation and very high temperatures in the D" zone. Under the effect of heat flow from the core, hot and dense material of the lower mantle could rise and produce superplumes, e.g., African or East Pacific. We also considered the possibility of magma generation for the Kola alkaline igneous province from an anomalous zone in the lower mantle. According to [42], the total volume of alkaline rocks of the Khibina and Lovozero massifs is about 11000 km^3. Using the average contents of rare elements in these rocks (e.g., Zr and Th), we can estimate the volume of the primitive or depleted mantle as sources of these rocks. According to our isotopic studies [31, 32], the giant alkaline intrusions of the Kola Peninsula were generated from a mantle source depleted in rare elements. The further calculation showed that the volume of mantle material for the formation of the Khibina and Lovozero massifs should be about 3.7×10^6 km^3 for primitive mantle and 3.8×10^7 km^3 for depleted mantle. These volumes are not realistic and are comparable or even exceed the volumes of the largest mantle plumes. For example, the volume of the Siberian superplume is about 7×10^6 km^3 [43]. Moreover, there is no geological evidence of the relation of the Kola alkaline province with any large plume. Thus, our calculations may indicate a much more enriched source of alkaline magmas for the Kola alkaline province. They could be derived from an anomalous layer of the lower mantle. General regularities in fractionation of rare and radio active elements in mantle reservoirs allow us to estimate the distribution of incompatible elements in the anomalous layer. As was mentioned above, U content in the anomalous zone is 25.6 ppm [40]. If all incompatible elements are proportionally fractionated in mantle processes, their contents should be (like the U content) by about two orders of magnitude higher than in the primitive mantle. These estimates are semi quantitative. However, such concentrations at 15% melting degree of anomalous lower mantle domains (this melting degree is normally accepted for superplumes producing flood basalts) and the addition of 5–30% of metasomatizing fluid could provide for the generation of alkaline magmas with concentrations of rare elements close to their average concentrations in the ultrabasic alkaline magmas parental for the ultrabasic alkaline rock associations in Polar Siberia and the Kola Peninsula. The calculations suggested Nb contents of 0.7 ppm in the primitive mantle, 70 ppm in the anomalous mantle, 490 ppm (calculated value) in a metasomatizing melt, and 131 ppm in parental magma [42] and assumed 5% melting degree of a metasomatically modified substrate. At these Nb concentrations, the calculated contribution of metasomatic material is 1.3%. At lower melting degree of 1%, the amount of the metasomatizing melt-fluid will be very low (about 0.013%). Our studies, together with literature data [44] demonstrate that the concentrations of metasomatic minerals (amphibole, mica, apatite, etc.) in the mantle are low (a few fractions of percent). Metasomatic processes in the mantle frequently do not cause the formation of specific minerals. The addition of rare elements by metasomatizing melts can be low and only increases their concentrations in rock-forming minerals.

Therefore, the evolution of alkaline magmatism in the Earth's history was determined by the intensity of mantle metasomatism causing the formation of enriched mantle reservoirs as sources of alkaline magmas. The global development of metasomatism in the mantle was related to the subduction of oceanic crust initiated at the Archean–Proterozoic boundary. The genesis of alkaline and carbonatitic magmas and related giant raremetal deposits may have been caused by mantle metasomatism induced by the partial melting of the anomalous substrate in the lower mantle. Such anomalous zones are revealed in recent studies using seismic tomography [40, 41].

5 Conclusions

The appearance of alkaline rocks and carbonatites at the Archean–Proterozoic boundary coincided with a number of major events on Earth. Most authors associate this boundary with the change in the geodynamic regime of our planet–plate tectonics joined the plume tectonics. At this stage of the Earth's evolution, an oxygen atmosphere appeared, and ocean sediments were oxidized. As a geochemical consequence of global degassing of subducted material and the release of oxidized fluid (water and carbon dioxide), the mantle was oxidized and large-scale metasomatosis of the mantle began, leading to the appearance of source reservoirs enriched with rare and ore elements, alkaline rocks, carbonatites of super-large deposits, and later kimberlites. The later appearance of the deepest alkaline rocks-kimberlites in the history of the Earth is most likely due to the formation of continental cratons and a thickened Mature lithosphere (necessary conditions for the generation of kimberlites) at later stages of our planet's development.

Acknowledgements This work was supported by the Russian Federation represented by the Ministry of education and science of Russia No. 075-15-2020-802.

References

1. Campbell, J., Griffiths, R.: The changing nature of the mantle hotspots through time: implica-tions for the chemical evolution of the mantle. J. Geol. **99**, 497–523 (1990)
2. Hofmann, A.: Mantle geochemistry: the message from oceanic volcanism. Nature **385**, 219–229 (1997)
3. Kogarko, L., Henderson, I., Pacheco, A.: Primary carich carbonatite magma and carbonate–silicate–sulfide liquid immiscibility in the upper mantle. Contrib. Mineral Petrol **121**, 267–275 (1995)
4. Logachev, N.: Volcanic and Sedimentary Rock Associations in Rifts of Eastern Africa. Nauka, Moscow (1977). [in Russian]
5. Lazarenkov, V.: Analysis of Associations of Continental and Oceanic Alkaline Rocks. Nedra, Leningrad (1988). [in Russian]
6. Khain, V.: Top Problems of Modern Geology. Nauka, Moscow (1994). [in Russian]
7. Holland, H.: The Chemical Evolution of the Atmosphere and Ocean. Princeton Univ. Press, Princeton (1984)
8. Cloud, P.: Atmospheric and hydrospheric evolution on the primitive earth. J. Geol. **100**, 729–736 (1992)
9. Walker, J.: Evolution of the Atmosphere. Macmillan Publ, New York (1977)
10. Arculus, R., Delano, J.: Implications for the primitive atmosphere of the oxidation state of earth's upper mantle. Nature **288**, 72–74 (1980)
11. Haggerty, S., Tompkin, S.: Redox state of earth's upper mantle from kimberlitic ilmenites. Nature **303**, 295–300 (1983)
12. Richardson, S.: Latter day origin for diamonds of eclogitic paragenesis. Nature **322**, 623–626 (1986)
13. Kramers, S.: Lead, uranium, strontium, potassium, and rubidium in inclusionbearing diamonds and man-tle- derived xenoliths from Southern Africa. Earth. Planet. Sci. Lett. **42**, 58–70 (1979)

14. Ozima, M., Zashu, S., Nitoh, O.: Primitive helium in diamonds. Geochim. Cosmochim. Acta **47**, 2217–2224 (1983)
15. Sobolev, N., Efimova, E., Pospelova, L.: Dia-mondiferous peridotite xenoliths in kimberlites and the problem of diamond nature. Geol. Geofiz. **12**, 25–29 (1981)
16. Meyer, H.: Inclusions in Diamonds. In: Nixon, P.H. (ed.) Mantle Xenoliths, pp. 501–522. Wiley, Chichester (1987)
17. Giardini, A., Melton, C.: The nature of cloudlike inclusions in two arkansas diamonds. Am. Mineral. **60**, 931–933 (1975)
18. Shervais, J.: Ti–V plots and the petrogenesis of modern and ophiolitic lavas. Earth Planet. Sci. Lett. **59**, 101–118 (1982)
19. Ryabchikov, I., Kogarko, L., Ntaflos, T., Kurat, G.: Native metals in mantle xenoliths: dokl. Akad. Nauk **338**(1), 95–98 (1994)
20. O`Neil, H.: The origin of the moon and the early history of the earth: a chemical model. Geochim. Cosmochim. Acta **55**(2), 1159–1172 (1991)
21. Bullhaus, C.: Redox states of lithospheric and asthenospheric upper mantle. Contrib. Mineral Petrol **114**, 331–348 (1993)
22. Kogarko, L., Ryabchikov, I.: Redox equilibria in alkaline lavas from trindade island, Brazil: int. Geol. Rev. **36**, 473–483 (1991)
23. Galimov, E.: Redox evolution of the earth caused by a multistage formation of its core: earth planet. Sci. Lett. **233**, 263–276 (2005)
24. Green, D., Falloon, T., Taylor, W.: Mantlederived magmas: roles of variable source peridotite and variable C–H–O fluid comparisons. In: Mysen, B. (eds.) Magmatic Processes: Physicochemical Principles, vol. 1, pp. 158–172. Geochem. Soc. Spec. Publ. (1987)
25. Kasting, J., Eggler, D., Raeburn, S.P.: Mantle redox evolution and the oxidation state of the archean atmosphere. J. Geol. **101**(2), 245–257 (1993)
26. Bell, K.: Carbonatite Genesis and Evolution. Unwin Hyman, London (1989)
27. Bunge, H., Grand, S.: Nature **405**, 337–340 (2000)
28. Ringwood, A.: Slab–mantle interactions: 3. petrogenesis of intraplate magmas and structure of the upper mantle. Chem. Geol. **82**, 187–207 (1990)
29. Kearey, P., Vine, F.: Global Tectonics. Blackwell, Boston (1996)
30. Wendland, R., Harrison, W.: Rare earth partitioning between immiscible carbonate and silicate liquids and CO vapor: results and implication for the formation of light rare earthenriched rocks. Contrib. Mineral. Petrol. **69**, 409–419 (1979)
31. Kogarko, L.: Problems in studies of the origin of giant rare metal deposits in the kola peninsula. in russian arctic zone: geological history, metallogeny, and geoecology, , pp. 773–788. St. Petersburg (2002) [in Russian]
32. Kogarko, L., Kramm, U., Grauert, B.: New data on the age and origin of alkaline rocks of the lovozero pluton: constraints from rubidium and strontium isotopic geochemistry: dokl. Akad. Nauk SSSR **260**(4), 1000–1005 (1981)
33. Meen, J., Ayers, J., Fregeau, E.: A Model of Mantle Metasomatism by Carbonated Alkaline Melts: Traceelement and Isotopic Compositions of Mantle Source Regions of Carbonatite and Other Continental Igneous Rocks, pp. 464–500. Unwin Hyman, London (1989)
34. Kogarko, L.: The Upper Mantle Heterogeneity and the Magmatism of Oceanic Islands. in Oceanic Magmatism, pp. 73–80. Nauka, Moscow (1986) [in Russian]
35. Fransis, D.: Some implications of xenolith glasses for the mantle sources of alkaline mafic magmas. Contrib. Mineral Petrol **108**, 175–180 (1991)
36. Edgar, A., Lloyd, F., Forsyth, D., Barnett, R.: Origin of glass in upper mantle xenoliths from the quaternary volcanics of gees, west eifel. Germany Contrib. Mineral Petrol **103**, 277–286 (1989)
37. Kogarko, L., Kurat, G., Ntaflos, T.: Carbonate metasomatism of the oceanic mantle beneath fernando de noronha island. Brazil. Contrib. Mineral Petrol **140**, 577–587 (2001)

38. Rivalenti, G., Mazzucchelli, M., Girardi, V., et al.: Composition and processes of the mantle lithosphere in northeastern brazil and fernando de noronha: evidence from mantle xenoliths. Contrib. Mineral Petrol **138**, 308–325 (2000)
39. Adam, J., Green, T.: Experimentally determined partition coefficients for minor and trace elements in peridotite minerals and carbonatitic melt, and their relevance to natural carbonatites. Eur. J. Mineral **13**, 815–827 (2001)
40. van der Hilst, R., Karason, H.: Compositional heterogeneity in the bottom 1000 kilometers of earth's mantle: toward a hybrid convection model. Science **283**, 1885–1888 (1999)
41. Kellogg, L., Hager, B., van der Hilst, R.: Compositional stratification of the deep mantle. Science **283**, 1881–1884 (1999)
42. Arzamastsev, A., Bea, F., Glaznev, V., et al.: The kola alkaline province in paleozoic: the composition of parental magmas and the magma generation conditions. Ross. Zh. Nauko Zemle **3**(1), 1–35 (2001)
43. Milanovsky, E.: Rift zones of the geologic past and their associated formations: report 2. Int. Geol. Rev. **18**, 619–639 (1976)
44. Menzies, M., Hawkesworth, C.: Mantle Metasomatism. Academic, London (1987)

Problems of Mantle Structure and Compositions of Various Terranes of Siberian Craton

I. V. Ashchepkov, N. V. Vladykin, A. Ivanov, S. Babushkina, M. Vavilov, and N. Medvedev

Abstract

Structure of mantle keel beneath Siberian craton was determined for mantle terranes using monomineral thermobarometry and database of EPMA analyses (>80000). Geochemical features of the minerals have been determined by LA ICP MS. Mantle sections consist of 6–7 paleo-subducted slab separated by pyroxenite, eclogite and dunite lenses, disturbed by later plume melt intrusion accompanied by metasomatism and melting. The cross-sections and long transects show the differences between terranes compiled by microplates in Archean. Extended sub-meridional structures of crust tectonic terranes are not always confirmed on the mantle level. Under the Anabar and Aldan shields mantle sections are coarser and consist of 3–4 large horizons of dunites, with pyroxenite and metasomatic lenses. Suture zones between protocratons are

I. V. Ashchepkov (✉) · M. Vavilov
V.S. Sobolev Institute of Geology and Mineralogy, Siberian Branch of RAS, Novosibirsk, Russia
e-mail: igor.ashchepkov@igm.nsc.ru

N. V. Vladykin
Vinogradov Institute of Geochemistry, Siberian Branch of RAS, Irkutsk, Russia
e-mail: vlad@igc.irk.ru

A. Ivanov
Scientific Research Geological Enterprise, ALROSA Company, Mirny, Russia
e-mail: IvanovAS@alrosa.ru

S. Babushkina
Institute of Diamond and Noble Metals Geology, Siberian Branch of Russian Academy of Sciences, Yakutsk, Russia
e-mail: sa.babushkina@s-vfu.ru

N. Medvedev
A.V. Nikolaev Institute of Inorganic Chemistry SB RAS, Novosibirsk, Russia
e-mail: medvedev@niic.nsc.ru

N. Vladykin (ed.), *Alkaline Rocks, Kimberlites and Carbonatites: Geochemistry and Genesis*, Springer Proceedings in Earth and Environmental Sciences,
https://doi.org/10.1007/978-3-030-69670-2_2

saturated with eclogites. Pyroxenite layer at the level of 3.5–4.5 GPa appeared in Early Archean when eclogites were melted at 130 km. In Early Archean protocratons: granite-greenstone terranes–Tunguska, Markha, Berekte and Sharyzhalgai mantle lithosphere is less depleted and often metasomatized. The Daldyn and Magan granulite-orthogneiss terranes have a layered structure of ML with folds, manifested in sections from north to south from Udachnaya to Krasnopresnenskaya. Metasomatism and alkalinity of pyroxenes and the amount of phlogopite are growing from Daldyn to Alakit fields. The most productive confined to the dunite cores. In Magan terrane, the thin layering of the craton keel is replaced by depleted productive horizon at the lithosphere base. In granite-greenstone Markha eclogites (metapelitic) assume subduction of continental sediments. Gentle tilting to the west dominate of mantle structures.

Keywords

Mantle lithosphere · Terranes · Siberian craton · Thermobarometry · Peridotite · Eclogites · Garnets · Kimberlites · Transects · Rare elements geochemistry · Layering · Oxidizing potential · Plume

1 Introduction

In geodynamics, the concept of terranes as structural units, including those within the Siberian craton [1, 2], formed at a certain time and geodynamic conditions with lithological features of the section structure, is widely used. The separation of mantle terranes as part of the Siberian craton was proposed [3] during the construction of a 1,000 km long cross section from the Mirny field to the Anabar shield, where 4100 analyses of pyrope from 65 kimberlites were used. This paper substantiates the isolation of individual mantle terranes within the Siberian craton, which mainly coincide with terranes established for the Earth's crust, based on 57000 electron microprobe analyses with reconstructions of PT conditions and layering under individual pipes and kimberlite fields, as well as geochemical features of mantle minerals and rocks. Such work using almost twice less analyses was published earlier [4]. This paper includes additional data for the northern regions and details the main studied fields of the Central part of the Yakut kimberlite province (YKP). Comparison of crust and mantle structures and compositions has been carried out earlier in different parts of the World [5], but not many materials for the mantle keels of individual cratons have been accumulated in earlier works. A study using geophysical methods showed that the Siberian craton has a mosaic structure [6]. Not only the depth of Moho and the structure of the mantle [7], but also the composition of granulites [8] vary significantly under certain areas and terranes. The mantle layering was determined rather roughly for some areas of the craton Slave under groups of kimberlite pipes [9], Finland [10] and other areas of the World. The traverses of the Kaapvaal Craton, the Dharwar Craton, [3, 11] as well as the Congo Craton [12] showed the mantle layering more clearly. They were constructed using garnet barometry [13]. The method of calculation of Mg' coexisting olivine [14] through garnet-orthopyroxene thermobarometry does not completely agree with natural, because Ni and Fe–Mg methods of thermobarometry are not always correlated. In our work we used the original methods of monomineral thermobarometry [4] and calculating the Mg' number or ferruginosity (Fe#) of olivine coexisting with garnet, ilmenite and chromite through regression equations. This work is based on >35000 electron microprobe analyses of the first author and additionally on the database (>22000 analyses) of analyses of the laboratory of X-ray methods of the Central Analytical Laboratory of ALROSA Company in Mirny (analyst A.S.Ivanov), as well as S.A. Babushkina analyses (Western Ukukit, Ogoner-Yuryakh, Orto-Yargyn, Toluopka and Biriginda, and partially Ebelakh and Kuranakh fields), A.P.Smelov data (Manchary and Carnian collector) and K.S. Ivanov.N. Egorov (Muro-Kova field, lamprites of the Eastern Sayan), as well as data from plume testing of the Morkoka, Horbosuonka, and Tumanshet river basins–according to M.A. Vavilov and colleagues. Diamond inclusions analyses were also used [15].

2 Methods of Research

The mineral compositions have been analyzed in the Collective Analytic Center of multi-element and isotope studies (CACMIS) of the Siberian Branch of the Russian Academy of Sciences on the CamebaxMicro X-ray microanalyzer (using standard conditions), 15 kV, 20 mA current, and the measurement time of 15 s per each circle. Natural samples and synthetic glasses were used as reference standards. Method of PAP data correction using the Karat program [16]. Relative standard deviation was <1.5%; accuracy ~3–5%. The composition of trace elements at first was analyzed in CACMIS by the of LA-ICP-MS method on the Finnigan Element mass spectrometer with Nd YAG laser attachment: UV 266 (analyst S.V.Palessky) and in the Nikolaev institute of Inorganic Chemistry of the Russian Academy of Sciences on iCAP Qc (Thermo Scientific) mass spectrometer with NWR 213 (New Wave Research), Nd YAG: UV 133 (WDS) laser attachment using NIST 610, 612, 614 SRM standards (analyst N.S. Medvedev).

P-T-FO2 values were calculated by monomineral methods [4] using the original TERA55 program [17], which allows calculating several pairs of PT parameters, recording them together with mineral compositions, and the resulting matrices are exported to Excel and further to Grapher, Surfer and other applications. Mantle sections and transects are calculated using Surfer8 software using the "contour map" diagram (as if they were wells in the mantle). The isolines were determined based on the following parameters: (1) pipe spacing; (2) calculated pressures, temperatures, and fO2 values for garnet [18] and ilmenite [19]; (3) and characteristics of mineral compositions, usually Fe# for olivines coexisting with minerals

(Fe#Ol). In this work, we mainly used Fe#Ol and fO2. The grids for calculations were obtained using the Surpher 8 software using kringing method with linear approximation between points [4, 20].

3 Tectonic Zoning of the Siberian Craton and Kimberlite Volcanism and Tectonic Structures Ratio

The zoning and scheme of terranes Siberian craton of Gladochub [1] with co-authors was taken as the basis for the terranes. The kimberlite fields trace the central part of the craton in the SSW-NNE direction, crossing the main structures of the Siberian craton [21] through Lower Lena, West and East Ukukit, Upper-Muna, Nakyn, Daldyn, Alakit, and Malo-Botuobinsky (Mirninsky) fields (Fig. 1). Kimberlites are dated as 350–360 Ma [2, 21–23] except for Toluopsky field (417–400 Ma). Further extensions are the Mura-Kova and Tumanshet placer areas (fields), probably formed by kimberlites or lamproites of Devonian age [24].

We have constructed global sections of the mantle–transects (MT) through the sub-craton lithospheric mantle (SCLM) in this direction from the Mirny field to Low Lena region, as well as across the Siberian Craton in the north from the Kharamai field to the lower reaches of the Olenek River and the Lena River mouth. In addition, in the south there is a section from the Ingashi lamproite field (Eastern Sayan) in the south through the Tumanshet and–Mura–Kova placer to the Tychan and Tyradak collectors and further through Mirny field in the west to the Manchary field [24] in the east. Compared to previous versions [4], the field includes new data for the Morkoka field [25], the Udzha river basin, the East-Ukukit, Orto-Yargyn, Molodo and Biriginda fields, and Khorbosuonka fields.

The main productive diamondiferous areas are concentrated within granulites–the orthogneiss Magansky and East and West Daldyn terranes. At the same time, the suture zones like Khapchan terrane [8] contain more eclogites and eclogite diamonds in placers [26] and kimberlite pipes. Apparently, the permeability of the zone served as a major factor in the abundance of kimberlite fields near the eastern part of the Anabar Shield. Within the granite-greenstone terranes, kimberlite fields are relatively rare. However, some of them are very productive, such as Nakyn [27, 28], where kimberlites carry specific eclogites that have a pelitic photoliths of continental type.

4 Layered Peridotite Mantle of Some Areas Near Siberian Craton

TPXFO2 diagrams for kimberlite pipes from the main areas of Siberian kimberlite magmatism have been described earlier in a number of works and are given in electronic supplements to them [29–36].

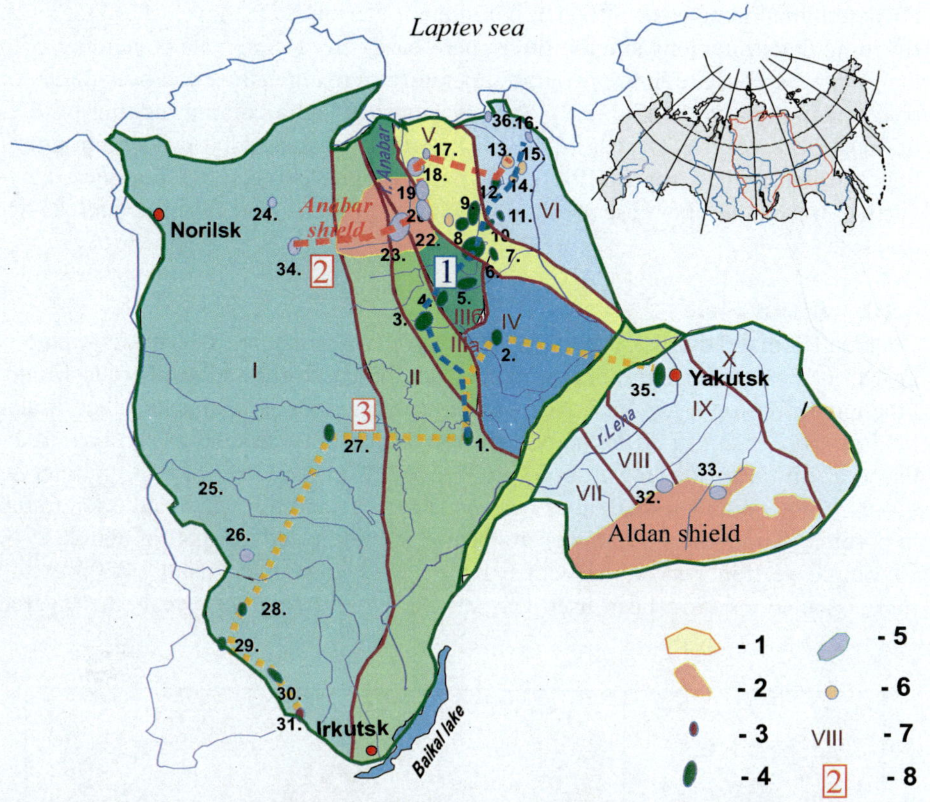

Fig. 1 Location of kimberlite fields on the Siberian Platform [21]. 1. Siberian Platform. 2. Shields. 3. Precambrian kimberlites. 4. Paleozoic kimberlites. 5. Triassic kimberlites. 6. Upper Jurassic kimberlites. 7. Tectonic terranes according to Gladkochub et al. [1]. 8. Direction lines of the lithosphere mantle cuts. Fields: 1–Kharamay–Anabar-Kuoyka–Lena Mouth (Carnian collector); 2–Malo Botuobinsky–Toluopsky field; 3-Ingashi (Sharyzhalgai)–Manchary field (W. Aldan). Fields: 1. Malo-Botuobinsky, 2. Nakyn; 3. Alakit, 4. Daldyn, 5. Upper Muna, 6. Chomurdakh, 7. Severnay, 8. Western Ukuit, 9. Eastern Ukukit, 10. Ogoner-Yuryakh, 11. Upper Motorchuna, 12. Merchimden, 13. Kuoyka, 14. Molodo, 15. Toluopsky, 16. Orto-Yargyn, 17. Ebelakh, 18. Starorechensky, 19. Ary-Mastakh, 20. Dyuken, 21. Luchakansky, 22. Kuranakh, 23. Middle Kuonapka, 24 Kotuisky, 25. Chadobetsky, 26. Taichikun-Nembinsky, 27. Tychansky, 28. Muro-Kovinsky, 29. Tumanshet, 30. Beloziminsky, 31. Ingashinsky, 32. Chompolinsky, 33. Tobuk-Khatystyrsky, 34. Kharamaisky, 35. Manchary; 36. Lower Lensky

4.1 Devonian (to Silurian) Stage of Kimberlite Magmatism

4.1.1 Daldyn Field

Mantle section beneath Udachnaya pipe was constructed using ~2000 deep rock associations have been used [22, 29–32, 37–47] (Fig. 2) reveals the layered structure of the mantle, which is evident in the alternation of sub-calcium (sub-Ca) and enriched pyrope differences, in the fluctuation of ferruginosity (Fe#), and in the

PT determination consists of 6 (12) individual plates up to the level of 7.0 GPa [4]. Ilmenitic determinations and the lithosphere base refer to intervals coinciding with the dunite-harzburgite lenses. High-temperature pyroxenes trace the boundaries of paleosubduction plates (Fig. 2A). Sharply bended mantle column judging by definitions for xenoliths Zarnitsa [4, 29–31], in this version added by definitions for high chrome garnets (up to 19% Cr_2O_3) for eclogite garnets and xenolites. Convective branch at the base continues with definitions for ilmenite and Ti–Cr diopsides.

4.1.2 Alakit Field

Detailed PT diagram for Aykhal [29] detects a sharp structure. It detects continuous growth of Fe# grants and other minerals from the base of the lithosphere to Moho. Fractional division by P–Fe% by 6–7 pressure intervals is similar to other Alakit field pipes (Fig. 3A) For the mantle section under tr. Aykhal (Fig. 2B) has established a substantial heating from 38 to 45 mW/m^2 in the lower part in the interval 0.8–1.0 GPa within garnet dunites horizon [34, 38], xenolites which are often found in porphyry kimberlites. Eclogites are located in middle and low part of mantle [54].

Mantle section beneath Udachnaya pipe was constructed using ~2000 deep rock associations have been used [22, 29–32, 37–47] (Fig. 2) reveals the layered

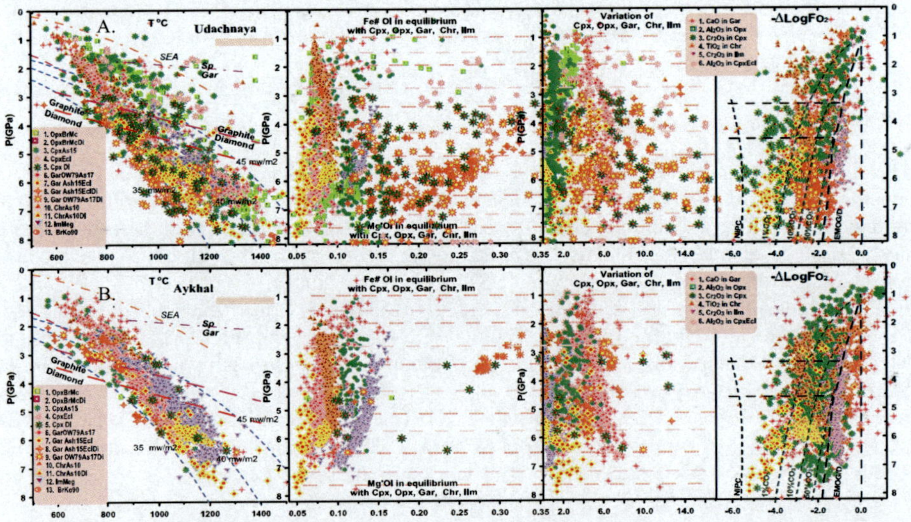

Fig. 2 **A** PXFO2 chart for all data on Udachnaya. Symbols: 1. Opx: T °C-[48]–P(GPa) [49], 2. The same for inclusions in diamonds 3. Cpx: T °C [50] and P(GPa) [34]. 4. °C and P(GPa) (Ashchepkov et al. [29] Un eclogue) 5. The same for diamond inclusions; 6. Gar: T°C [51]–P (GPa) [36] for peridotites; 7. The same for diamond inclusions; 8. Krogh [52], and P(GPa) [34], The same for diamond inclusions; 10. Chromite T °C [53]–P(GPa) [36]; 11. The same for diamond inclusions; 12. Ilmenite [19]–P(GPa). 13. T °C–P(GPa) [48]. **B** PXFO2 diagram for minerals from Aykhal pipe

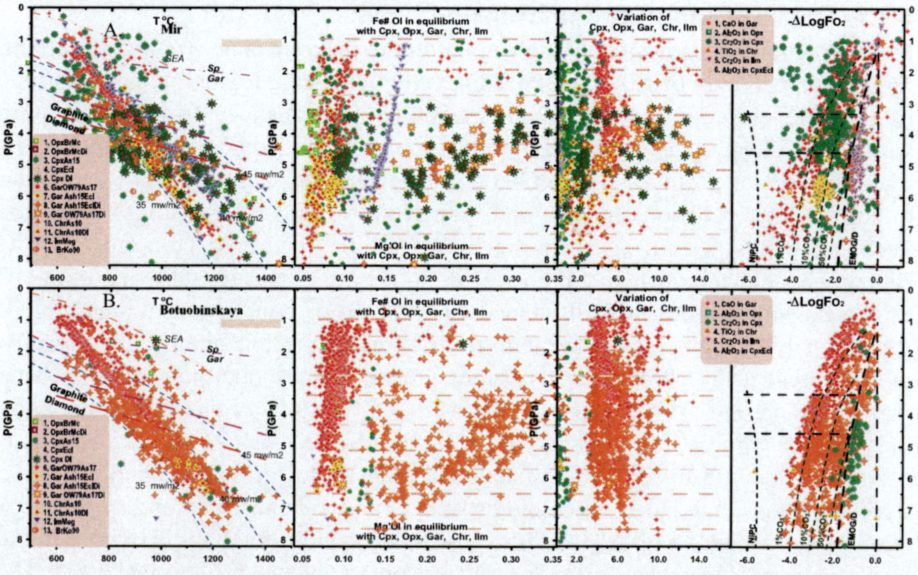

Fig. 3 PXFO2 diagram for minerals from: **A** Mir pipe (Malo-Botuobinsky field) and **B** the same for Botuobinskaya pipe (Nakyn field)

structure of the mantle, which is evident in the alternation of sub-calcium (sub-Ca) and enriched pyrope differences, in the fluctuation of ferruginosity (Fe#), and in the PT determination consists of 6 (12) individual plates up to the level of 7.0 GPa [4]. Ilmenitic PT estimates and the lithosphere base refer to intervals coinciding with the dunite-harzburgite lenses. High-temperature pyroxenes trace the boundaries of paleo-subduction plates (Fig. 2A). Sharply bended mantle column judging by definitions for xenoliths Zarnitsa [4, 29–31], in this version added by definitions for high chrome garnets (up to 19% Cr_2O_3) for eclogite garnets and xenolites. Convective branch at the base continues with definitions for ilmenite and Ti–Cr diopsides.

4.1.3 Alakit Field

Detailed PT diagram for Aykhal [29] detects a sharp structure. It detects continuous growth of Fe# grants and other minerals from the base of the lithosphere to Moho. Fractional division by P–Fe% by 6–7 pressure intervals is similar to other Alakit field pipes (Fig. 3A) For the mantle section under tr. Aykhal (Fig. 2B) has established a substantial heating from 38 to 45 mW/m^2 in the lower part in the interval 0.8–1.0 GPa within garnet dunites horizon [34, 38], xenolites which are often found in porphyry kimberlites. Eclogites are located in middle and low part of mantle [54].

4.1.4 Malo-Botuobinsky (Mirninsky) Field

Mantle section beneath Mir pipe in shown on the diagram (Fig. 3A) [4, 20, 30, 55, 56]. The upper level of 1.0–3.0 GPa is unevenly warmed, judging by the definitions for ilmenite, chromite, and pyroxenite garnets. In the range of 3.0–4.0 GPa, a more substantial warming is established for clinopyroxenes from diamond inclusions. The next interval from 4.5 to 5.0 GPa corresponds to PT definitions for garnet and orthopyroxene inclusions in diamond, which lie on the cold peridotite section of geotherm ~800 °C, and pyroxene inclusions in diamond give To up to 1250 °C. Below the inclined convective branch from 6.0 to 8.0 GPa is traced by chromite PT estimates with temperature difference of 300° from conductive gradient of 35 mW/m^2, which is given by eclogite pyroxenes and sub-calcic garnets to convective branch obtained by chromites, pyroxenes ilmenite. Under the pipe Mir–exhausted garnets are present from the base of SCLM to 4.5 GPa. Cut of mantle column under the pipe–the garnets are present from the base of SCLM to 4.5 GPa. The section of the mantle column under Internationalnaya pipe is close (Fig. 4B) and is characterized by the abundance of eclogites and clinopyroxenes with high content of Na_2O, Cr_2O_3, FeO, and Al_2O_3 [20], which are typical for hybrid (with eclogite material) peridotite rocks. From the asthenosphere level (6.5 GPa) up to 5.0 GPa, the mantle is heated.

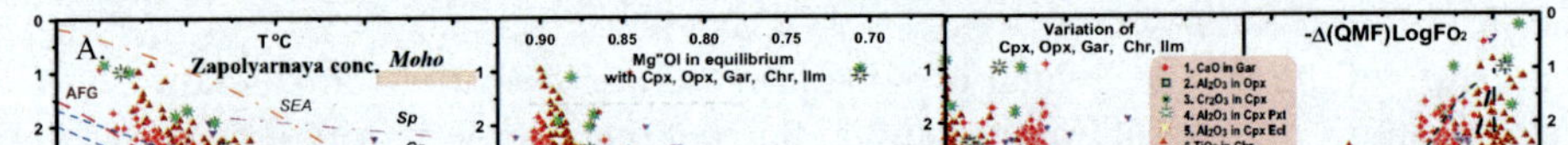

Fig. 4 PXFO2 diagram for minerals from: **A** Mir pipe (Malo-Botuobinsky field) and **B** the same for Botuobinskaya pipe (Nakyn field)

4.1.5 Nakyn Field

Mantle incisions beneath the Botuobinskaya (Fig. 4B) and Mayskaya [27] pipes body are similar to those previously published for Nyurbinskaya pipe [57, 58]. The division of the garnet geotherm for the Botuobinskaya pipe into two branches implies heating, which coincides with the development of ilmenite metasomatites. Peridotite column under Nakyn pipes is more heated, than in Daldyn-Alakit field, judging by garnet thermobarometry, to the 40 mW/m^2 geotherm. Pyroxene geotherm, constructed for the mantle column beneath Botuobinskaya pipe is high-temperature up to 4.8 GPa and only in shallow level approaches the normal of 40–35 mW/m^2 geotherm. The main part of the chromite points in the middle part determines the fracture at the geotherm about 5.0 GPa with heating lower. The column is composed of 12 horizons, significantly lherzolite and harzburgite type. Clinopyroxenes, typical for deformed peridotite, were found in both pipes. Eclogite definitions give gradual stages of Fe# growth. According to chromites, it is possible to assume intercalation peridotites and eclogite horizons, judging by discreteness of the chromite trend of Botuobinskaya pipe chromite trend (Fig. 3B). In the column of the mantle beneath Mayskaya pipe [27], definitions for sub-calcic garnets, demonstrate even sharper layering.

4.1.6 Upper Muna Field

The field contains several diamondiferous pipes [2, 21]. The Zapolyarnaya [59] and Novinka pipes [60] (Fig. 4A), Komsomolskaya-Magnitnaya pipe [61], have a close structure of mantle columns, which consists of 4 large intervals. In Zapolyarnaya, pyroxenite rocks are sporadically found, ranging from the base of the section 6.5 GPa to 3.0 GPa. A high-temperature garnet branch for enriched Ti associations near the protokimberlite mantle bodies, which formed the PT melt trajectories formed due to the melts, is also noted. Under the Novinka pipe [60] the presence of pyroxene-rich rocks in the whole pressure range with minimum definitions near 35 mW/m^2 and irregular heating to 40 and 45 mW/m^2 at the base of the section is characteristic. The presence of a convective branch below 7.5 GPa indicates that these pipes have a deep embedding and should be productive.

4.1.7 Morkoka Field

In this Morkoka field [25] there are many sub-calcic garnets which compose 4 intervals, starting from 4 to 8 GPa, together with garnets with increased CaO, which are also numerous in the upper part of the mantle section, where chromite PT determination also drops. The interval of increased Fe# at 3.8–4.5 GPa, apparently, marks the pyroxenite horizon. The geotherm in the lower part up to 4.5 GPa is relatively low-temperature, and above it is high-temperature (40–45 mW/m^2) and formed jointly by pyroxenite and ilmenite determinations. The eclogite garnets form an interval of 6–7 GPa and are found in the middle part of the interval (Fig. 4B).

4.1.8 Toluopsky Field

The largest in the North of YKP Aerogeologichaskaya pipe [21, 62] (Fig. 5A) has a deeply metasomatized manty column, especially in the lower part of the interval, where complex step trends for ilmenites are manifested, which means a multistage introduction of protokimberlite melts. The peridotite part consists of 6 main intervals, close to vertical and the slope of the trends imply washing out by the evolving melts. The eclogites are mainly of ferrous type.

4.1.9 West Ukukit Field

For the Leningrad pipe it was reconstructed (Fig. 5B) the layered section with more ferrous upper part up to 3.5 GPa and high-magnesium peridotite at the base of 5 GPa determined by pyrope compositions. Peridotite inclusions of garnets, Cr-diopsides and chromites in Svetlana and Lorik pipes [63] gives PT estimates forming a stepwise trend divided into 3 large intervals, among the pyropes there are many calcium-type garnets locating in the middle part of the section. The lower part is subject to Ti metasomatism, although ilmenite is practically absent and is replaced by chromite with high content close to ulvospinel.

4.1.10 Prisayanye

Ingashinsky lamproites is the 1st diamond manifestation on the Eastern Sayan, the Iskra vein contains pyrope and pyrope–almandine garnets and Cr-diopsides. On lherzolite pyropes [64] the pressure variations of 2–7 GPa are determined. The

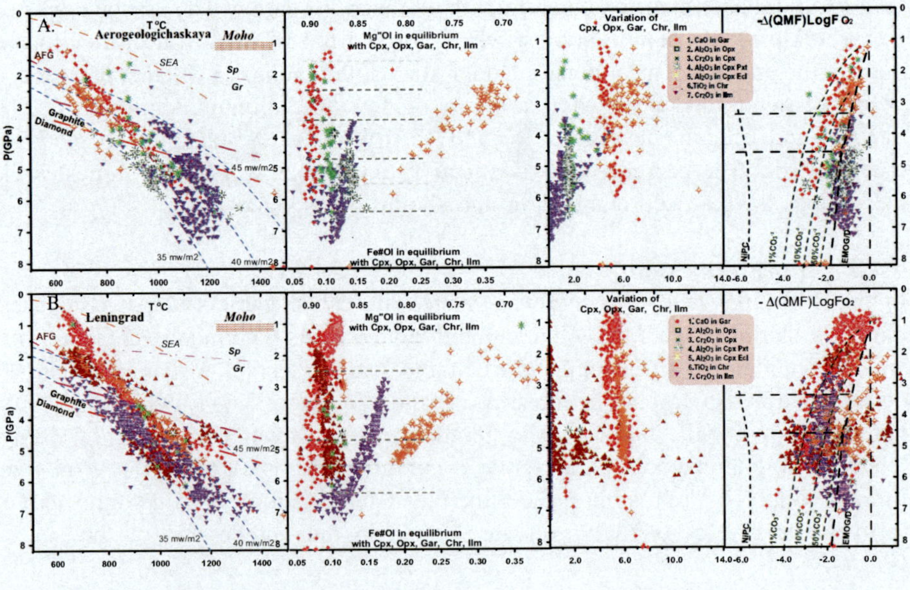

Fig. 5 PXFO2 diagram for minerals from: **A** Aerologicheskaya pipe (Toluopskoe Field) and **B** Leningrad pipe (Western Ukukit field)

average geotherm coincides with convective branch at the base of SCLM, where the definitions on eclogite garnets (Fe# = 0.25) (5–7 GPa) also are present (Fig. 6A). Placer of the Tumanshet River basin–the left tributary of the Biryusa River, in Sayan Foothils (Fig. 6B), contain pyrope referring to diamond facies. There are 3 large intervals with the heating in the 5–7.5 GPa interval. Oscillations of Ca assume the presence of several horizons of pyroxenites in the mantle section.

The PT for pyropes of the Mura-Kovinsky collector reveal close type of geothermal section, but wide variations in Fe# probably mean a large-scale interaction with melts in the peridotite column. The trends of these southern locations are all different, assuming individual sources (Fig. 6B).

4.2 Mantle Under Triassic Kimberlites

4.2.1 Kharamai Field

Early Triassic kimberlites located in northern part of YKP were dated by Zaitsev [65]. The mantle sections beneath the Prianabarie kimberlite pipes from the Ary-Mastakh to the Kuranakh fields are sharply different [66]. In the western part of the Kharamai field [66, 67], using the example of Evenkiyskaya (Fig. 7A) pipe the garnet-pyroxene–containing associations represent deep lithospheric roots starting from 6.5 GPa.

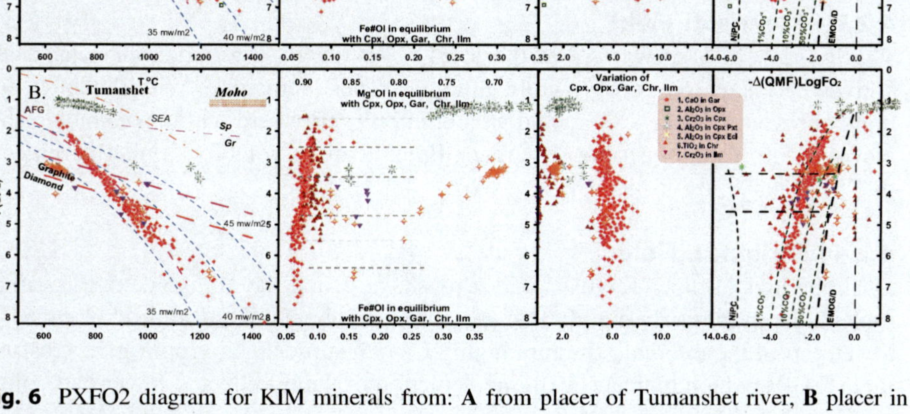

Fig. 6 PXFO2 diagram for KIM minerals from: **A** from placer of Tumanshet river, **B** placer in Mura-Kovinsky field

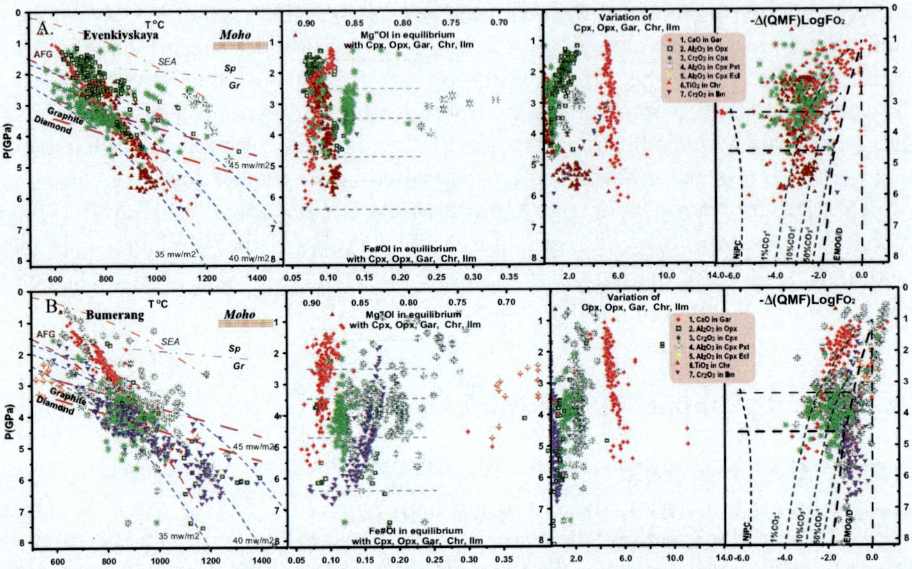

Fig. 7 PXFO2 diagram for minerals from: **A** for the Evenkiyskaya pipe from Kharamai field. **B** Khardakh and Boomerang pipes, Ary Mastakh field

4.2.2 Ary-Mastakh Field

Close conditions were determined for the western part of Anabar Shield under Khardakh pipe [66] (Fig. 7B). The mantle of Prianabarie judging by xenoliths is mainly represented by giant grained dunites with pyroxene-garnet slices generated during regeneration by melts which created also glimmerite metasomatites. Garnets of lherzolite type in Boomerang are carried out mainly from the upper parts of the section (Fig. 7B).

4.2.3 Kuranakh Field

The lithosphere base was set at the level of <6.0 GPa for Malokonamskaya and Universiteteskaya pipes [65], while clinopyroxenes (sometimes with ilmenite) give higher pressures (up to 7 GPa) and are common everywhere within the shield. They are more ferrous than peridotite garnets. Ilmenite definitions for kimberlite pipes in the Prianabarye are usually limited to 7.5–4.0 GPa,

4.2.4 Biriginda Field

In kimberlites of pipes Ksyusha, Chekhovskaya, Blind the orange eclogitic garnets prevail, the pyropes are mainly low-pressure, and abundant chromites belong to the lower part of the interval. The rare highly Cr-rich sub-calcic pyropes give pressures up to 7 GPa, which indicates extreme depletion of the middle and lower parts of the section.

4.2.5 Manchary Field

Under this pipe, garnet peridotite are mainly related to the upper and lower parts of the interval [24]. On middle part the chromite rocks (harzburgites) subjected to Ti-metasomatism are mainly found, which compose three intervals from 3 to 5.5 GPa. In this interval, dunite pyropes are rarely found.

4.2.6 Orto-Yargyn Field

Among garnets from this field in the majority of pipes predominate eclogite almandines, which give together with the pyropes a split common conductive geotherms (Fig. 8A). The usual advective-type branch is close to 35 mW/m^2 at the base of SCLM and 50 mW/m^2 at the top is defined for the mantle under the Arkhangelsk pipes [68] is installed for many pipes of Siberia and the World. The higher temperature geotherm is close to the conductive 45 mW/m^2. Pyropes give a jugged P-Fe# trend.

4.2.7 Laptev Sea and Lena River Mouth, Carnian Collector

The diagram for kimberlite indicator minerals (KIM) from the highly diamondiferous placers of the Carnian collector at the Lena River mouth is based on the published materials [69] and unpublished data of A.P. Smelov and S.A. Babushkina and original analyses (Fig. 8B). Wide distribution of eclogite garnets of diamond association and absence of ilmenites in the placers indicates a source close in type to the Ebelyakh placers.

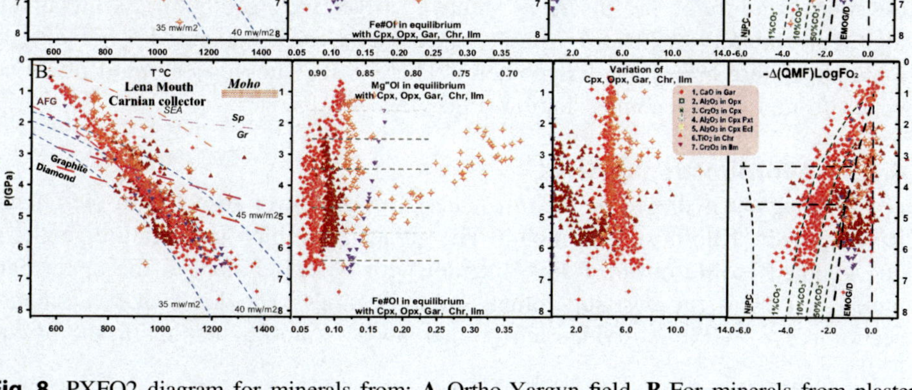

Fig. 8 PXFO2 diagram for minerals from: **A** Ortho-Yargyn field. **B**-For minerals from plaster minerals of the Carnian collector near the Lena River mouth

4.2.8 Middle and Lower Stretches of the Anabar River

Ebelyakh field. Placer areas along Mayat and Kholomolokh rivers near Ebelyakh [26] and by inclusions in diamonds allow reconstructing a section of mantle consisting of 4 intervals, with sub vertical P-Fe# trends that are marked by the appearance of sub-calcic garnets and oscillations of their Fe#. The eclogites give an inclined upward trend of Fe# at pressure decrease [34]. The mantle is unevenly heated, that is especially characteristic for eclogites, which on the PT diagram give "cloud" from 35 to 50 mW/m^2. The ilmenite step trend of 7–2.5 GPa implies 4–5 horizons at the bottom of the section.

In the middle stretches of the Udzha River (Tomtor field) pyrope and eclogite garnets give rather high-temperature geotherm with wide temperature dispersion. The ilmenite trend, more extended than under the Ebelyakh field on the PT diagram, divides into the high-temperature part, which coincides with the pyrope PT trend, and the low-temperature part with the eclogitic almandine garnet field. Layering is noticeable in the lower part of the section by Fe# and CaO oscillations.

4.3 Middle-Late Jurassic Stage of Kimberlite Magmatism

4.3.1 Kuoyka Field

This field forms a dense population of Jurassic pipes, approximately 175–150 Ma, but some of them seem to belong to the earlier Devonian and Triassic stages [65]. For Obnazhennaya pipe with fresh xenoliths of pyroxenites and eclogites [44, 70, 71] mineral thermobarometry pressure give estimates ~4 GPa. However, within the field, there are the diamond placers, and in some pipes, the amount of high-chromium pyropes is significant (Fig. 9A). Pyropes and other minerals from Djanga, Vodorasdelnaya, Zaozernaya defined the pressure up to 7 GPa, and contain sub-calcium garnets.

4.3.2 Molodo Field

Mary's pipe minerals gives a fairly long section from 7.5 to 2 GPa, which in the lower part differs in discrete values. Advective geotherm is relatively high-temperature. Inclusions from the eclogite type diamonds lie on the low-temperature branch of the geotherm. There is the sub-vertical trend of P-Fe#, which from 4 GPa becomes sloping (Fig. 9B).

4.3.3 Khorobosuonka Field

A PTX diagram is drawn on the dirt material in the lower stretches of the Olenek River near the Khorbosuonka mouth. The garnets give high-temperature geotherm and inclined profile on the P-Fe# diagram with Fe enrichment in the upper part, which is typical for post-superplume Jurassic pipes. The lower part of mantle section (P. > 4 GPa) is divided into 5 intervals by CaO fluctuations in the pyrope [20].

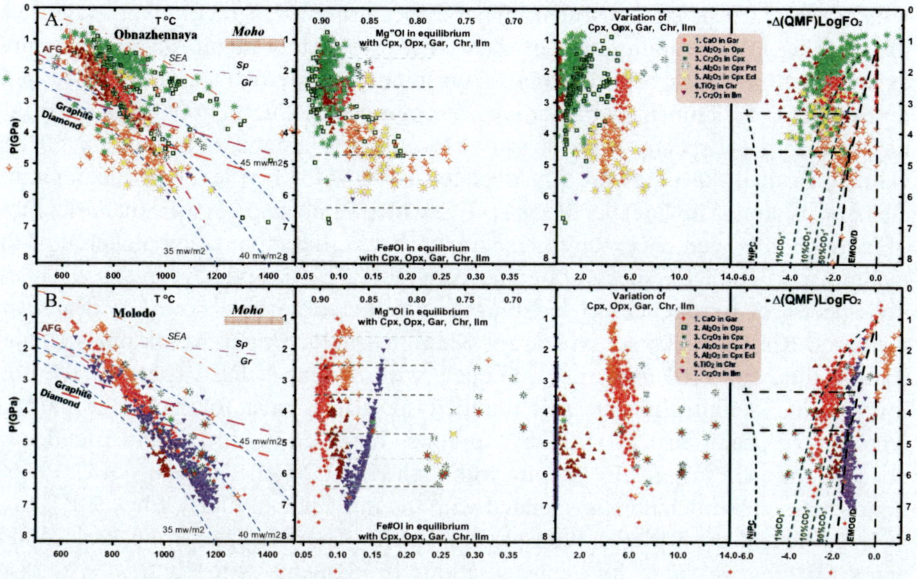

Fig. 9 PXFO2 diagram for minerals from: **A** from Obnazhennaya pipe (Kuoyka field); **B** for (Molodo field)

5 Geochemical Characteristics of Minerals of Various Mantle Terranes

Peridotite minerals of the Siberian platform have a very wide range of compositions in comparison with African, North American and other cratons [70–76]. The variations of TRE garnet compositions complement the characteristics of mantle columns. Garnets often have peaks of Pb and U, which are more expressed in garnets with S-shaped distributions of rare-earth elements (REE), typical for rocks of depleted type, sometimes with enrichment of light (LREE) and incoherent elements as a whole, and often with minima of highly charged elements (HFSE), that reflects characteristics of rocks formed in subduction environment [77]. At the same time, protokimberlite metasomatism leads to enrichment with HFSE and U, Th (Fig. 10). Peridotite garnets from kimberlites of the Daldyn field often have S–type distributions (Udachnaya and Zarnitsa), which is especially characteristic for large pipes, which contain deeply depleted mantle substrate at the base of the mantle column. Garnets from Udachnaya are diverse in their REE configuration, sometimes quite complex. Many garnets are characterized by peaks of U and often an increased level of incoherent elements. Garnets from Alakit pipes differ in the level of heavy REE (HREE). Pyropes from the Sytakanskaya [33] and Komsomolskaya pipes [54] often have S-spectra and have higher concentrations of lithophiolitic elements–they have deeper minimums of HFSE. For garnets of Komsomolskaya Aykhal, pipes have typical small Ce peaks and often-higher concentrations of incompatible rare elements (TRE), deep minimums of Zr, which are better manifested in the most exhausted differences of Komsomolskaya pipe. The pyrope garnets of Mir pipe reveal LREE enrichment, while among the pyropes of Internationalnaya pipe only rare grains have spectra with sharply elevated LREE and minimum in HREE. Garnets from these pipes are not depleted by HFSE. The level of incompatible elements is almost at the level of the LREE, which indicates a significant enrichment of incompatible elements with fluids and, probably, the input of the material of the mature crust. The REE spectra of garnets are more diverse and have peaks in Th and Pb. Spectra of pyroxenite, of harzburgite and lherzolite type reveal U peaks and increased Rb, Cs peaks are typical for Kharamai field. Upper–Muna pipes contain sharply different TRE of pyropes. Garnets with S–shaped REE spectra have high level of incoherent elements and rounded spectra, typical for lherzolite garnets, which have peaks also in U. Nakyn pyropes have Ce-minimum and rounded or concave from Gd to Tm REE spectra with high HREE level. The peaks of U, Pb and high Rb are noted, which is associated with the influence of fluids. Elevated Th and U levels reveal the varieties with enriched REE. The analyzed Prianabarie differences (Trudovaya) have low concentrations of elements with contrast spectra or enriched characteristics with smoothed spectra. Typical for most garnets from the Udzha river are flattened REE spectra, sometimes with bends and increased Ta, Nb, U, Th levels associated with protokimberlite metasomatism. Garnets from the Carnian collector have a high peak U, often Rb and spectra with an inflection in the LREE part, which is usually due to the influence of fluids. Inclined REE spectra are typical for pyrope of Tumanshet area. The spider-diagram shows U peaks less than Th and almost undifferentiated HFSE, along with increased Rb, Cs.

6 Mantle Cross-Sections

Using reconstructed mantle columns under individual pipes, we constructed sections subject to regeneration with newly formed clinopyroxenes in the lower part of the mantle cross-sections (MC) enriched with Fe# [4, 29–34]. The presence of sub-rock garnets marks the depletion levels. To see the details of the structure, some sections were prepared by the compositions of several individual minerals.

The cut through Daldyn field is the most detailed (Fig. 11). Dark (blue) areas correspond to Ilm- and Cpx-megacrystals formed by protokimberlite melts. Judging by the dark region at the level of 4 GPa, the level of metasomatism in the eastern part in the area of Zarnitsa pipe is growing in comparison with the western part near Udachnaya pipe. High concentration of small pipes in Zarnitsa group probably reflects high concentration of melts in the permeable zone. At least 3 units are distinguished in the upper part of the MC on the fO2 diagram, which is comparable with 3 peaks on the histograms for ilmenites in Daldyn district [32]. In the lower part of SCLM, the double zone of garnet depletion indicates the slope of the structure to the west. The area of increased Fe# near 4.0 GPa refers to the

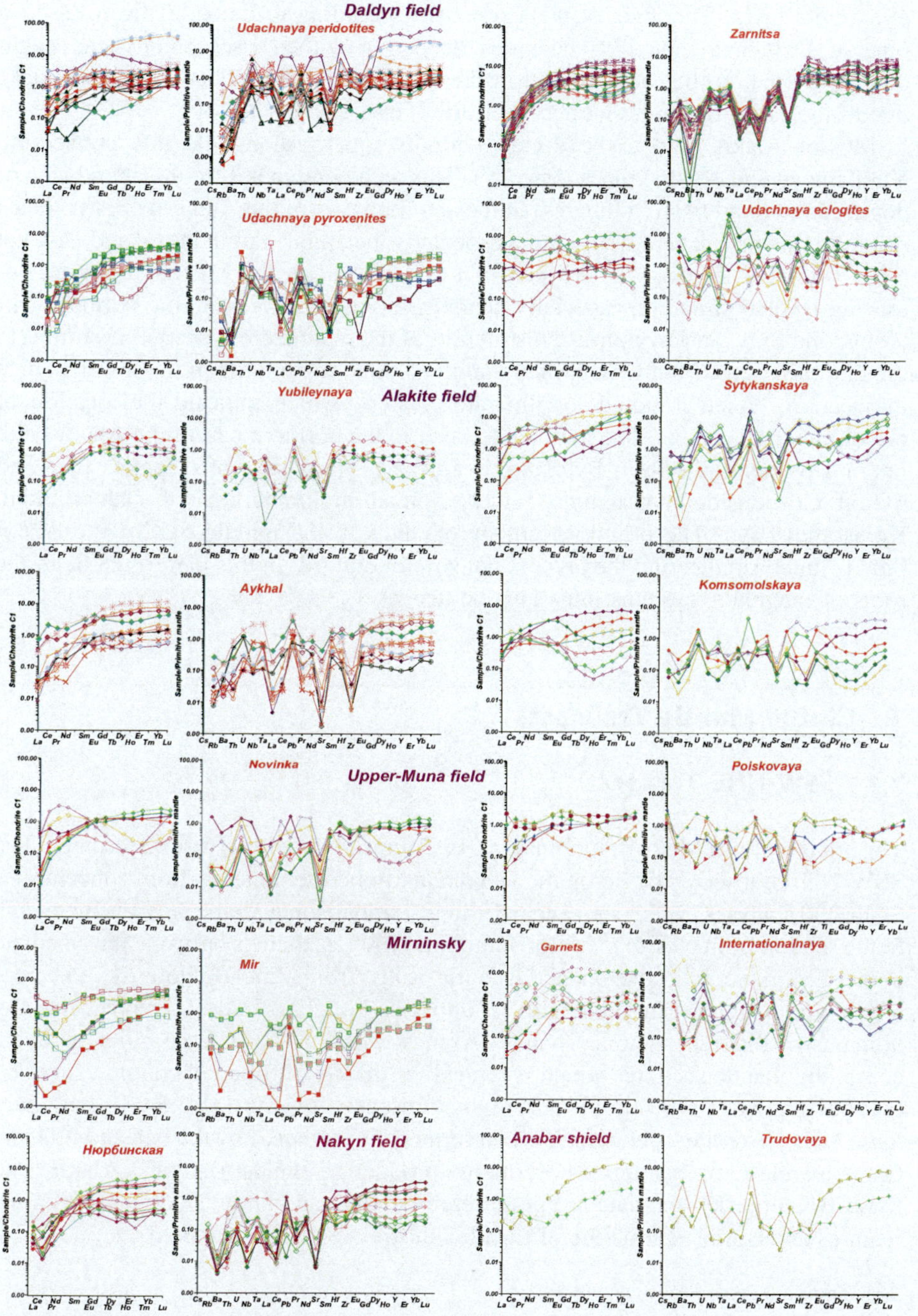

Fig. 10 Normalized rare earth element distributions (REE) [78] and multicomponent spiderdiagrams normalized to primitive mantle [79] for garnets from different fields of the Yakut kimberlite province

pyroxenite layer. The level of metasomatism according to ilmenites from Zarnitsa pipe to Dalnyaya pipe [29] becomes deeper, and then rises again near Zagadochnaya pipe. In the southern part of the Daldyn field the MC are less contrasting due to the rarity of dunite garnets with low Fe# values.

MC of Alakit field to be divided into two parts along the line connecting Yubileinaya and Aykhal pipes (Fig. 11). This area corresponds to the abundance of depleted diamond bearing dunites. In the southern part of the Alakit field, MC has a more contrasting layered structure, especially near the Yubileinaya and Aykhal pipes (Fig. 11). In the southern part of the Alakit field, the MC has a more contrasting layered structure, especially near Krasnopresnenskaya. In the northern part around the Sytykanskaya pipe, the structure of the mantle cross-section is similar to the northern part of the Daldyn field, and has three main units in the lower part of the section, which is noted for ilmenite clusters, which indicate the location of protokimberlite centers. The pyroxenite layer in the northern part of the Alakit field, judging by xenoliths from Sytykanskaya pipe [33] consists of eclogites [54] and hybrid Cr-diopside pyroxenites, widely spread in kimberlites of Daldyn field. Komsomolskaya. The proportion of pyroxenites in the middle part of the Alakit field is much smaller and the layer is not so pronounced. In this part of the field, the roots of magmatic systems must also be deeper.

7 Global Mantle Transects

7.1 SSW-NNE Transect

This transect from Malo-Botuobinskoye to Tolupskoye fields crosses 5 terranes. The SSW-NNE transect is based on the Devonian kimberlite minerals from concentrates ($\sim$40000 analyses) with a more dense grid in comparison with the previous work [3]. In the section from Nakyn to Malo-Botuobinsky fields, there is only one intermediate point of Morkoka pipe, the extended mantle profile from Malo-Botuobinsky to Upper Muna and Ogoner–Yuryakh and Toluopsky (Fig. 12) fields is represented by domains of different tectonic mantle terranes. There are 6–7 large units or layers compiling mantle sections, which is typical for mantle of Siberian craton, visible by fO2 fluctuations. In the middle part of the transect below the 4.0–3.5 GPa boundary there is a pyroxenite layer, which is determined in most areas by the Fe# and fO2 and CaO increase in garnets. For more productive fields–Daldyn, Alakit and Malo-Botuobinsky–the data are more representative and allow studying individual features and complex structure of mantle lithosphere under each field.

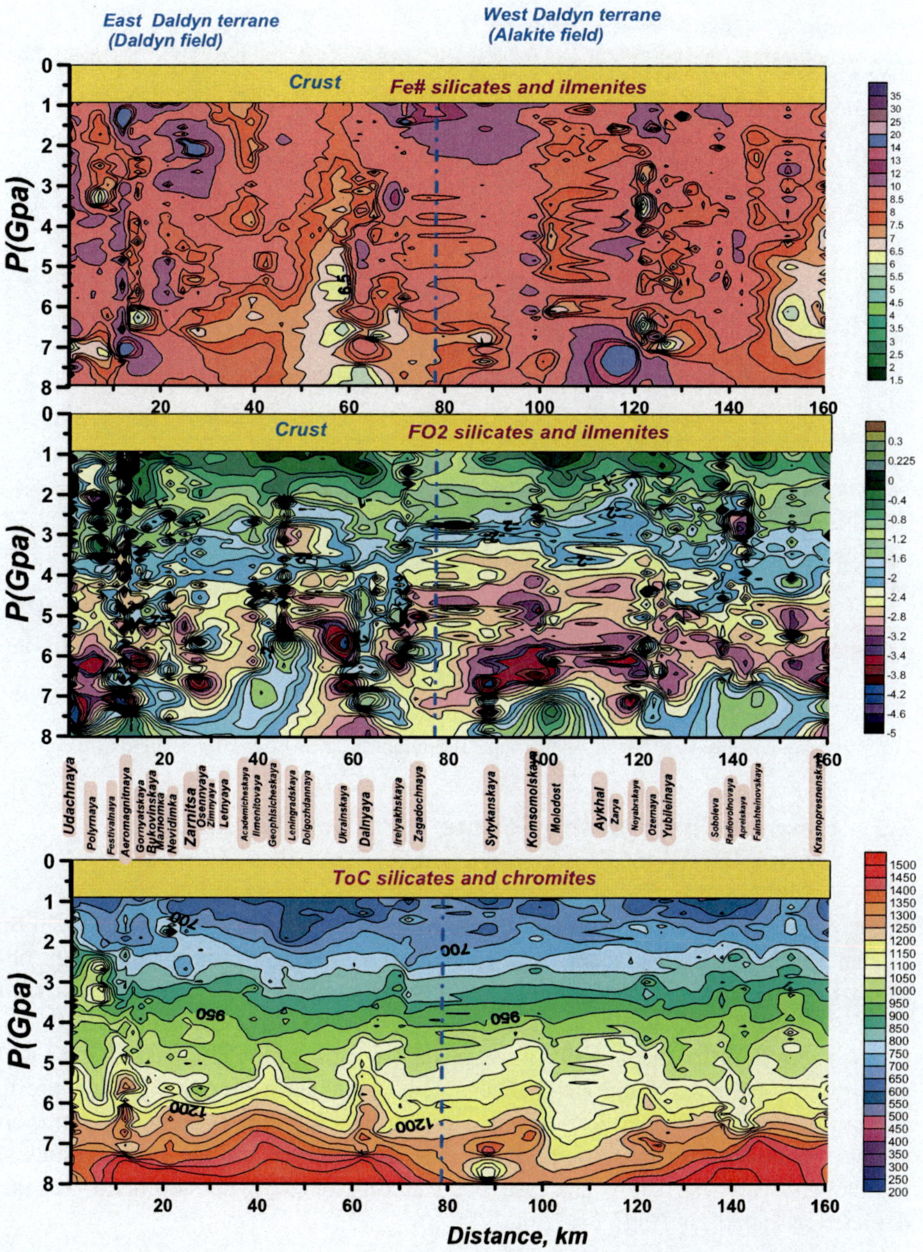

Fig. 11 Mantle transect through Daldyn–Alakit fields, from Udachnaya to Krasnopresnenskaya pipes

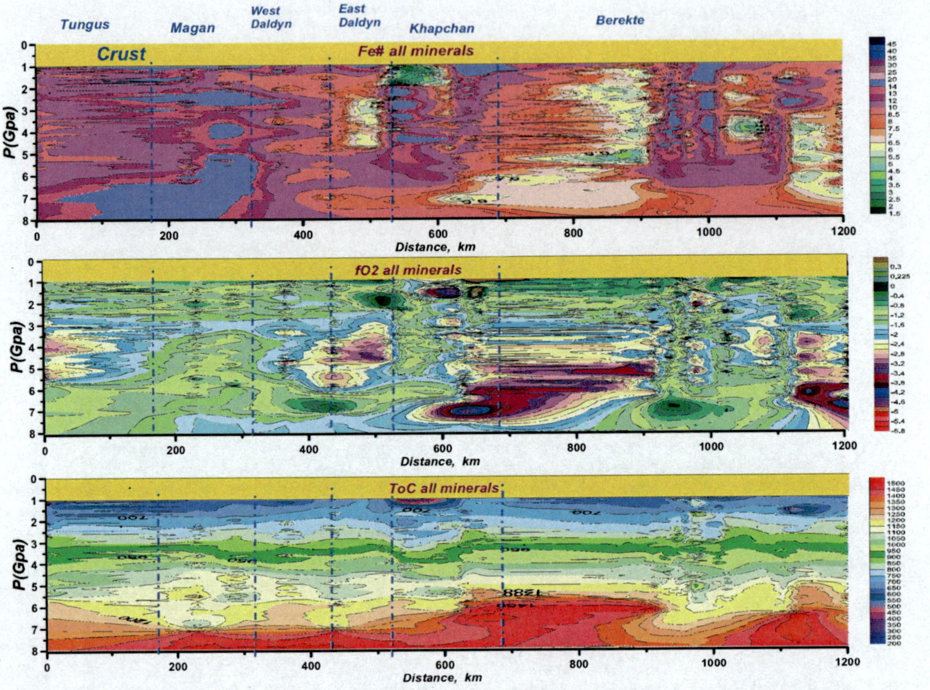

Fig. 12 Mantle transect SSW-NNV from the Kharamai field to Lower Lena region

7.2 Transect Through the Southern Part of the Siberian Craton (NNW–SSE)

In the peripheral areas, a profile was built on the compositions of garnets from the Tychan and Tyradak placers, as well as pyropes from the Tumanshet and Muro-Kovinsky fields from alluvial placers, as well as the KIM minerals from Nakyn field [27, 28] repeats the general features of the previous one (Fig. 13). The southeastern fields of the Tunguska terrane show signs of flat layered structure with moderately restored conditions in each area. All mantle domains in this profile contain predominantly peridotite minerals with low ilmenite content and represent typical continental mantle transects. In the eastern part of the Aldan block, a multilevel-layered structure has also been reconstructed, with absence of deeply depleted associations (Fig. 14) [30].

7.3 SW-SE Transect in the North of the Siberian Craton

The transect from the Kharamai Field (Ashchepkov et al. 2016) through the Orto-Yargyn and Kuoyka fields [70] and the Lena mouth [80] as well as the profile through the Anabar shield to the Laptev Sea coast (Fig. 13) and extends further to

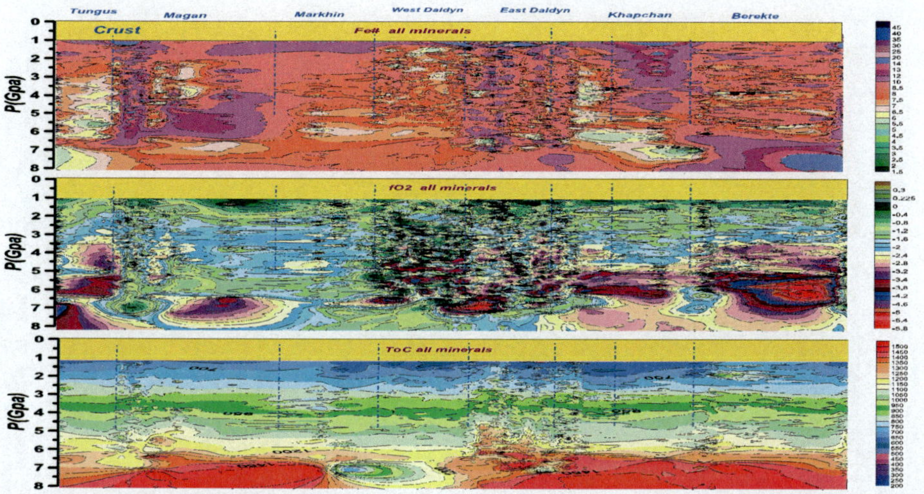

Fig. 13 Mantle transect SSW-NNV from the Little Botuobinskoe field to the West-Ukukite region

the mouth of the Lena River, where diamondiferous placers of Carnian age were found along the Laptev Sea coast [69, 80]. This transect crosses the Ary-Mastakh, Starorechenskoye, Dyuken, Kuranakh and Ogoner-Yuryakh fields, as well as the Kuoyka and Toluopsky fields.

At least 3 large parts are divided in SCLM under the Anabar Shield and the lower plate is sloped southwest, in the same direction as in Daldyn Field. The structure of the mantle sections of the West Anabar, Kuranakh, Dyuken, and Orto-Yargyn fields is divided into 4 large units and differs from the layering under the Ary-Mastakh and Starorechensky fields, which consist of 3 large units. A much more complex layering of the mantle of the Kuoyka field is emphasized on the P–fO_2 diagrams by marking permeable zones with metasomatic ilmenite. In the northeastern part of the mantle section of the Birekte terrane there is a flattened and simple structure with low fO_2 values in the northern part of the terrane, which is favorable for diamondiferous [20] (Fig. 13).

8 Discussion

8.1 Comparisons with Geophysical Data

Our data and calculations show that in most cases the lithosphere thickness in Devonian time was about 250–270 km [20], which is supported by geophysical models [79, 80] as well as by findings of high chromium up to 16% Cr_2O_3 in garnets. Evidence of high lithosphere thickness in the northern part of the Siberian

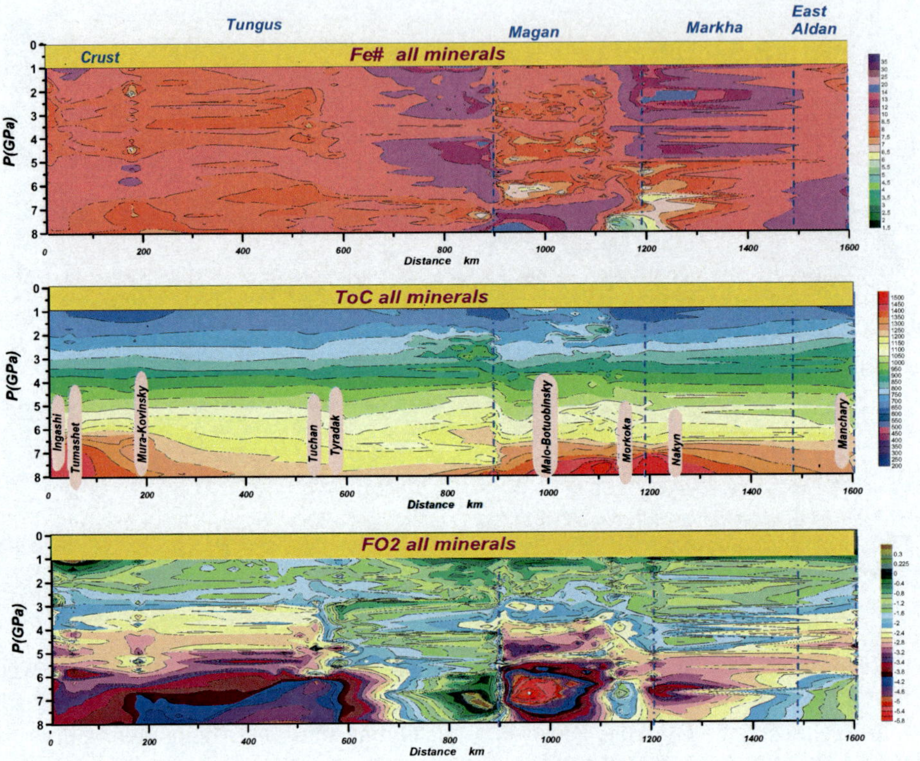

Fig. 14 Mantle transect SW-SE from Tumanshet to Chompolo field

craton, according to I.Yu. Kulakov [7] is not very obvious. The mantle cut built on the minerals of the Carnian collector gives a lithosphere thickness of up to 8 GPa–270 km. Lower SCLM values are found under the Kuoyka Field, although there are also high chrome garnets (SCLM up to 250 km).

Geophysical models refer to the present time and do not reflect the heat associated with the Devonian plume. Most geophysical models, both seismic models [81, 82] and those using the thermodynamic approach based on the Gibbs method of free energy minimization [82], show similar structures of the lithospheric mantle. The velocities of seismic waves gradually increase with depth, in the eastern part and near the basin of the Vilyui river they capture the Moho uprise. The depth of local velocity decrease of the zone varies from 260 km in the West to 220 km in the eastern part of the craton.

A mantle model built with the help of seismic tomography [83] shows that the thickness of the lithosphere under the central part of the Siberian craton is 250 km and a little less under the Anabar shield.

Thermal structure of lithospheric mantle on seismic profiles defined in works of O. Kuskov [82] generally corresponds to the structures defined in our work, but it is colder by 50 ~ 75 °C. This may be explained by this difference in cooling since the Upper Devonian time.

8.2 Variations in Structure and Composition of Mantle Lithosphere of Different Terranes

Selected tectonic terranes within the central and northern part of the Siberian craton have sub meridional extension, as well as kimberlite fields. Archean granite-greenstone and tonalite–trondjemite terranes: Tungus, Markha, West Aldan, and Sharyzhalgai, which are ~3.8–2.5 billion years old [8, 84] include relatively few kimberlite fields. East Yakutian protocraton Birekte with an age of ~2.7–2.1 billion years contain a large number of kimberlites, often barren, but the placers suggest rich sources in the north-east of the craton. In the mantle under the granite-greenstone terrane Birekte (Kuoyka, Molodo, Orto-Yargyn, Toluopsky, Lower Lensky and other fields), contains moderately depleted peridotite along with hybrid eclogites [70] of about 4.0 GPa boundary. The eclogites should be melted at this level in the Late Archean under high temperatures [85] with formation of pyroxenites. This may be one of the reasons for the division of the lithospheric mantle into two parts at the level of 3.5–4 GPa and the origin of pyroxenite lens. Although recent events correspond to the Proterozoic activation [71]. The abundant metasomatism [72] does not contribute to the diamond grade. The Toluopsky field is characterized by a thick lithospheric keel and a sharp division into the upper and lower parts of the SCLM, as for the Priazovie mantle [20]. The Markhinsky granite-greenstone terrane have specific composition of SCLM. Here in the mantle under the Nakyn field, the peridotite section contains abundant interlayers of various eclogites mainly of highly aluminous type and garnet- mica bearing rocks (metapelites in the protolith) [58] (, as well as peridotite xenoliths and minerals [27]. The mantle under the Magansky granulite-orthogneiss terrane under Tr.

The Mir pipe in Magan terrane is composed of relatively fertile [84] and metasomatically changed peridotite in the upper and middle parts of the section [4, 20, 55]. Depletion begins at 5.0 GPa beneath the Mir pipe, and under International pipe a significant number of eclogites and hybrid peridotites was found in the lower part of the section, starting from the base of the lithosphere, which is a promising factor ability of diamond. The general low Fe# and other basalt components (CaO, Na_2O, TiO_2) are characteristic of the lower parts of the Magan terrane (Malobotuobinsky and Morkoka fields) have a sharply depleted lower part of the SCLM but this is not confirmed in the northern part in the terrane in the Kharamaisky and Ary–Mastakhsky field, probably here the mantle terrane represents a collage of smaller plates.

Early–Proterozoic Daldynsky granulite-orthogneiss terrane includes the most numerous and productive kimberlites. According to this scheme, kimberlites are located mainly within the Paleoproterozoic accretion zone [8], which has an age of ~1.8 or 1.9 billion years [86]. The corresponding correlating peaks of isochronous Re/Os ages for silicate minerals in mantle xenoliths [71], sulfides in pyropes [87] and zircons from gneisses [8] were found. Zircons [88] of the lower crust xenoliths record the events of magmatic underplating of basite intrusions at the time of plume magmatism. Daldynsky accretion area [8] represents a structure permeable for plume melts.

Under the Western Daldynsky terrane, in the mantle lenses of dunites are localized near the Aykhal and Yubileinaya pipe. The eclogites are common in Komsomolskaya [54] and Sytykanskaya [33] pipes, where peridotites are fertile and richer in pyroxenes. In the East Daldyn terrane beneath the Udachnaya pipe, the layered section consists of harzburgites, close to abyssal peridotite of subduction type in composition [46, 47] with lenses of diamond dunites and eclogites. The eclogites in this terrane often belong to Mg type like beneath the Komsomolskaya pipe [54], which is associated with tonalite-trondyemite restitis (cumulates). The structure of the Upper Moon field SCLM contains sub-calcium garnets starting from 3 GPa and reveals a tendency to increase the number of the pyroxenes in peridotites with depth. Common feature of all silicate minerals in this field is rather high degree of oxidation associated with metasomatism.

The general inclination of the mantle layers to the east for Daldyn-Alakit district coincides with the one installed on the geophysical profiles of the mantle [7]. The kimberlite fields are located close to the edges of tectonic terranes. We cannot determine the slope angles of terrane boundaries, but we can characterize in general the composition and some features of such mantle terranes.

Within the limits of collision Khapchan terrane there are concentrated the fields in which the share of eclogite matter is essential. In the richest placers of Ebelyakh with middle current of the Anabar river eclogite inclusions in diamonds sharply prevail as well as in the area of the Udzha river. The considerable quantity of eclogite garnets is characteristic both for the Kuranakh field and is established in the East and West Ukukite, and Ogoner-Yuryakh fields. In the western part under the Khapchan terrane (accretion complex), the SCLM mantle is extremely depleted in Al and Ca, but dunites are more Fe- rich than Marian peridotite [89]. The thick eclogite pyroxenite lens is perspective for diamond. Less contrasting layers have been diagnosed in the mantle sections under the northern fields of Ogoner-Yuryakh field of the Khapchan terrane than in the central regions of the YKP. The mantle beneath the Anabar shield and Prianabarie is depleted in FeO, CaO and Al_2O_3 domain of the mantle as compared to most other northern fields. The layered structure of SCLM transects under individual fields is sometimes similar [20, 36]. However, there are certain differences in mantle layering beneath the Alakit, Daldyn, and Upper-Muna districts and much more significant with SCLM under the Malo-Botuobinsky field [20].

Continuous trends with the splitting of the CaO in pyrope garnet compositions with the increasing depth have been determined for the Siberian craton and in many regions including South Africa [90, 91]. In most cases, there the number of dunites and pyroxenites increases simultaneously to near the base of the lithosphere.

Comparison of profiles shows that granulite-orthogneiss terranes have a finer layered structure of the lithospheric mantle as a whole. The detailed section of the

Daldyn terrane reveals alternation of rather heterogeneous layers of the mantle and the folded structure of the mantle sections. The geochemical data indicate that within the granulite orthogneiss terrane there are large variations in the REE and TRE spectra and ultra-depleted compositions as well as the subduction-related geochemical characteristics. For significantly silicic terranes, the influence of subducted continental sediments is also evident.

We can state that Magan and Daldyn terranes, as well as Khapchan terranes should be divided into several units by their extension, and Anabar shield should be singled out as a separate structure with the close mantle characteristics and structure. For detailed partitioning, we need not only thermobarometric works using microprobe analyses but also geochemical data +and age determination for the rocks and minerals.

8.3 Lithosphere Layering

Formation of the layered lithosphere is considered by most researchers to be the result of repeated subduction and joining the slabs [92–94]. In the early Archean heat regime [95] did not allow deep subduction [96] and primary continents did not have deep roots. Only after the cooling the deep subduction accompanied by superplume events [97] was formed deep continental roots, which were considered may reached the boundary of 410 km [90]. Ultra depleted craton nuclei like Zimbabwe [97], the Anabar Shield could be the result of high melting of early plumes or boredom and accumulation of early essentially olivine plates of the ancient mantle. It seems wrong to stretch the mantle terrane from south to north through the Anabar Shield, although there are signs and models of the so-called "stacked slabs" [91] with a steep slope [98]) but such structures are mainly characteristic of suture zones. Kimberlites, which were rising along the steeply tilting eclogite part slab, should significantly melted it (mainly pyroxenes), and capture many almandine grants.

During the subduction processes, eclogites and peridotites could be essentially melted mainly during the superplume events [99], when the rocks of the plunging plate were melted to acquire buoyancy. For peridotites, the melting lines calculated from experimental data by Herzberg [100] are mostly near 5–6 GPa [46, 71].

In classical works of mantle petrology, all geotherms are considered to be conductive [45], but verification has shown that this phenomenon is quite rare. Geothermes by pyrope garnets [34–47, 51, 54, 86–104] and calculated by mineral thermobarometry the most reliable methods [50, 48] give inclined geotherms with a sub-adiabatic inclination, which cross the conductive [105] due to the fact that during the superplume period there was a washing of mantle columns with melts saturated with volatiles, which reduce the melting temperature [106]. Therefore, after superplumes, the P-Fe# trends for garnets became smoothed out and the tilt often changes and became close to sub vertical. At the same time, melt percolation waves often form a secondary layering regulated by phase equilibrium and temperature regime, as well as permeability of the mantle, which even questioned the primary subduction genesis of the craton keel [91]. In fact, garnet geotherms slide along the minimum on solidus peridotite in the presence of CO_2–H_2O [107]. The oscillations of Mg' or Fe# for olivines in the mantle section give possibility to determine the layering mainly beneath the pipes, where the roots were not affected by significant pervasive metasomatism.

Garnet geothermal geotherms based on eclogite garnets using a thermometer [52] are often lower temperature and more close to conductive. But after the super plumes like beneath Premier pipe after Bushveld event the eclogites may be more high temperature than peridotites [108].

8.4 The Evolution of Mantle Incisions in Time

The lower boundary in the early Archean was regulated by the olivine flotation level in the melt ~8.0 GPa [109]. The kimberlite related plumes and super plumes often form the melted lenses at the boundary ~6 GPa, because there the minimum of olivine strength in the presence of the melt locates [110]. Under stress, the melting increases. In the presence of oxy melt–and carbonatite melt, the boundary shifted upwards [111].

Comparison of mantle section reconstructions for Late Devonian, Early Triassic and Late Jurassic kimberlites shows the evolution of mantle sections over time.

Layered thick lithospheric keels and heating at the base of the lithosphere at 6.5–7.5 GPa are characteristic of the Devonian period. Cr-hornblende and pargasite amphiboles (and richterites) are typical for the mantle at this time especially in the marginal craton regions. In the Early Triassic period, the lithosphere basement was at the level between 6.0 and 5.5 GPa, as under the Kuranakh field. Here low chrome garnets in peridotite and pyroxenite at the base and middle part of the section are typical, as well as significant heating up to the level of ~5.0 GPa and abundant phlogopite and Fe–Ti–metasomatism. At this time, metasomatism in the middle and upper levels of mantle incisions is widespread in Prianabarie. Many pyroxenite associations are found at the top of mantle sections, starting mainly from 4.0 GPa.

The conclusions about delamination and reduction of lithosphere thickness to the north after Perm-Triassic superplume [67] are not confirmed [66]. Some decrease in thickness and the asthenosphere boundary was probably related to the melt concentration in the lower parts of craton.

It is quite probable that outside the direct influence of Siberian Permo–Triassic superplume, the lithosphere remained as powerful as in Devonian time in the central part of the craton, for example, under the northern part of the craton 600 west of the Lena river mouth. Similar placers of pyrope belonging to the Upper Triassic (Carnian time) are widely distributed from the mouth of the Lena River [69] along the coast of the Olenek Gulf of the Laptev Sea [80] in the northern part of Siberia.

The thickness of the lithospheric mantle in Upper Jurassic was reduced to 180 km, and sporadically to 130 km locally in some areas, mainly under the Kuoyka field [70]. These may be rift structures that are permeable at the mantle level. The decrease in the level at which peridotite xenoliths were captured may be explained by the presence of an intermediate magmatic chamber located mainly at the pyroxenite lens on the border between the upper part of the SCLM ~4.0 GPa, where the abundance of eclogites, at least in the north of the Siberian craton, is the general rule. The level of magma generation for all kimberlites is much deeper than 6.0 GPa.

9 Conclusions

1. The structures beneath the Siberian craton are very diverse and vary greatly between different terranes, which are also not very stable in composition. In the Archean, the protocratons such as Tungussky, Markhinsky, and Birektinsky terranes have a sharply thinly layered unfolded structure of SCLM relatively homogeneous and often metasomatized mantle with phlogopites possibly with the participation of continental sediments. The collision terranes have folded structures with abundant eclogites than the SCLM under the orthogneiss Daldyn collision terrane, which consists of more contrasting layers.
2. The thickness of the lithosphere slightly decreased with time from 270–250 km in the Upper Devonian time to 250–220 km in the Lower Triassic and up to 130–180 km in the Upper Jurassic time, due to the interaction with the super plume melts. This may be the result of local ascent of melt and creation of upper asthenospheric lenses.
3. Within the Siberian craton there are no signs of detachment and delamination of mantle lithosphere under the influence of P-T plume.
4. The essential dunitic composition of the Prianabarie mantle in the lower part of the section (having positive buoyancy with respect to pyroxenites and eclogites) is probably related to the high crust position in the area of the shield.
5. The impact of a concentrated flow of fluid melts, possibly locally, significantly affected the composition and mantle, but did not completely change the primary structure of the mantle.

Acknowledgements The research was done with the financial support of RFBR, grant project No. 19-05-00788, 20-05-00261, as well as the Integration Project of the Research Center of the Russian Academy of Sciences, Block 1.4. and by the state order of IGM SB RAS. and the study was performed by the governmental assignment in terms of Project IX. 129.1.4., RNF-21-17-00015.

References

1. Gladkochub, D.P., Pisarevsky, S.A., Donskaya, T.V., Natapov, L.M., Mazukabzov, A.M., Stanevich, A.M., Sklyarov, E.V.: Siberian Craton and its evolution in terms of Rodinia hypothesis. Episodes **29**, 169–174 (2006). https://doi.org/10.1134/S0016852112040024

2. Smelov, A.P., Zaitsev, A.I.: The age and localization of kimberlite magmatism in the yakutian kimberlite province: constraints from isotope geochronology—an overview. In: Pearson, D. et al., (eds.) Proceedings of 10th International Kimberlite Conference. Springer, New Delhi, pp 225–234 (2013). https://doi.org/10.1007/978-81-322-1170-9_14
3. Griffin, W.L., Ryan, C.G., Kaminsky, F.V., O'Reilly, S.Y., Natapov, L.M., Win, T.T., Kinny, P.D., Ilupin, I.P.: The Siberian lithosphere traverse: mantle terranes and the assembly of the Siberian Craton. Tectonophysics **310**, 1–35 (1999). https://doi.org/10.1016/S0040-1951(99)00156-0
4. Ashchepkov, I.V., Vladykin, N.N., Ntaflos, T., Kostrovitsky, S.I., Prokopiev, S.A., Downes, H., Smelov, A.P., Agashev, A.M., Logvinova, A.M., Kuligin, S.S., Tychkov, N.S., Salikhov, R.F., Stegnitsky, YuB., Alymova, N.V., Vavilov, M.A., Minin, V.A., Babushkina, S.A., Ovchinnikov, YuI., Karpenko, M.A., Tolstov, A.V., Shmarov, G.P.: Layering of the lithospheric mantle beneath the Siberian Craton: modeling using thermobarometry of mantle xenolith and xenocrysts. Tectonophysics **634**, 55–75 (2014). https://doi.org/10.1016/jtecto200911013
5. Helmstaedt, H.: Crust–mantle coupling revisited: the Archean Slave craton, NWT. Canada Lithos **112**(S2), 1055–1068 (2009). https://doi.org/10.1016/jepsl201104034
6. Artemieva, I.M., Thybo, H., Cherepanova, Y.: Isopycnicity of cratonic mantle restricted to kimberlite provinces. Earth Planet. Sci. Lett. **505**, 13–19 (2019). https://doi.org/10.1016/jepsl201809034
7. Koulakov, I., Bushenkova, N.: Upper mantle structure beneath the Siberian craton and surrounding areas based on fieldal tomographic inversion of P and PP travel times. Tectonophysics **486**, 81–100 (2010). https://doi.org/10.1016/jtecto201002011
8. Rosen, O.M., Levskii, L.K., Zhuravlev, D.Z., Rotman, A.Y., Spetsius, Z.V., Makeev, A.F., Zinchuk, N.N., Manakov. A.V., Serenko, V.P.: Paleoproterozoic accretion in the northeast siberian craton: isotopic dating of the anabar collision system. Stratigrafiya Geologicheskaya Korrelyatsiya **14**, 3–24 (2006). https://doi.org/10.1134/S0869593806060013
9. Aulbach, S., Griffin, W., Pearson, N., O'Reilly, S.Y.: Nature and timing of metasomatism in the stratified mantle lithosphere beneath the central Slave craton (Canada). Chem. Geol. **352**, 153–169 (2013). https://doi.org/10.1016/jchemgeo201305037
10. Lehtonen, M.L., O'Brien, H.E., Peltonen, P., Johanson, B.S., Pakkanen, L.K.: Layered mantle at the Karelian Craton margin: P-T of mantle xenocrysts and xenoliths from the Kaavi-Kuopio kimberlites, Finland. Lithos **77**, 593–608 (2004). https://doi.org/10.1016/jlithos200404026
11. Griffin, W.L., Kobussen, A.F., Babu, E.V.S.S.K., O'Reilly, S.Y., Norris, R., Sengupta, P.: A translithospheric suture in the vanished 1-Ga lithospheric root of South India: evidence from contrasting lithosphere sections in the Dharwar Craton. Lithos **112**(2), 1109–1119 (2009). https://doi.org/10.1016/jlithos200905015
12. Batumike, J.M., Griffin, W.L., O'Reilly, S.Y.: Lithospheric mantle structure and the diamond potential of kimberlites in southern DR Congo. Lithos **112**(1), 166–176 (2009). https://doi.org/10.1016/jlithos200904020
13. Ryan, C.G., Griffin, W.L., Pearson, N.J.: Garnet geotherms: pressure-temperature data from Cr-pyrope garnet xenocrysts in volcanic rocks. J. Geophys. Res. Solid Earth **101**, 5611–5625 (1996)
14. Gaul, O.F., Griffin, W.L., O'Reilly, S.Y., Pearson, N.J.: Mapping olivine composition in the lithospheric mantle. Earth Planet Sci. Lett. **182**, 223–235 (2000). https://doi.org/10.1016/S0012-821X(00)00243-0
15. Logvinova, A.M., Taylor, L.A., Floss, C., Sobolev, N.V.: Geochemistry of multiple diamond inclusions of harzburgitic garnets as examined in situ. Int. Geol. Rev. **47**, 1223–1233 (2005). https://doi.org/10.2747/0020-681447121223
16. Lavrent'ev Yu, G., Korolyuk, V.N., Usova, L.V., Nigmatulina, E.N.: Electron probe microanalysis of rock-forming minerals with a JXA-8100 electron probe microanalyzer. Russian Geol. Geophys. **56**, 1428–1436 (2015). https://doi.org/10.1016/jrgg201509005

17. Ashchepkov, I.V.: Program of the mantle thermometers and barometers: usage for reconstructions and calibration of PT methods. Vestnik Otdelenia nauko Zemle RAN **3**, NZ6008 (2011). https://doi.org/10.2205/2011NZ000138
18. Gudmundsson, G., Wood, B.J.: Experimental tests of garnet peridotite oxygen barometry. Contrib. Mineral. Petrol. **119**, 56–67 (1995). https://doi.org/10.1007/BF00310717
19. Taylor, W.R., Kammerman, M., Hamilton, R.: New thermometer and oxygen fugacity sensor calibrations for ilmenite and chromium spinel-bearing peridotitic assemblages. In: 7th International Kimberlite Conference Extended abstracts. Cape town, pp. 891–901 (1998)
20. Ashchepkov, I.V., Ivanov, A.S., Kostrovitsky, S.I., Vavilov, M.A., Babushkina, S.A., Vladykin, N.V., Tychkov, N.S., Medvedev, N.S.: Mantle terranes of the Siberian craton: their interaction with plume melts based on thermobarometry and geochemistry of mantle xenocrysts. Geodyn. Tectonophys. **10**(2), 197–245 (2019). https://doi.org/10.5800/GT-2019-10-2-0412
21. Kostrovitsky, S.I., Morikiyo, T., Serov, I.V., Yakovlev, D.A., Amirzhanov, A.A.: Isotope-geochemical systematics of kimberlites and related rocks from the Siberian Platform. Russian Geol. Geophys. **48**, 272–290 (2007). https://doi.org/10.1016/jrgg200702011
22. Agashev, A.M., Ionov, D.A., Pokhilenko, N.P., Golovin, A.V., Cherepanova, Yu., Sharygin, I.S.: Metasomatism in lithospheric mantle roots: constraints from whole-rock and mineral chemical composition of deformed peridotite xenoliths from kimberlite pipe Udachnaya. Lithos **160–161**, 201–215 (2013). https://doi.org/10.1016/jlithos201211014
23. Egorov, K.N., Solov'eva, L.V., Kovach, V.P., Men'shagin Yu, V., Maslovskaya, M.N., Sekerin, A.P., Bankovskaya, E.V.: Petrological features of olivine-phlogopite lamproites of the Sayan region: evidence from Sr-Nd isotope and ICP-MS trace-element data. Geochem. Int. **44**, 729–735 (2006). https://doi.org/10.1134/S0016702906070093
24. Smelov, A.P., Andreev, A.P., Altukhova, Z.A., Babushkina, S.A., Bekrenev, K.A., Zaitsev, A.I., Izbekov, E.D., Koroleva, O.V., Mishnin, V.M., Okrugin, A.V., Oleinikov, O.B., Surnin, A.A.: Kimberlites of the Manchary pipe: a new kimberlite field in Central Yakutia. Russ. Geol. Geophys. **51**, 121–126 (2010). https://doi.org/10.1016/jrgg200912012
25. Garanin, V.K., Zvezdin, A.B., Okrugin, G.V.: Mineralogy of oxide minerals from kimberlites of the Morkoka pipe in connection with the assessment of its diamond content (Yakutian diamondiferous province). Bulletin of the Moscow University Series 4. Geology **4**, 39–46 (1998)
26. Shatsky, V.S., Zedgenizov, D.A., Ragozin, A.L., Kalinina, V.V.: Diamondiferous subcontinental lithospheric mantle of the northeastern Siberian Craton: Evidence from mineral inclusions in alluvial diamonds. Gondwana Res. **28**, 106–120 (2015). https://doi.org/10.1016/jgr201403018
27. Tolstov, A.V., Minin, V.A., Vasilenko, V.B., Kuznetsova, L.G., Razumov, A.N.: A new body of highly diamondiferous kimberlites in the Nakyn field of the Yakutian kimberlite province. Russ. Geol. Geophys. **50**, 162–173 (2009). https://doi.org/10.1016/jrgg200800901
28. Spetsius, Z.V.: Petrology of highly aluminous xenoliths from kimberlites of Yakutia. Lithos **77**, 525–538 (2004)
29. Ashchepkov, I.V., Logvinova, A.M., Ntaflos, T., Vladykin, N.V., Downes, H.: Alakit and Daldyn kimberlite fields, Siberia, Russia: Two types of mantle sub-terranes beneath central Yakutia? Geosci. Front. **8**, 671–692 (2017). https://doi.org/10.1016/jgsf201608004
30. Ashchepkov, I.V., Vladykin, N.V., Ntaflos, T., Downes, H., Mitchell, R., Smelov, A.P., Alymova, N.V., Kostrovitsky, S.I., Rotman, A.Y., Smarov, G.P., Makovchuk, I.V., Stegnitsky, Y.B., Nigmatulina, E.N., Khmelnikova, O.S.: Regularities and mechanism of formation of the mantle lithosphere structure beneath the Siberian Craton in comparison with other cratons. Gondwana Res. **23**, 4–24 (2013). https://doi.org/10.1016/jgr201203009
31. Ashchepkov, I.V., Ntaflos, T., Kuligin, S.S., Malygina, L.V., Mityukhin, S.I., Vladykin, N. V., Palessky, S.V., Khmelnikova, O.S.: Deep seated xenoliths in Udachnaya pipe from the brown breccia. In: Pearson, D. et al., (eds.) Proceedings of 10th International Kimberlite

Conference. Springer, New Delhi, pp. 59–73 (2013). https://doi.org/10.1007/978-81-322-1170-9_5

32. Ashchepkov, I.V., Alymova, N.V., Logvinova, A.M., Vladykin, N.V., Kuligin, S.S., Mityukhin, S.I., Downes, H., Stegnitsky, Y.B., Prokopiev, S.A., Salikhov, R.F., Palessky, S. V., Khmelnikova, O.S.: Picroilmenites in Yakutian kimberlites: Variations and genetic models. Solid Earth **5**, 915–938 (2014). https://doi.org/10.5194/se-5-915-2014
33. Ashchepkov, I.V., Logvinova, A.M., Reimers, L.F., Ntaflos, T., Khmel'nikova, O.S.: The Sytykanskaya kimberlite pipe: evidence from deep-seated xenoliths and xenocrysts for the evolution of the mantle beneath Alakit Yakutia. Russia Geosci. Front. **6**, 687–714 (2013). https://doi.org/10.1016/jgsf201408005
34. Ashchepkov, I.V., Ntaflos, T., Logvinova, A.M., Spetsius, Z.V., Vladykin, N.V.: Monomineral universal clinopyroxene and garnet barometers for peridotitic, eclogitic and basaltic systems. Geosci. Front. **8**, 775–795 (2017a). https://doi.org/10.1016/jgsf201606012
35. Ashchepkov, I.V., Ntaflos, T., Spetsius, Z.V., Salikhov, R.F., Downes, H.: Interaction between protokimberlite melts and mantle lithosphere: Evidence from mantle xenoliths from the Dalnyaya kimberlite pipe, Yakutia (Russia). Geosci. Front. **8**, 693–710 (2017b). https://doi.org/10.1016/jgsf201605008
36. Ashchepkov, I.V., Pokhilenko, N.P., Vladykin, N.V., Logvinova, A.M., Kostrovitsky, S.I., Afanasiev, V.P., Pokhilenko, L.N., Kuligin, S.S., Malygina, L.V., Alymova, N.V., Khmelnikova, O.S., Palessky, S.V., Nikolaeva, I.V., Karpenko, M.A., Stagnitsky, Y.B.: Structure and evolution of the lithospheric mantle beneath Siberian craton, thermobarometric study. Tectonophysics **485**, 17–41 (2010). https://doi.org/10.1016/jtecto200911013
37. Sobolev, N.V.: Deep-seated inclusions in kimberlites and the problem of the composi-tion of the upper mantle American Geophysscal Union, p. 422. Washington, DC (1977)
38. Pokhilenko, N.P., Pearson, D.G., Boyd, F.R., Sobolev, N.V.: Megacrystalline dunites: sources of Siberian diamonds Carnegie Institute Washington Yearbook **90**, 11–18 (1991)
39. Pokhilenko, L.N., Mal'kovets, V.G., Kuz'min, D.V., Pokhilenko, N.P.: New data on the mineralogy of megacrystalline pyrope peridotite from the Udachnaya kimberlite pipe, Siberian Craton, Yakutian diamondiferous province Doklady. Earth Sci. **454**, 179–184 (2014). https://doi.org/10.1134/S1028334X14020159
40. Pokhilenko, N.P., Sobolev, N.V., Kuligin, S.S., Shimizu, N.: Peculiarities of distribution of pyroxenite paragenesis garnets in Yakutian kimberlites and some aspects of the evolution of the Siberian craton lithospheric mantle. In: Proceedings of the VII International Kimberlite Conference The PH Nixon volume, pp. 690–707 (1999)
41. Pokhilenko, N.P., Agashev, A.M., Litasov, K.D., Pokhilenko, L.N.: Carbonatite metasomatism of peridotite lithospheric mantle: implications for diamond formation and carbonatite-kimberlite magmatism. Russ. Geol. Geophys. **56**, 280–295 (2015). https://doi.org/10.1016/jrgg201501020
42. Jagoutz, E., Lowry, D., Mattey, D., Kudrjavtseva, G.: Diamondiferous eclogites from Siberia: Remnants of Archean oceanic crust. Geochim. Cosmochim. Acta **58**, 5195–5207 (1994). https://doi.org/10.1016/0016-7037(94)90304-2
43. Bascou, J., Doucet, L.S., Saumet, S., Ionov, D.A., Ashchepkov, I.V., Golovin, A.V.: Seismic velocities, anisotropy and deformation in Siberian cratonic mantle: EBSD data on xenoliths from the Udachnaya kimberlite. Earth Planet. Sci. Lett. **304**, 71–84 (2011). https://doi.org/10.1016/jepsl201101016
44. Ionov, D.A., Doucet, L.S., Xu, Y., Golovin, A.V., Oleinikov, O.B.: Reworking of archean mantle in the NE Siberian craton by carbonatite and silicate melt metasomatism: evidence from a carbonate-bearing, dunite-to-websterite xenolith suite from the Obnazhennaya kimberlite Geochimica et Cosmochimica. Acta **224**, 132–153 (2018). https://doi.org/10.1016/jgca201712028
45. Boyd, F.R., Pokhilenko, N.P., Pearson, D.G., Mertzman, S.A., Sobolev, N.V., Finger, L.W.: Composition of the Siberian cratonic mantle: evidence from Udachnaya peridotite xenoliths. Contrib. Mineral Petrol **128**, 228–246 (1997). https://doi.org/10.1007/s004100050305

46. Ionov, D.A., Doucet, L.S., Ashchepkov, I.V.: Composition of the Lithospheric Mantle in the Siberian Craton: new Constraints from Fresh Peridotites in the Udachnaya-East Kimberlite. J. Petrol. **51**, 2177–2210 (2010). https://doi.org/10.1093/petrology/egq053
47. Doucet, L.S., Ionov, D.A., Golovin, A.V., Pokhilenko, N.P.: Depth, degrees and tectonic settings of mantle melting during craton formation: inferences from major and trace element compositions of spinel harzburgite xenoliths from the Udachnaya kimberlite, central Siberia. Earth Planet. Sci. Lett. **359–360**, 206–218 (2012). https://doi.org/10.1016/jepsl201210001
48. Brey, G.P., Kohler, T.: Geothermobarometry in four-phase lherzolites II New thermobarometers, and practical assessment of existing thermobarometers. J. Petrol. **31**, 1353–1378 (1990). https://doi.org/10.1093/petrology/3161353
49. McGregor I.D.: The system MgO-SiO2-Al2O3: solubility of Al2O3 in enstatite for spinel and garnet peridotite compositions. Am. Mineral. **59**, 110–119 (1974). http://www.minsocam.org/ammin/AM59/AM59_110.pdf
50. Nimis, P., Taylor, W.: Single clinopyroxene thermobarometry for garnet peridotites Part I Calibration and testing of a Cr-in-Cpx barometer and an enstatite-in-Cpx thermometer. Contrib. Mineral. Petrol. **139**, 541–554 (2000). https://doi.org/10.1007/s004100000156
51. O'Neill, H.S.T.C., Wood, B.J.: An experimental study of Fe-Mg-partitioning between garnet and olivine and its calibration as a geothermometer. Contrib. Mineral. Petrol. **70**, 59–70 (1979)
52. Krogh, E.J.: The garnet-clinopyroxene Fe-Mg geothermometers: a reinterpretation of existing experimental data. Contrib. Mineral. Petrol. **99**, 44–48 (1988). https://doi.org/10.1007/BF00399364
53. O'Neill, H.S.C., Wall, V.J.: The olivine orthopyroxene-spinel oxygen geobarometer, the nickel precipitation curve, and the oxygen fugacity of the Earth's upper mantle. J. Petrol. **8**, 1169–1191 (1987). https://doi.org/10.1093/petrology/28.6.1169
54. Pernet-Fisher, J.F., Howarth, G.H., Liu, Y., Barry, P.H., Carmody, L., Valley, J.W., Bodnar, R.J., Spetsius, Z.V., Taylor, L.A.: Komsomolskaya diamondiferous eclogites: evidence for oceanic crustal protoliths. Contrib. Mineral. Petrol. **167**, 981 (2014). https://doi.org/10.1007/s00410-014-0981-y
55. Sobolev, N.V., Pustyntsev, V.I., Kuznetsova, I.K., Khar'kiv, A.D.: New data on the mineralogy of the diamond-bearing eclogites from the "Mir" pipe (Yakutia). Int. Geol. Rev. **12**, 657–659 (1970)
56. Spetsius, Z.V., Serenko, V.P.: Composition of the Continental Mantle and Low Crust Beneath the Siberian Platform, p. 271. Nauka, Moscow (1990)
57. Riches, A.J.V., Liu, Y., Day, J.M.D., Spetsius, Z.V., Taylor, L.A.: Subducted oceanic crust as diamond hosts revealed by garnets of mantle xenoliths from Nyurbinskaya, Siberia. Lithos **120**, 368–378 (2010). https://doi.org/10.1016/jlithos201009006
58. Spetsius, Z.V., Taylor, L.A., Valley, J.W., Ivanov, A.S., Banzeruk, V.I.: Diamondiferous xenoliths from crustal subduction: Garnet oxygen isotopes from the Nyurbinskaya pipe, Yakutia. Eur. J. Mineral. **20**, 375–385 (2008). https://doi.org/10.1127/0935-1221/2008/0020-1828
59. Ashchepkov, I., Medvedev, N., Vladykin, N., Ivanov, A., Downes, H.: Thermobarometry and geochemistry of mantle xenoliths from Zapolyarnaya pipe, Upper Muna field, Yakutia: implications for mantle layering, interaction with plume melts and diamond grade. Minerals **10**(9), 740 (2020). https://doi.org/10.3390/min10090755
60. Nimis, P., Kuzmin, D.V., Malkovets, V.: Error sources in single-clinopyroxene thermobarometry and a mantle geotherm for the Novinka kimberlite. Yakutia. Am. Mineral. **101** (10), 2222–2232 (2016). https://doi.org/10.2138/am-2016-5540
61. Dymshits, A.M., Sharygin, I.S., Malkovets, V.G., Yakovlev, I.V., Gibsher, A.A., Alifirova, T.A., Vorobei, S.S., Potapov, S.V., Garanin, V.K.: Thermal state, thickness, and composition of the lithospheric mantle beneath the upper muna kimberlite field (siberian craton) constrained by clinopyroxene xenocrysts and comparison with daldyn and mirny fields. Minerals **10**, 549 (2020). https://doi.org/10.3390/min10060549

62. Sun, J., Tappe, S., Kostrovitsky, S.I., Liu, C-Z., Skuzovatov, S.Y., Wu, F-Y.: Mantle sources of kimberlites through time: a U-Pb and Lu-Hf isotope study of zircon megacrysts from the Siberian diamond fields. Chem. Geol. **479**, 228–240 (2018). https://doi.org/10.1016/jchemgeo201801013
63. Kornilova, V.P., Spetsius, Z.V., Pomazanskiy, B.S.: Petrographic-mineralogical peculiarities and feasibility of kimberlite pipes Lorik and Svetlana diamond grade re-estimation (West Ukukit field, Yakutia). Reg. Geol. Metal. **68**, 92–99 (2016)
64. Egorov KN, Kiselev AI, Men'shagin YuV, Minaeva YuA (2010) Lamproite and kimberlite of the Sayany area: Composition, sources, and diamond potential. Doklady Earth Sci. https://doi.org/10.1134/S1028334X10120251
65. Zaitsev, A.I., Smelov, A.P.: Isotope geochronology of kimberlite formation rocks of Yakut province. Institute of Diamond and Noble Metals Geology, Siberian Branch of Russian Academy of Sciences, p. 105
66. Ashchepkov, I.V., Kuligin, S.S., Vladykin, N.V., Downes, H., Vavilov, M.A., Nigmatulina, E.N., Babushkina, S.A., Tychkov, N.S., Khmelnikova, O.S.: Comparison of mantle lithosphere beneath early Triassic kimberlite fields in Siberian craton reconstructed from deep-seated xenocrysts. Geosci. Front. **7**, 639–662 (2016). https://doi.org/10.1016/jgsf201506004
67. Griffin, W.L., Natapov, L.M., O'Reilly, S.Y., van Achterbergh, E., Cherenkova, A.F., Cherenkov, V.G.: The Kharamai kimberlite field, Siberia: modification of the lithospheric mantle by the Siberian Trap event. Lithos **81**, 167–187 (2005). https://doi.org/10.1016/jlithos200410001
68. Afanasiev, V.P., Ashchepkov, I.V., Verzhak, V.V., O' Brien, H., Palessky, S.V.: PT conditions and trace element variations of picroilmenites and pyropes from the Arkhangelsk region. J. Asian Earth Sci. **70–71**, 45–63 (2013). https://doi.org/10.1016/jjseaes201303002
69. Sobolev, N.V., Logvinova, A.M., Nikolenko, E.I., Lobanov, S.S.: Mineralogical criteria for the diamond potential of Upper Triassic placers on the northeastern margin of the Siberian Platform. Russ. Geol. Geophys. **54**, 903–916 (2013). https://doi.org/10.1016/jlithos201607030
70. Taylor, L.A., Gregory, A., Keller, S.R., Remley, D.A., Anand, M., Wiesli, R., Valley, J., Sobolev, N.V.: Petrogenesis of group A eclogites and websterites: evidence from the Obnazhennaya kimberlite, Yakutia. Contrib. Mineral. Petrol. **145**, 424–443 (2003). https://doi.org/10.1007/s00410-003-0465-y
71. Ionov, D.A., Doucet, L.S., Carlson, R.W., Golovin, A.V., Korsakov, A.V.: Post-Archean formation of the lithospheric mantle in the central Siberian craton: Re–Os and PGE study of peridotite xenoliths from the Udachnaya kimberlite. Geochim. Cosmochim. Acta **165**, 466–483 (2015). https://doi.org/10.1016/jgca201506035
72. Solov'eva, L.V., Kalashnikova, T.V., Kostrovitsky, S.I., Ivanov, A.V., Matsuk, S.S., Suvorova, L.F.: Metasomatic and magmatic processes in the mantle lithosphere of the Birekte terrain of the Siberian craton and their effect on the lithosphere evolution. Geodyn. Tectonophys. **6**(3), 311–344
73. Boyd, F.R., Nixon, P.H.: Ultramafic nodules from the Kimberley pipes. South Africa Geochim. Cosmochim. Acta **42**, 1367–1382 (1978). https://doi.org/10.1016/0016-7037(78)90042-X
74. Dawson, J.B.: Kimberlites and their Xenoliths. Springer-Verlag, Berlin, New York (1980)
75. Lazarov, M., Brey, G.P., Weyer, S.: Evolution of the South African mantle—A case study of garnet peridotites from the Finsch diamond mine (Kaapvaal craton); part 1: inter-mineral trace element and isotopic equilibrium. Lithos **154**, 193–209 (2012). https://doi.org/10.1016/jlithos201207013
76. Stachel, T., Viljoen, K.S., McDade, P., Harris, J.W.: Diamondiferous lithospheric roots along the western margin of the Kalahari Craton the peridotitic inclusion suite in diamonds from Orapa and Jwaneng. Contrib. Mineral. Petrol **147**, 32–47 (2004). https://doi.org/10.1007/s00410-003-0535-1

77. Manning, C.E.: The chemistry of subduction-zone fluids. Earth Planet Sci. Lett. **223**, 1–16 (2004). https://doi.org/10.1016/j.epsl.2004.04.030
78. Evensen, N.M., Hamilton, P.J., Onions, R.K.: Rare-earth abundances in chondritic meteorites. Geochim. Cosmochim. Acta **42**, 1199–1212 (1979). https://doi.org/10.1016/0016-7037(78)90114-X
79. McDonough, W.F., Sun, S.S.: The composition of the earth. Chem. Geol. **120**, 223–253 (1995). https://doi.org/10.1016/0009-2541(94)00140-4
80. Grakhanov, A.S., Zarubin, R.A., Bogush, I.N., Yadrenkin, A.B.: Discovery of upper triassic diamond placers in the olenek bay of the laptev sea. Otechestvennaya Geol. **1**, 53–61 (2009)
81. Pavlenkova, N.I.: Seismic structure of the upper mantle along the long-range PNE profiles —rheological implication. Tectonophysics **508**, 85–95 (2011). https://doi.org/10.1016/jtecto201011007
82. Kuskov, O.L., Kronrod, V.A., Prokofyev, A.A., Pavlenkova, N.I.: Thermo-chemical structure of the lithospheric mantle underneath the Siberian craton inferred from long-range seismic profiles. Tectonophysics **615–616**, 154–166 (2014). https://doi.org/10.1016/jtecto201401006
83. McKenzie, D., Priestley, K.: The influence of lithospheric thickness variations on continental evolution. Lithos **102**, 1–11 (2008). https://doi.org/10.1016/jlithos200705005
84. Gladkochub, D.P., Donskay, T.V., Stanevich, A.M., Pisarevsky, S.A., Zhang, S., Motov, Z. L., Mazukabzov, A.M., Li, H.: U-Pb detrital zircon geochronology and provenance of Neoproterozoic sedimentary rocks in southern Siberia: new insights into breakup of Rodinia and opening of Paleo-Asian Ocean. Gondwana Res. **65**, 1–16 (2019). https://doi.org/10.1016/j.gr.2018.07.007
85. Van Hunen, J., van den Berg, A.P.: Plate tectonics on the early Earth: Limitations imposed by strength and buoyancy of subducted lithosphere. Lithos **103**, 217–235 (2008). https://doi.org/10.1016/jlithos200709016
86. Smelov, A.P., Kotov, A.B., Sal'nikova, E.B., Kovach, V.P., Beryozkin, V.I., Kravchenko, A.A., Dobretsov, V.N., Velikoslavinskii, S.D., Yakovlev, S.Z.: Age and duration of the formation of the billyakh tectonic melange zone, anabar shield. Petrology **30**, 286–301 (2012). https://doi.org/10.1134/S0869591112030058
87. Malkovets, V.G., Griffin, W.L., O'Reilly, S.Y., Wood, B.J.: Diamond, subcalcic garnet, and mantle metasomatism: kimberlite sampling patterns define the link. Geology **35**, 339–342 (2007). https://doi.org/10.1130/G23092A1
88. Koreshkova, M., Yu, Downes H., Nikitina, L.P., Vladykin, N.V., Larionov, A.N., Sergeev, S.A.: Trace element and age characteristics of zircons in granulite xenoliths from the Udachnaya kimberlite pipe, Siberia. Precambr. Res. **168**, 197–212 (2009). https://doi.org/10.1016/jprecamres200809007
89. Parkinson, I.J., Pearce, J.A.: Peridotites from the Izu-Bonin-Mariana forearc (ODP Leg 125): evidence for mantle melting and melt-mantle interaction in a supra-subduction zone setting. J. Petrol. **39**, 1577–1618 (1998). https://doi.org/10.1093/petroj/3991577
90. O'Reilly, S.Y., Zhang, M., Griffin, W.L., Begg, G., Hronsky, J.: Ultradeep continental roots and their oceanic remnants: a solution to the geochemical "mantle reservoir" problem? Lithos **112**, 1043–1054 (2009)
91. Griffin, W.L., O'Reilly, S.Y., Abe, N., Aulbach, S., Davies, R.M., Pearson, N.J., Doyle, B.J., Kivi, K.: The origin and evolution of the Archean lithospheric mantle. Precambr. Res. **127**, 19–41 (2003). https://doi.org/10.1016/S0301-9268(03)00180-3
92. Pearson, D.G.: The age of continental roots. Lithos **48**, 171–194 (1999). https://doi.org/10.1016/S0024-4937(99)00026-2
93. Pearson, D.G., Snyder, G.A., Shirey, S.B., Taylor, L.A., Carlson, R.W., Sobolev, N.V.: Archaean Re-Os age for Siberian eclogites and constraints on Archaean tectonics. Nature **374**, 711–713 (1995). https://doi.org/10.1038/374711a0
94. Santosh, M., Maruyama, S., Yamamoto, A.: The making and breaking of supercontinents: Some speculations based on superplumes, super downwelling and the role of tectosphere. Gondwana Res. **15**, 324–341 (2009). https://doi.org/10.1016/jgr200811004

95. Perchuk, A.L., Safonov, O.G., Smit, C.A., van Reenen, D.D., Zakharov, V.S., Gerya, T.V.: Precambrian ultra-hot orogenic factory: making and reworking of continental crust. Tectonophysics **746**, 572–586 (2018)
96. Gerya, T.: Precambrian geodynamics: concepts and models Gondwana Research **25**, 442–463 (2014). https://doi.org/10.1016/jgr201211008
97. Smith, C.B., Pearson, D.G., Bulanova, G.P., Beard, A.D., Carlson, R.W., Wittig, N., Sims, K., Chimuka, L., Muchemwa, E.: Extremely depleted lithospheric mantle and diamonds beneath the southern Zimbabwe Craton. Lithos **112**, 1120–1132 (2009). https://doi.org/10.1016/jlithos200905013
98. Snyder, D.B.: Stacked uppermost mantle layers within the Slave craton of NW Canada as defined by anisotropic seismic discontinuities. Tectonics **27**, TC4006 (2008). https://doi.org/10.1029/2007TC002132
99. Condie, K.C.: Supercontinents and superplume events: distinguishing signals in the geologic record. Phys. Earth Planet. Inter. **146**, 319–332 (2004). https://doi.org/10.1016/jpepi200304002
100. Herzberg, C.: Geodynamic Information in Peridotite Petrology. J. Petrol. **45**, 2507–2530 (2004). https://doi.org/10.1093/petrology/egh039
101. Nimis, P., Zanetti, A., Dencker, I., Sobolev, N.V.: Major and trace element composition of chromian diopsides from the Zagadochnaya kimberlite (Yakutia, Russia): Metasomatic processes, thermobarometry and diamond potential. Lithos **112**, 397–412 (2009). https://doi.org/10.1016/jlithos200903038
102. Pernet-Fisher, J.F., Howarth, G.H., Pearson, D.G., Woodland, S., Barry, P.H., Pokhilenko, N.P., Pokhilenko, L.N., Agashev, A.M., Taylor, L.A.: Plume impingement on the Siberian SCLM: evidence from Re-Os isotope systematics. Lithos **218**, 141–154 (2015). https://doi.org/10.1016/jlithos201501010
103. Roden, M.F., Patiño-Douce, A.E., Jagoutz, E., Laz'ko, E.E.: High pressure petrogenesis of Mg-rich garnet pyroxenites from Mir kimberlite. Russia Lithos **90**(1–2), 77–91 (2006). https://doi.org/10.1016/jlithos200601005
104. Ernst, W.G.: Earth's thermal evolution, mantle convection, and Hadean onset of plate tectonics. J. Asian Earth Sci. **145**, 1334–1348 (2017). https://doi.org/10.1016/jjseaes201705037
105. Pollack, H.N., Chapman, D.S.: On the regional variation of heat flow, geotherms and lithospheric thickness. Tectonophysics **38**, 279–296 (1977)
106. Wyllie, P.J., Ryabchikov, I.D.: Volatile components, magmas and critical fluids in upwelling mantle. J. Petrol. **41**, 1195–1206 (2000). https://doi.org/10.1093/petrology/4171195
107. Foley, S.F., Yaxley, G.M., Rosenthal, A., Buhre, S., Kiseeva, E.S., Rapp, R.P., Jacob, D.E.: The composition of near-solidus melts of peridotite in the presence of CO_2 and H_2O between 40 and 60 kbar. Lithos **112**(S1), 274–283 (2009). https://doi.org/10.1016/jlithos200903020
108. Korolev, N.M., Melnik, A.E., Li, X.-H., Skublov, S.G.: The oxygen isotope composition of mantle eclogites as a proxy of their origin and evolution: a review. Earth Sci. Rev. **185**, 288–300 (2018)
109. Agee, C.B.: Crystal-liquid density inversions in terrestrial and lunar magmas. Phys. Earth Planet. Inter. **107**, 63–74 (1998). https://doi.org/10.1016/S0031-9201(97)00124-6
110. Karato, S.: Rheology of the Earth's mantle: a historical review. Gondwana Res. **18**, 17–45 (2010). https://doi.org/10.1016/jgr201003004
111. Tappe, S., Foley, S.F., Jenner, G.A., Heaman, L.M., Kjarsgaard, B.A., Romer, R.L., Stracke, A., Joyce, N., Hoefs, J.: Genesis of ultramafic lamprophyres and carbonatites at AillikBay, Labrador: a consequence of incipient lithospheric thinning beneath the North Atlantic craton. J. Petrol. **47**, 1261–1315 (2006). https://doi.org/10.1093/petrology/egl008

Zhidoy Alkali-Ultramafic Rock and Carbonatite Massif: Geochemical Features, Its Sources Aednd Ore-Bearing

N. V. Vladykin, I. A. Sotnikova, and N. V. Alymova

Abstract

The article considers geological position of the Zhidoy massif and its age. The scheme of magmatism of the massif has been developed. The graphs of paired correlations of petrogenic elements in massif rocks which had consistent trend in composition are given for validation purposes. The present article provides graphs of REE spectra and the spider diagram of rare elements concentration in the massif rocks. Pyroxenites are the early rocks of the massif, which are the ores for titanium. Titanium is concentrated in three minerals: titanium magnetite, ilmenite and perovskite. The main type of titanium ores, perovskite, is known only in the Zhidoy massif. A conclusion about the mantle sources of the primary magma of the massif is drawn based on the geochemistry of the isotopes Sr and Nd.

Keywords

Zhidoi massif · Carbonatites · Pyroxenites · Perovskite · Titanium deposit

N. V. Vladykin (✉) · I. A. Sotnikova · N. V. Alymova
Vinogradov Institute of Geochemistry SB RAS, 1A Favorsky Str., Irkutsk 664033, Russia
e-mail: vlad@igc.irk.ru

Irkutsk Scientific Center of Siberian Branch of Russian, 134 Lermontov St., Irkutsk 664033, Russia

N. Vladykin (ed.), *Alkaline Rocks, Kimberlites and Carbonatites: Geochemistry and Genesis*, Springer Proceedings in Earth and Environmental Sciences,
https://doi.org/10.1007/978-3-030-69670-2_3

1 Introduction

Titanium is a strategic element required by industry. Its concentrate used to be supplied to Russia from Ukraine, and there is now a large shortage of this raw material. In Russia, its deposits are represented by titanium-magnetite, from which it is difficult to extract titanium. In Irkutsk region the deposits of titanium on the Murun massif [1] have been discovered in recent years, where they are represented by quartz zones with pure TiO_2–brookite, anatase and rutile, and titanium deposit in the Zhidoy massif, represented by pyroxenites with perovskite, ilmenite and titanium-magnetite. The Murun massif is located at hard-to-reach area of the Irkutsk region, and the Zhidoy array is 80 km from Irkutsk, near the Taliyany village (20 km from Angarsk town) [2]. Geological work is currently being carried out on Zhidoysky massif for the reserves estimation by the company Joint-Stock Company « Tiamin Resource Baikal». The ores of the Zhidoy massif are the magmatic pyroxenites of the main phase, occupying 95% of the area of the massif and coming to the surface. Besides the pyrocenites, there are residential bodies of illites, alkali and nepheline sienites and carbonatites. In present a 60 m carbonatite body with rare-metal circon-pyrochlor mineralization has now been found during the drilling, as at Beloziminskiy carbonatite deposit, the province of which includes Zhidoy massif. Therefore, carbonatites may also be of industrial interest as sources of rare elements.

For unknown reason Zhidoy massif began to be called Zadoy in publication [3]. This leads to a great confusion, as there is also an ultramafic Zadoy massif in PreSayan. Thus, it is necessary to return the old name Zhidoy to the massif.

2 Geological Structure of the Zhidoy Massif

Zhidoy massif is located 80 km east of Irkutsk, on the Nephelinovy stream, at its outlet in Big Zhidoy River [2, 4]. Zhidoy massif occupies an area of 0.85 km^2 and has a horseshoe shape with a length at the center of the intrusion of 2 km at a width of 600 m. According to the geophysical data of the rock, the massif extends up to Big Zhidoy River. We have developed a new layout of the structure of the Zhidoy massif [5], presented in Fig. 1.

The massif is located in the early Precambrian Sharzhalgai block of the PreSayan protrusion of the foundation of the Siberian platform. It is breaking through rogue biotit nests. In the northern part of the massif in the exocontactant part, the phenytes are formed, reveled by a borehole drilled in the adjacent embedded rocks. The alkaline-ultramafic carbonate formation massif is up of 90% ore pyroxenites with different contents of titanomagnetite, ilmenite and perovskite. It is considered [2, 3] that four phases of implemention in the massif–pyroxenites, ijolites, sienites and calcite carbonatites, however the last three species of rocks form small thickness dykes and can be considered as dykes of additional intrusions. On the massif there are four varieties of pyroxene, with different forms of Ti-ore, which gradually

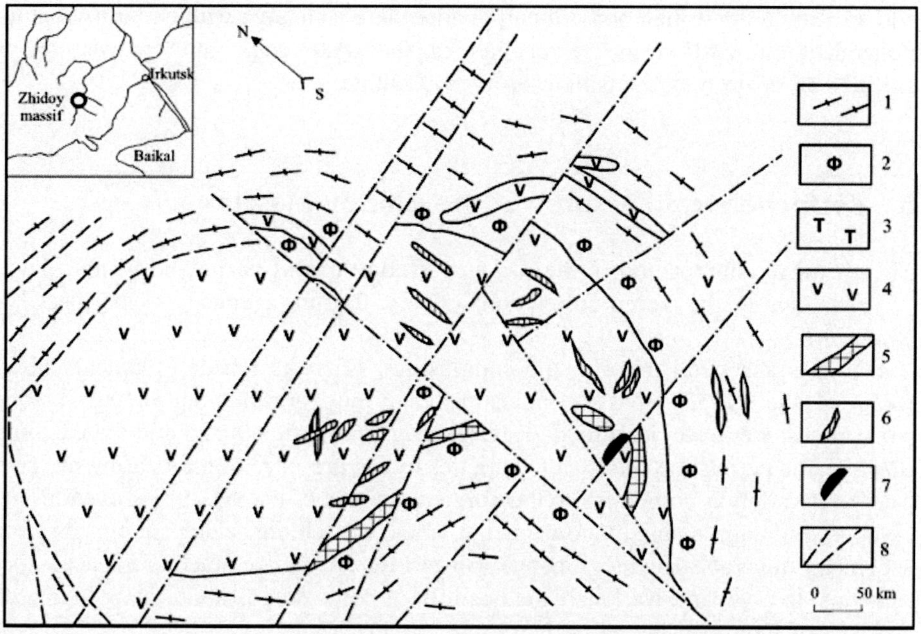

Fig. 1 New scheme of Zhidoy massif structure. Legend: 1−breed, 2−phenytes, 3−lamprophyre dykes, 4−pyroxenites, 5−ijolite bodies, 6−syenite dykes, 7−carbonatite dykes, 8−faults

alternate. These are the following: 1−pyroxenites with titanium-magnetite (divided into magnetite and ilmenite), without perovskite; 2−pyroxenites with titanium-magnetite (divided into magnetite and ilmenite) and individual secretions of ilmenite; 3−pyroxenites with titanium-magnetite, ilmenite, and enriched with perovskite up to 7%; 4−pyroxenite titanit containing with small amount of titanium-magnetite and without perovskite (not ore). Pyroxenites are fine-grained, indicating the hypabyssal position of the massif. In addition to pyroxene, up to 5% of pyroxenites contain biotite.

Dykes of alkaline and nepheline sienites are crossing the pyroxenite and located in different parts of the massif (core of boreholes 10 and 11).

The thickness of the dykes is 1–2 m and a length of up to 500 m. They are composed of microcline, nepheline with small amounts of pyroxene and biotite. Titanit, apatite, zircon and magnetite were found from the accessory minerals.

Dykes of carbonatites have been found in two places—in the center of the carbonatite massif (20 cm thickness) the digestion of sienite (1 m thickness), which in turn separates pyroxenite and in the southern, peripatetic part cross the ijollite vein body. Here the thickness of carbonatite is 2 m.

The fall of its contact is almost vertical. Near this point, the drilled borehole entered the 60 m carbonatite body. It may have been drilled by the fall of the dyke carbonatite body. Calcite carbonatites contain small amounts of biotite, pyroxene and apatite. In the drilled body of carbonatite there is an area with circon-pyrochlor mineralization, and in the lower part of the dyke carbonatite passes into a silicate-carbonate rock containing up to 50% albite.

3 Petrochemical Features of the Zhidoy Massif

The chemical composition of the rocks of Zhidoy massif is shown in Table 1.

A diagram of the paired correlations of rock forming elements of the massif is given in Fig. 2.

There is a general trend in the composition of rocks forming elements in all rocks of the Zhidoy massif, indicating that they are homodromous. Among pyroxenites we have identified 4 facies–containing titanomagnetite, containing ilmenite and perovskite and titanite varieties by the presence of ore-bearing mineral. The latter variety is coinciding to the northern contact of the massif. Pyroxenites are composed of augitte, and titanomagnetite, mica and seldom calcite are present from secondary minerals. Titanite, ilmenite, perovskite, apatite are noted of the accessory minerals. In pegmatite varieties, the quantity of perovskite is abnormally increased and it is of industrial interest [5].

4 Geochemical Features of the Zhidoy Massif Rocks

The concentrations of rare elements in the rocks of the massif were studied by the ICP-MC method (IGC SB RAS, analyst Mitrofanova A.Yu.), presented in Table 2. The spectra of rare earth elements are shown in Fig. 2.

As can be seen from the diagram, spectra of rare-earth elements for rocks of the massif are identical, which also indicates their genetic connection. The highest concentrations are observed in perovskite pyroxenites and the lowest in sienites. In sienites, beginning with holmium, concentrations of heavy lanthanides have increased. Europia fractionation is almost no observed (Fig. 3).

Spider-diagrams are shown at Fig. 4. The behavior of rare elements in pyroxenites and ijolites is similar. They have negative anomalies of lead, strontium and zirconium with hafnium and rather gentle curves of rare earth elements. There are other trends in sienites–sharp positive anomalies of strontium, lead and zirconium with hafnium, which is associated with the process of differentiation and accumulation of these elements in later rocks, the sienites. The distribution of rare elements in phenites differs from their distribution in magmatic rocks.

Table 1 Chemical Composition of Zidoy massif rock

	SiO_2	TiO_2	Al_2O_3	Fe_2O_3	MnO	MgO	CaO	Na_2O	K_2O	P_2O_5
Pyroxenite										
11/45	26,97	8,6	3,37	16,26	0,17	8,95	20,26	0,9	0,86	3,73
11/72	27,36	9,04	4,78	15,86	0,19	5,88	16,28	0,9	1,76	2,32
11/285	28,99	6,54	7,06	16,01	0,17	9,49	14,95	1,1	2,77	3,38
11/80	27,57	8,49	5,42	17,79	0,18	8,61	16,01	0,8	2,04	3,09
10/35a	22,02	11,6	1,88	37,96	0,29	6,77	10,28	0,4	0,28	1,05
11/270	35,78	5,33	7,94	17,86	0,3	5,66	13,31	2,2	1,48	1,56
11/228	30,01	7,6	3,75	21,89	0,25	5,69	15,58	1	0,78	2,48
10/300	31,66	5,51	9,89	15,62	0,27	5,12	15,41	3,3	2,24	1,86
11/559	37,92	5,39	9,89	14,02	0,3	3,29	12,83	2,2	2,8	1,15
11/698	28,9	3,88	7,79	15,07	0,37	7,19	17,94	1,8	0,72	1,61
10/350	30,84	5,63	8,89	16,32	0,29	4,38	15,92	3,3	1,3	2,05
10/417v	31,97	4,89	6,61	13,44	0,24	5,7	19,4	2,2	0,16	2,12
10/126a	36,44	2,25	21,3	10,32	0,22	2,4	10,21	12	3,89	0,28
10/630	35,75	3,55	14,5	12,25	0,23	4,26	16,04	6,2	0,11	1,08
10/870	18,23	8,1	0,85	28,81	0,25	6,71	20,92	0,4	0,08	8,45
10/417г	31,97	6,56	2,45	15,83	0,3	6,19	20,28	0,7	0,03	1,81
10/756	33,28	4,05	6,29	17,61	0,25	9,19	14,31	1,3	1,15	0,83
10/520	31,89	5,93	1,93	20,65	0,2	11,6	19,24	0,7	0,2	2,59
10/128	28,89	7,54	4,22	15,94	0,19	11,7	18,04	0,8	1,58	3,46
10/950	30,86	6,56	2,64	17,81	0,18	11,3	20,04	0,6	0,28	3,79
11/87	26,53	11,9	0,56	16,97	0,23	4,98	16,91	0,7	0,14	3,17
10/35	26,72	9,4	1,83	16,89	0,15	8,24	21,14	0,4	0,16	3,87
10/114	28,31	8,22	3,81	17,73	0,22	10,5	18,86	0,8	1,36	3,56
11/200	28,89	8,57	3,53	18,63	0,2	7,02	16,83	1	0,58	2,97
11/240	30,76	7,3	5,47	17,99	0,2	7,67	14,69	1,2	1,66	2,7
316/7	39,02	4,22	9,48	10,04	0,42	3,86	15,9	2,5	0,72	0,99
10/420	33,73	5,64	4,02	13,09	0,27	6,8	21,6	1,5	0,1	2,29
10/550	32,07	2,36	12,5	7,78	0,14	10	25,75	0,8	0,33	6,61
10/730	36	3,81	6,94	17,88	0,25	10,3	14,89	1,8	0,91	1,05
328	29,85	11,2	4,72	18,38	0,23	9,64	22,54	0,8	0,04	2,33
330	36,38	0,45	0,28	52	0,16	4,57	4,72	2,5	0,16	0,29
334	33,53	5,87	4,19	21,18	0,22	10,9	20,29	0,9	0,24	2,77
335	34,01	5,63	4,12	21,01	0,22	11,1	20,3	0,9	0,18	2,57
315/10	45,4	0,84	7,45	12,81	0,18	23,6	6,31	1	0,71	0,27
10/450	33,46	5,54	4,46	21,45	0,23	10,6	20,01	1	0,36	2,88
10/733	39,66	3,68	4,86	19,89	0,29	9,88	18,79	1,3	0,37	1,05
Ijolite										
10/375	43,48	1,98	10,7	14,26	0,32	8,78	14,03	3,2	1,36	0,66
315/22	43,39	4,65	13,5	13,97	0,31	4,34	11,12	3,8	2,28	1
331	43,6	3,76	14,2	12,18	0,38	3,13	10,77	4,9	3,51	0,59
11/460	41,15	3,38	9,02	17,69	0,38	4,72	11,98	2,7	3,02	1,51
326	38,79	3,12	15,2	13,08	0,27	4,16	12,17	7,6	2,87	1,13
315/11	48,95	1,33	12,8	14,35	0,19	10,9	8	2,2	0,89	0,5

(continued)

Table 1 (continued)

	SiO_2	TiO_2	Al_2O_3	Fe_2O_3	MnO	MgO	CaO	Na_2O	K_2O	P_2O_5
315/17	48,03	0,15	27,7	2,81	0,04	2,6	12,69	2,1	3,65	0,02
10/126	39	3,13	18,2	10,92	0,21	2,27	10,05	10	0,72	0,38
10/417	38,28	2,97	17,9	11,29	0,19	3,01	11,8	4,6	2,38	0,84
11/496a	45,21	3,91	13,1	13,25	0,39	2,68	10,98	4,8	1,95	0,72
11/407	47,48	0,39	13,8	2,89	0,02	0,9	9,37	4,5	5	0,11
Syenite										
10/325	53,45	0,91	15,4	7,82	0,12	5,68	5,74	4,3	3,01	0,31
11/405	52,72	0,98	15,2	6,05	0,07	2,32	5,76	5,3	4,89	0,34
11/513	51,52	2,23	13,5	11,82	0,14	3,98	6,9	4,5	3,19	0,13
11/700	51,48	1,19	14,3	9,71	0,14	6,64	6,65	3,7	2,9	0,41
11/390	52,97	1,44	15,4	9,28	0,18	3,73	3,77	6,2	2,88	0,56
11/525	53	0,58	17,4	6,76	0,16	2,24	4,39	3,7	8,2	0,22
11/532	54,04	0,26	23,5	3,49	0,09	0,16	1,19	7,6	6,91	0,08
11/330	54,84	0,73	16,5	6,3	0,07	2,39	4,67	6,4	3,37	0,24
11/539	57,07	0,71	18,5	4,54	0,12	1,18	3,37	5,5	6,12	0,09
11/580	59,77	0,22	16,9	4,72	0,03	1,28	2,9	4,3	8,29	0,01
10/600	54,06	0,33	19,1	3,75	0,1	0,79	3,26	11	1,03	0,37
11/435	50,6	1,08	12,9	12,69	0,19	6,46	8,45	4,5	1,84	0,1
11/539	57,58	0,7	18,6	4,3	0,11	0,8	3,14	6	5,99	0,09
11/532	53,34	0,26	22,9	3,58	0,08	0,56	1,36	7,9	6,81	0,07
10/163	49,23	0,36	22,9	5,94	0,14	0,85	3,2	9,7	0,43	0,12
10/164	53,21	0,21	19,6	5,79	0,15	1,12	3,7	11	1,23	0,1
10/36	48,31	0,38	22,6	5,42	0,1	0,53	1,36	11	3,57	0,1
10/36a	50,66	0,66	19,8	4,07	0,09	0,93	4,28	6,1	6,32	0,11
10/185	49,9	0,75	18,6	4,92	0,13	1,23	5,27	8,9	2,97	0,5
315/4	53,03	1,46	17,8	6,19	0,13	2,15	6	2	8,22	0,18
315/18	51,26	0,25	23,3	4,64	0,07	3,52	10,71	2,6	2,36	0,04
315/20	55,93	2,71	15,1	14,83	0,22	3,49	7,94	4	0,05	0,94
317/9	55,54	0,90	18,8	5,56	0,16	0,59	3,02	6,6	6,64	0,13
329	53,36	0,50	22,5	5,29	0,08	0,3	1,28	7,1	7,69	0,03
332	55,89	0,02	20,9	4,16	0,09	0,22	1,16	7,8	7,07	0,03
315/2	50,22	1,70	18,5	6,77	0,21	1,42	4,74	5,5	5,09	0,24
316/4	57,1	0,10	23,2	3,02	0,13	0,29	1,32	5,8	7,32	0,03
16/175	52,58	0,22	20,0	2,84	0,12	0,69	3,95	12	1,27	0,27
317/8	63,12	0,23	17,9	2,61	0,07	0,62	2,24	5,6	6,84	0,05
317/17	55,23	0,84	18,9	5,54	0,13	0,68	6,5	4,1	5,41	0,14
Phenite										
336	53,64	1,37	15,5	10,09	0,13	4,63	7,11	3,4	2,48	0,59

Note The XRF analyses data IGC SB RAS, 2018, total iron Fe_2O_3

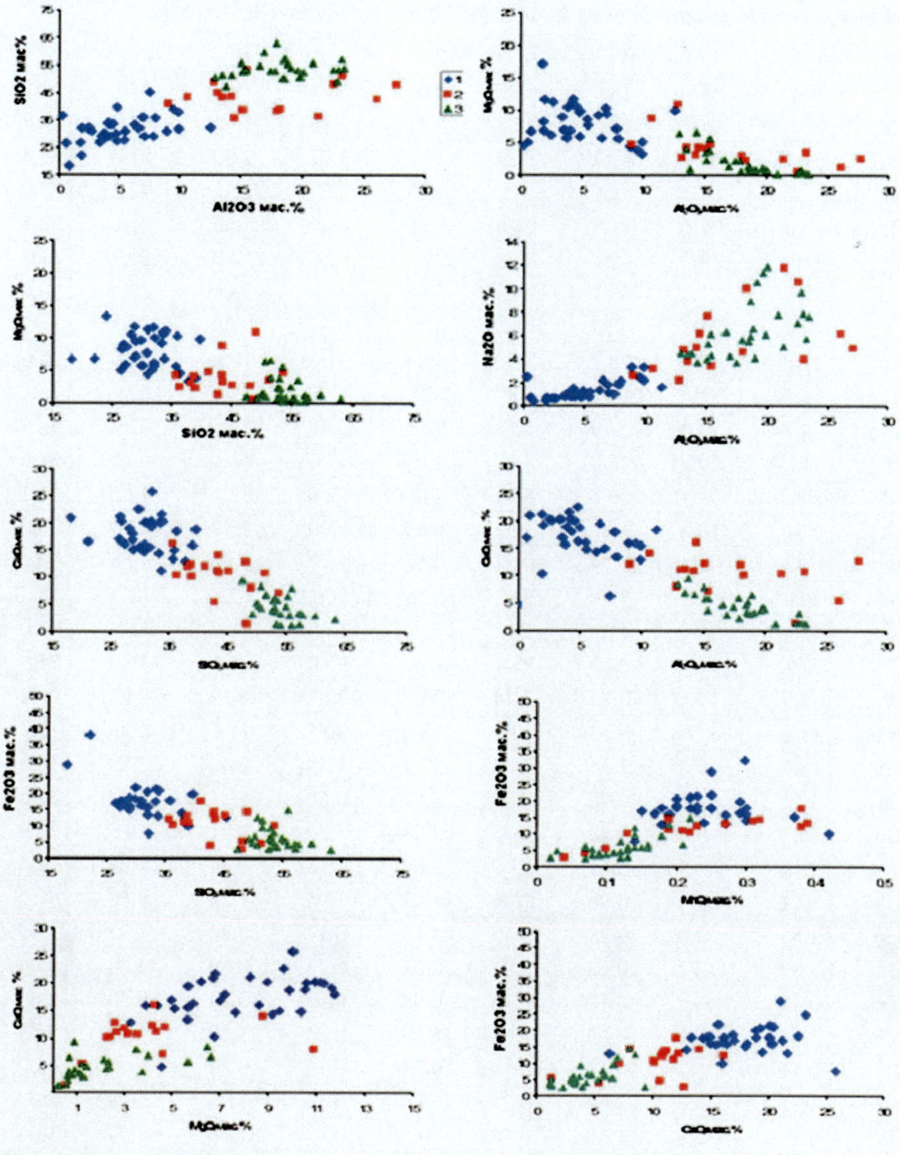

Fig. 2 Paired correlations of rock-forming elements in the rocks of the Zhidoy massif: 1—pyroxenites, 2—ijolites, 3—syenites

Table 2 Contents of rare elements in the rocks of the Zhidoy massif

No.	1	2	3	4	5	6	7	8	9
V	434	654	186	68.2	323	66.4	103	446	189
Cr	6.78	1.0	1.0	43.7	4.47	2.84	1.94	64.3	139
Co	37.7	47.1	4.12	13.9	15.2	3.77	2.36	87.0	33.4
Ni	8.47	14.2	5.56	24.9	9.08	3.57	2.53	42.2	33.0
Cu	84.3	94.1	11.8	21.9	20.0	7.94	6.55	216	28.7
Zn	170	297	62.0	74.5	218	63.6	54.0	205	111
Ga	30.5	27.1	43.1	9.55	34.6	44.8	38.60	20.5	25.0
Ge	1.64	2.93	0.72	4.56	2.27	0.84	0.89	1.98	1.83
Rb	27.8	1.92	98.9	3.00	56.4	91.6	105	4.78	75.6
Sr	1438	979	1909	100	3841	1865	1202	628	1601
Y	44.24	78.4	6.19	12.8	80.6	5.68	5.85	31.4	42.0
Zr	913.4	606	268	3.08	1954	475	680	342	376
Nb	271.8	349	66.6	8.38	446	52.0	82.1	17.5	21.2
Sn	5.32	10.2	1.81	1.55	8.69	1.09	2.37	4.71	2.80
Cs	0.43	0.74	1.24	0.82	0.53	1.77	1.96	1.06	0.59
Ba	244	57.2	3265	94.9	3224	1660	725	81.5	1850
La	130	360	11.9	11.6	326	13.8	12.3	63.5	134
Ce	276	943	23.9	25.1	764	28.0	31.6	163	257
Pr	32.5	100	2.84	3.48	74.8	2.87	4.14	24.1	29.7
Nd	113	377	9.52	13.2	254	9.85	15.8	108	105
Sm	18.7	66.0	1.34	2.62	37.7	1.58	2.85	20.9	16.8
Eu	5.51	19.0	0.78	0.76	10.7	0.68	0.81	5.57	3.66
Gd	15.2	50.5	1.48	2.73	27.8	1.30	1.81	16.5	12.9
Tb	2.13	6.13	0.21	0.40	3.64	0.18	0.34	1.98	1.66
Dy	10.1	25.4	0.96	1.86	17.3	1.05	1.23	8.60	8.12
Ho	1.55	3.80	0.28	0.40	3.15	0.17	0.20	1.22	1.47
Er	3.87	7.77	0.62	0.91	8.04	0.73	0.69	2.64	4.04
Tm	0.45	0.79	0.12	0.16	0.98	0.11	0.14	0.30	0.59
Yb	2.64	3.59	0.69	1.08	5.80	0.81	1.19	1.77	3.58
Lu	0.41	0.47	0.11	0.18	0.74	0.13	0.24	0.22	0.47
Hf	16.3	19.6	3.69	0.21	29.4	6.54	11.2	10.8	7.00
Ta	8.12	39.3	1.06	0.30	20.3	0.46	1.36	1.77	1.09
W	0.56	2.37	0.17	0.94	8.00	1.64	0.69	0.24	1.41
Tl	0.02	0.03	0.08	0.05	0.01	0.07	0.17	0.04	0.08
Pb	6.15	13.8	3.37	0.78	8.97	10.7	3.95	0.73	22.8
Th	9.49	25.3	2.91	0.39	19.5	10.9	4.95	1.87	16.0
U	1.69	8.16	0.97	0.14	3.47	1.85	1.31	0.31	1.69

Note 1, 4, 6–pyroxenites, 2, 5–ijolites, 3, 7, 8–syenites, 9–phenite. The samples were ana-lyzed by ICP-MS CUC of the IGC SB RAS, analyst Mitrofanova A.Yu. 2018

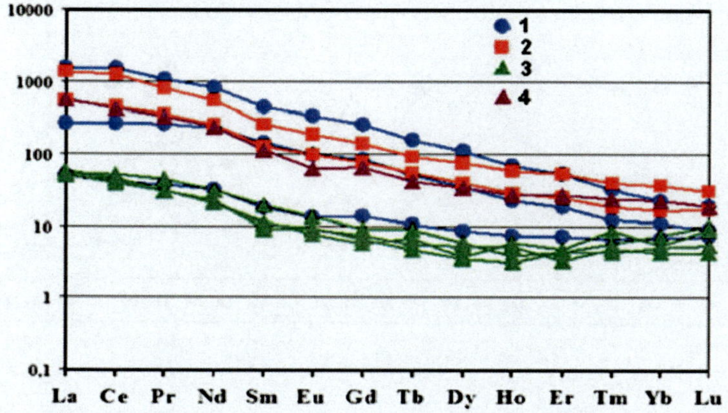

Fig. 3 Spectra of rare-earth elements in rocks of the Zhidoy massif (chondrite-normalized). Symbols: 1—pyroxenite, 2—ijolite, 3—syenite, 4—phenite

5 Mineralogical Features of Zhidoy Massif Rocks

The mineral compositions based on microprobe analysis data are given in Table 2. Both nepheline and cancrenite are found in nepheline sienites.

Pyroxene in them has an augite composition. The alkaline sienites are characterized by microcline-pertites, albite and pyroxene of aegirine-augite composition. In ijolites, pyroxene has titanium-augite composition, mica has magnesian biotite, titanium garnet-melanite, apatite and titanite contain iron admixture. In pyroxenites, titanite is silicone-free and perovskite is free of impurities (Table 3).

Please try to avoid rasterized images for line-art diagrams and schemas. Whenever possible, use vector graphics instead (see Fig. 1).

Table 4 shows the concentrations of yttrium and rare-earth elements in titanite and apatite (according to atomic emission analysis). The highest concentrations of rare-earth elements are found in pyroxenite titanites, while these concentrations are higher than in apatites, and in ijolites the reverse is true. The highest concentrations of rare-earth elements are observed in apatite from titanites.

Symbols: blue titanites from rocks: 1—pyroxenite, 2—biotite pyroxenites, 3—alkaline syenites, 4—nepheline syenites, 5—pegmatites of nepheline syenites; red apatites: 6—pyroxenites, 7—biotite pyroxenites, 8—phenytes.

The ratios of rare-earth element groups to yttrium in minerals of the Zhidoy massifs in titanites and apatites are given in Fig. 5. The red line connects paragenic titanite and apatite from pyroxenites.

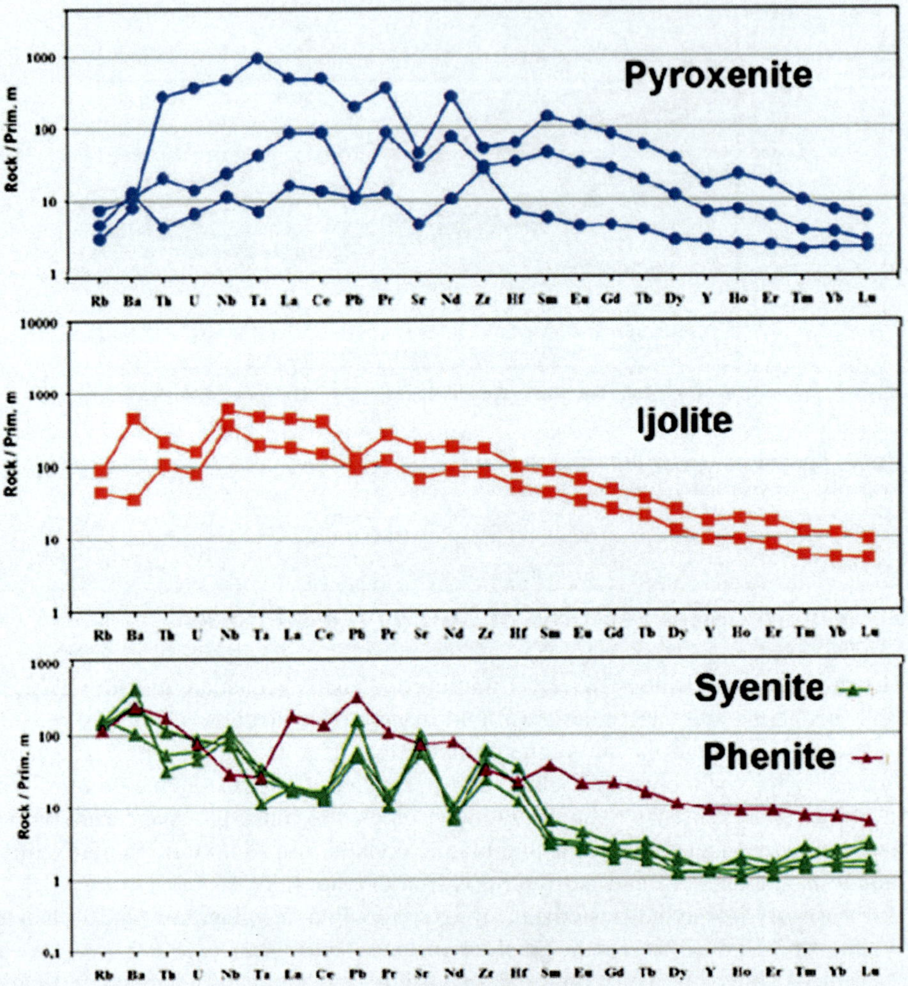

Fig. 4 Spider-diagram of the Zhidoy massif rocks

6 Isotopic Studies

Determination of C, O, Nd and Sr isotopes in carbonatites of the Zhidoy massif was carried out in the laboratory of Shinshy University, Japan [6], presented in Tables 5 and 6.

The ratios of oxygen and carbon isotopes are shown on the graph (Fig. 6). The values of these isotopes in carbonatites of the Zhidoy massif fall into the mantle square, which testifies to the mantle origin of the primary magma of the Zhidoy massif.

Table 3 Chemical composition minerals of the Zhidoy massif

No.	SiO_2	TiO_2	Al_2O_3	Fe_2O_3	FeO	MgO	CaO	Na_2O	K_2O	Total
1	42,9	0,17	33	0,85	0	0,91	0,24	14,9	7,06	100
2	46	0	34,5	0	0	0	0	13,1	6,41	100
3	43,17	0	32,92	0	0	0	2,99	17,9	0	97
4	41,07	0	31,47	0	0	0	8,41	15,5	0	96,4
5	64,14	0	18,71	0	0	0	0,55	2,62	13,98	100
6	66,22	0	21,63	0	0	0	0	12,2	0	100
7	53,68	0	2,53	0	13,39	9,41	19,5	0	0	98,51
8	50,38	0	2,15	27,43	0	2,51	8,74	8,8	0	100
9	43,23	5,25	7,03	0	12,72	0	29,56	0,8	0	98,59
10	35,26	3,25	13,64	0	25,61	10,76	0	0	8,63	97,15
11	38,39	6,68	0	10,29	0	10,78	33,87	0	0	100
12	2,89	0	0	0	0	0	54,12	0	0	97,01[a]
13	3,17	0	0,52	2,88	0	0	53,54	0,13	0	97,24*
14	31,47	35,24	1,47	3,07	0	0	28,74	0	0	99,99
15	0	58,25	0	0	0	0	41,75	0	0	100

Note 1 and 2–nepheline, 3 and 4–cankrenite, 5–microcline, 6–albite, 7–avgite, 8–egyrine-avgite, 9– titaniumavgite, 10–biotite, 11–granate melanite, 12 and 13–apatite, 14–titanite, 15–perovskite. Minerals from: nepheline syenites–1, 2, 3, 4, 7; alkaline syenites–5, 6, 8; ijolites–9, 10, 11, 13, 14; pyroxenites–12, 15. Microprobe analysis, Shinshi University, Japan, 2018.
[a]No 11–P_2O_5 = 39,99, No–P_2O_5 = 37.11

Table 4 Rare-earth elements and yttrium (in ppm) in titanites and apatites

No.	1	2	3	4	5	6	7	8	9	10
	329	326	331	333	328	334	335	326a	328a	336
Y	210	370	760	520	640	210	240	430	230	1200
La	510	620	2000	860	4000	640	810	1800	520	3000
Ce	1000	1400	4000	2200	8400	1300	1600	3600	800	6900
Nd	540	800	2000	1100	4100	910	1000	1300	360	3800
Sm	120	140	590	330	790	300	24	340	77	890
Eu	50	31	42	90	200	64	40	120	20	140
Gd	100	100	380	100	900	150	130	140	90	600
Dy	70	96	230	220	210	90	100	150	70	370
Ho	1	2	10	10	20	14	16	24	13	90
Yb	24	36	83	57	34	14	11	25	8	100
Lu	24	36	83	57	34	14	11	25	8	100

Note Minerals from rocks: 1, 4–syenites; 2, 3, 8–ijolites; 5, 6, 7, 9–pyroxenites; 10–phenites

Massifs: 1−Chadobets, 2−Yessey, 3−Guli, 4−Ingili, 5−Arbarastah, 6−Gornoye Ozero, 7−Kiisky, 8−Tatarsky, 9−Belaya Zima, 10−Srednyaya Zeema, 11−Tagna, 12−Zhidoy; B−isotopes of oxygen and carbon in the rocks of the Zhidoy massif: 1−carbonatite, 2−ijolites.

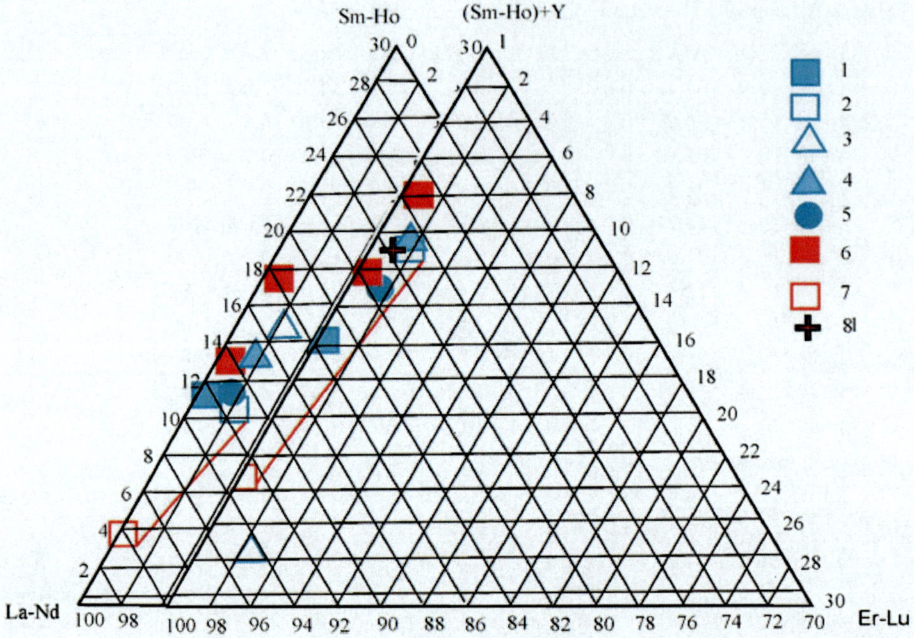

Fig. 5 Relations between groups of rare-earth elements and yttrium in minerals of the Zhidoy massif

Table 5 Values of oxygen and carbon isotopes in carbonatite of the Zhidoy massif

No. п/п	Mineral	d^{13}C	d^{18}O
1	Calcite	−5,8	7,2
2	Calcite	−5,6	7,5
3	Calcite	−5,5	7,1
4	Calcite	−5,2	7,5
5	Dolomite	−4,9	7,3

Table 6 Values of Sr and Nd isotopes in rocks of the Zhidoy massif

No.	Rb (ppm)	Sr (ppm)	^{87}Rb/^{86}Sr	^{87}Sr/^{86}Sr	^{147}Sm/^{144}Nd	^{143}Nd/^{144}Nd	eNd
1	0,67	768,04	0,00251	0,702489	0,107504	0,512436	2,8
2	33,61	1237,86	0,078525	0,704144	0,091018	0,512271	0,9
3	105,36	881,52	0,345757	0,706185	0,100649	0,512341	1,5
4	1,03	6050,42	0,000491	0,703351	0,082711	0,512263	1,4
5	1,79	4658,95	0,001109	0,703389	0,081792	0,512304	2,2
6	1,08	5908	–	0,70332	–	0,51206	3,8
7	1,34	5437	–	0,70317	–	0,51201	2,8

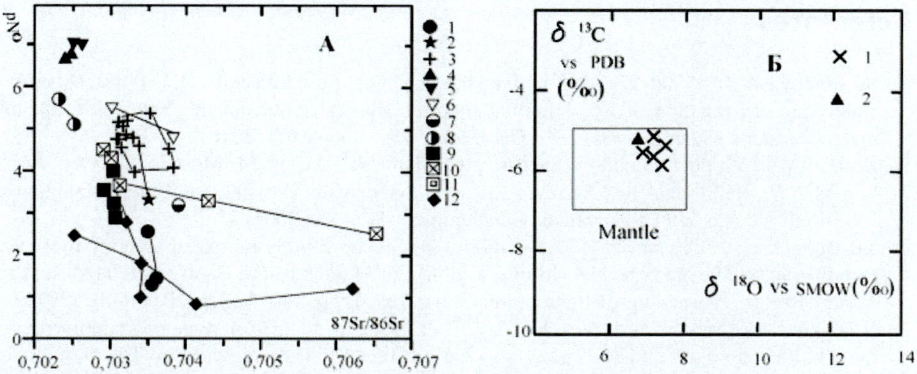

Fig. 6 Isotopes of Sr-Nd in carbonatite of Siberian platform framing [7]

Figure 6 shows data on ratios of epsilon Nd and ratios of strontium isotopes in carbonatite framing of the Siberian platform at the age of 650 million years. The carbonatites of the Zhidoy massif by these values are the closest to the carbonatites of the Sayan group (Belaya-, Srednyaya Zeema, and Tagna) to which they belong [5]. The isotope values for all carbonatites of the Siberian platform frame fall into the depleted mantle area, which can be explained by subduction under the platform of depleted oceanic basalts. The isotope data confirm the mantle source for the primary alkaline-ultramafic magma of the Zhidoy massif, the final differential of which is carbonatite.

7 Conclusions

1. Three varieties of ore pyroxenites have been defined−titanium-magnetite, ilmenite and perovskite ore.
2. The petrochemical diagrams show a common trend in the composition of rock forming elements, indicating the homomorphism of the rocks and their crystallization from a single primary magma.
3. Geochemical data also confirm the genetic relation of the Zhidoy massif.
4. Mantle source, the depurated mantle for the primary magma of the Zhidoy massif, has been determined on the basis of isotope data.

Acknowledgments The research was done with the financial support of the RFBR, grant project No. 20-05-00261, as well as the Integration Project of the Research Center of the Russian Academy of Sciences, Block 1.4. and the study was performed by the governmental assignment in terms of Project IX. 129.1.4., RNF-21-17-00015.

References

1. Borovikov, A.A., Vladykin, N.V., Tretiakova, I.G., Yu Dokuchits, E.: Physicochemical conditions of formation of hydrothermal titanium mineralization on the Murunskiy alkaline massif, western Aldan (Russia). Ore Geol. Rev. **95**, 1066–1075 (2018)
2. Konev, A.A.: Zhidoic alkaline ultrabasic plutonium. Science, p. 84. Moscow (1970)
3. Stories, S.V., Ilyasova, A.M., Konev, A.A., et al.: Geochemical evolution of the Zadoi alkaline and ultrabasic massif of Prisayanya. Geochemistry **1**, 3–18 (2007)
4. Yu Dokuchits, E., Alymova, N.V., Vladykin, N.V.: The Zhidoy massif of alkaline rocks and carbonatites and its ore potencial (Irkutsk region). In: Magmatism of Earth and related strategic metal deposits. Proseeding of International Conference, pp. 74–76. Saint Peterburg (2019)
5. Vladykin, N.V., Alymova, N.V., Dmitriev, R.P., Bobrov, V.A.: New type of titanium ores in the Zhidoi massif of alkaline rocks and carbonatites (Irkutsk region). In: Bychkov, A.L., Kazakov, V.B. (eds.) Actual problems of Baikal science, Issue 2, pp. 67–72. Publishing House of the Sochava Institute of Geography SB RAS, Irkutsk (2017)
6. Morikiyo, T., Takano, K., Miyazaki, T., et al.: Sr, Nd, C and O isotopic compositions of carbonatite and peralkaline silicate rocks from the Zhidoy complex, Russia). J. Mineral. Petrol. Sci. **95**, 162–172 (2000)
7. Vladykin, N.V., Morikiyo, T., Miyazaki, T.: Sr and Nd isotopes geochemistry of alkaline and carbonatite complexes of Siberia and Mongolia and some geodynamic consequences. Irkutsk Probl. Sources Deep Mag. Plumes **1**, 19–37 (2005)

Geochemistry, Mineralogy and Ore Content of Alkaline Granite Magmatism of East Sayan Zone (On the Example of Zashikhinsky Deposit)

N. V. Alymova and N. V. Vladykin

Abstract

The Zashikhinsky deposit (Irkutsk region) is confined to the Late Paleozoic East Sayan zone composed of rare-metal alkaline granites. In terms of differentiation degree and ore potential, the rare-metal granites are divided into three facies varieties: early facies involving microcline-albite amphibole-bearing granites, middle facies represented by leucocratic quartz-albite granites, and late facies including essentially pure albite varieties. Pure albite (albitites) is the most ore-bearing variety; REE contents can reach as high as 10000 ppm, in cases even 15440 ppm. In the plots of paired correlations of rock forming and trace element, all facies of the rare-metal granites define clear compositional trends that suggest their derivation from a single magma during its differentiation (from early facies to late ones). All facies of the alkaline granites of the massif show similar shapes of REE pattern and are different only in total REE contents. In addition, they also show similar REE spidergrams, which, however, differ only in small variations of anomalies sizes. Obtained geological, petrographic, and geochemical data suggest a common magmatic genesis of the rare-metal granites and albitites. Accessory minerals in rocks and ores are columbite, Nb-rutile, zircon, xenotime and gagaranite. It should be noted, that rocks of the complex show no metasomatic zoning and metasomatic alteration of minerals by other minerals, characteristic of metasomatic genesis.

N. V. Alymova (✉) · N. V. Vladykin
Vinogradov Institute of Geochemistry SB RAS, 1A Favorsky Str., Irkutsk 664033, Russia
e-mail: alymova@igc.irk.ru

N. V. Alymova · N. V. Vladykin
Irkutsk Scientific Center of Siberian Branch of Russian, 134 Lermontov St., Irkutsk 664033, Russia

N. Vladykin (ed.), *Alkaline Rocks, Kimberlites and Carbonatites: Geochemistry and Genesis*, Springer Proceedings in Earth and Environmental Sciences,
https://doi.org/10.1007/978-3-030-69670-2_4

Keywords

Zashikhinsky deposit · Rare-metal granites · Columbite · Mineral concentrators

1 Introduction

Rare-metal alkaline granites are typically related to complex rare-metal mineralization (tantalum-niobium, zirconium, rare earth, beryllium, lithium, and others). In terms of scales of the accumulation of Ta, Nb, Zr, Y and REE, these occurrences of alkaline granitoid magmatism are sufficiently large, having important practical significance in the mineral raw base of different countries, including Russia. The Zashikhinsky intrusive complex (Irkutsk Region) located within the Late Paleozoic East Sayan zone of the rare-metal alkaline granitic magmatism is ascribed to this group of deposits. This zone formed in the periphery of the Barguzin magmatic province [1], which includes the largest Angaro-Vitim granitoid batholith surrounded by the rift zones (Uda-Vitim, Synnyr, East Sayan and Saizhen) [2]. The formation of the Barguzin magmatic province and the East Sayan magmatic zone is related to the mantle plume activity in the late Carbonaceous-Early Permian [1].

The zone forms a NE-trending belt extended along the western boundary of the Tuva-Mongolian microcontinent. The unique rare-metal Zashikhinsky deposit with the highest tantalum content in Russia [3] is located at the continuation of this belt, northwards the Main Sayan Fault. Its ores are ascribed to the tantalum-niobium economic type. In addition, they are enriched with rare earth elements of yttrium group. The origin of the Zashikhinsky Deposit is still under debate. Many earlier researchers traditionally regarded the rocks of the complex and the associated mineralization to metasomatites [4–6]. According to other viewpoints, the rocks of the massif represent typical peralkaline (agpaitic) rare-metal granites, whose mineralogical and petrochemical features are characteristic of magmatic rocks [1, 7–9]. The leading role in the formation of the unique rare-metal mineralization is assigned to the processes of prolonged fractional crystallization of the melt leading to the accumulation of incompatible elements in the final products of the magmatic process [2, 8, 10, 11].

2 Analytical Research Methods

The analytical studies of the rocks and minerals of the Zashikhinsky complex were accomplished at the certified analytical center of the collective use for "Isotopic-geochemical investigations" at the Institute of Geochemistry, SB RAS and at Baikal Center of nanotechnologies located at Irkutsk National Research University.

The samples were analyzed for major oxides using XRF silicate analysis on a multichannel SRM-25 XRF spectrometer (Nauchpribor, Orel, Russia). The samples were ground and fused with $LiBO_2$ in induction oven in glassy-carbon crucibles at the temperature of 1100 °C. The sample: flux ratio was 1:2; sample weight was 1.25 gr.

Trace and rare earth elements were measured by ICP-MS spectroscopy under the standard conditions with Element 2 Thermo Finnigan. The samples were obtained by fusion with $LiBO_2$.

Alkaline elements (K, Na, Li, Rb, and Cs) were analyzed by flame photometry on the atomic emission flame spectrophotometer DFS-12. Powder subsamples, preliminarily decomposed by a mixture of HF and $HClO_4$ acids were used. The estimated uncertainties comply with those for this method (STP IGKh-009-97 technique).

In this study, mineral analysis was done on the JEOL JXA-8200 equipped with EDS (energy dispersive spectrometer) with Si(Li) detector, 133 eV resolution, and 5 WDS (wavelength dispersive spectrometers).

3 Geological Position and Structure of the Massif

The Zashikhinsky intrusive complex of alkaline rare-metal granitoids is located in the Nizhnyaya Uda area of the Irkutsk Region, the East Sayan Range. It is confined to the Main Sayan Fault being the marginal suture of the Siberian platform. In this area, the fault is traced by the exposures of sedimentary, volcanogenic, intrusive, and lithotectonic complexes of wide age range. One of the most important structural elements of this zone is the belt of moderate-alkaline and alkaline granitoids and comagmatic Paleozoic volcanic rocks known as the Ognitsky intrusive complex and the Verkhnyaya Nerkha Formation, respectively.

The massif is composed of alkaline rare-metal granites rich in trace elements, which form a separate body intruded in the Khailama granitoid massif of the Middle Paleozoic Ognitsky Complex [4]. The massif has NW-trending ellipsoid shape and covers an area of about 1.3 km^2 (see Fig. 1). The depth of the erosion section is around 300 m.

The Zashikhinsky granites are medium-grained rocks frequently displaying the massive structure and hypidiomorphous texture. The rock-forming minerals of the Zashikhinsky granites are quartz, microcline, and albite, whose variations in different facies varieties are significant. Subordinate minerals are arfvedsonite, aegirine, lepidomelane, tainiolite, polylithionite and fluorite. Accessory minerals are columbite, pyrochlore, Nb-rutile, xenotime, gagarinite, cryolite, zircon, thorite, carbonatite, and sulfide. The main ore minerals, which accumulate trace and rare earth elements in the Zashikhinsky alkaline granites, include columbite, Nb-rutile, zircon, xenotime and gagarinite. It should be noted, that the alkaline rocks of the massif show no metasomatic zoning and metasomatic alteration of minerals by other minerals, characteristic of metasomatic genesis [7].

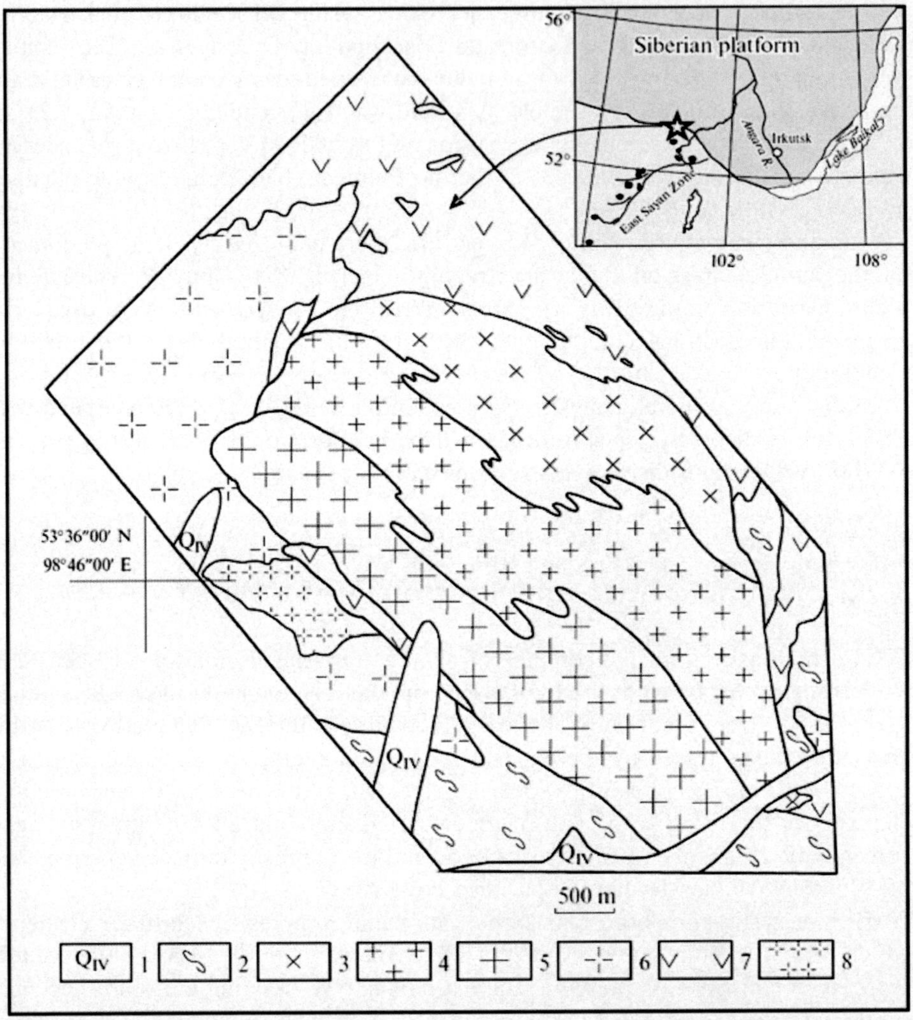

Fig. 1 Geological scheme of Zashikhinsky massif [4] with clarifications of authors. Symbols: 1 −Quaternary deposits (Q_{IV}). Zashikhinsky massif: 3−quartz-albite granites and albite massif (faction III), 4−leucocratic quartz-albit-microcline granites (faction II), 5−amphibole-containing quartz-microcline-albite granites (faction I). Middle Paleozoic intrusive rocks: 6−pegmatoid amphibole granites, 7−granites, syenites, granienites of the ognite complex (D_2og), 8−diorites of the Hoyto-okino complex (Pz_1ho). Early Proterozoic intrusive rocks: 2−shales, microgneisses, amphibolites of turquoise formation (PR_1br_2)

As regard to the degree of differentiation and ore potential, three rock facies are distinguished based on the mineral-petrographic composition and the contents of valuable components (see Fig. 1).

Amphibole-bearing quartz-microcline-albite granites include rocks with porphyritic pisolitic quartz and elongated prisms of black amphibole (arfvedsonite). This facies is traced as a single body in the southwestern near-contact part of the Zashikhinsky complex and is regarded as the oldest granite facies.

Rare-metal leucocratic quartz-albite-microcline granites, which contain both pisolitic quartz phenocrysts and its fine grains in the granite groundmass. This facies is observed in the center of the deposit.

Leucocratic quartz-albite granites grading into albitites, characterized by the domains of monomineral albitites observed between quartz-albite rocks. In terms of the chemical composition, the albite includes the sodic variety with less than 8% of anorthite. This facies variety occurs in the northeastern part of the Zashikhinsky intrusion and represents the main ore site of the deposit.

The high-grade ores of the deposit (average content of Nb_2O_5 = 0.03 wt% and Ta_2O_5 = 0.027 wt%) occur over an area of 30 × 1000 m^2. They compose complex strata body with an average thickness of about 150 m, steeply dipping (at an angle over 400–500) to the north. At the depth, they are replaced by with lower Ta and Nb concentrations, which contain lenses and blocks of high-grade ores [5].

The zones demonstrate gradual transitions between facies. We believe that the distinguished facies mark the magmatic differentiation of fluid-rich alkali-granite melt [7, 8].

4 Petrochemical and Geochemical Characteristicks of the Rocks

The Zashikhinsky rare-metal granites belong to the alkaline agpaitic granites and demonstrate higher concentrations of a great number of incompatible elements. The granites are characterized by high alkali content (up to 12.68 wt% $Na_2O + K_2O$), and agpaitic index of 1.0–1.2 and present sodium type of alkalinity (K_2O/Na_2O <1). In addition, they exhibit lower Ca, Mg, Al, P, Ba, and Sr contents (Table 1). In accordance with the alphabet granite classification, the Zashikhinsky rocks have the mineralogical and petrochemical characteristics of A-type granites [12, 13]. Moreover, the granites demonstrate the elevated contents of Rb, Li, Y, Zr, Hf, Ta, Nb, Th, U, Zn, Ga, REE relative to the Clarke levels [14]. Clear linear dependences in the plots of paired correlations of rock-forming and trace elements point to a common origin during the differentiation of alkaline-granite magma and the subsequent enrichment of granites with the corresponding rare elements from the first to the third facies group.

All facies of the Zashikhinsky granites show similar shapes of REE pattern. The REE patterns demonstrate a profound Eu negative anomaly that serves as an evidence for formation of the whole rock series through magmatic differentiation spidergrams (see Fig. 2a).

Table 1 Contents composition of granites in the massif (mas. %)

Oxide	Faction								
	I	I	I	II	II	II	III	III	III
SiO_2	76.13	76.08	73.81	72.51	68.33	73.18	70.40	67,16	70.33
TiO_2	0.08	0.01	0.02	0.04	0.05	0.04	0.06	0,25	0.08
Al_2O_3	12.30	12.70	12.9	13.58	16.92	13.42	15.60	16,23	14.18
FeO	0.58	1.95	2.65	2.36	2.26	2.47	2,06	3.00	3.14
MnO	0.01	0.09	0.17	0.07	0.11	0.16	0.08	0.05	0.06
MgO	0.05	0.05	0.06	0.05	0.05	0.05	0.05	0.30	0.11
CaO	0.18	0.12	0.43	1.21	0.21	1.32	0.72	1.18	0.83
Na_2O	2.4	5.29	5.48	5.72	11.12	4.85	7.32	9.11	6.84
K_2O	6.81	3.00	3.85	2.38	0.44	2.53	2.10	0.54	2.98
P_2O_5	0.03	0.01	0.01	0.02	0.14	0.03	0.05	0.14	0.03
BaO	0.03	0.02	0.02	0.03	0.03	0.02	0.02	0.02	0.02
SrO	0.47	0.22	0.19	0.83	0.47	0.71	0.42	0.46	1.09
ZrO	0.4	0.19	0.16	0.70	0.40	0.60	0.36	0.39	0.93
H_2O	1.14	0.52	0.46	0.14	0.14	0.12	0.67	0.96	0.10
F	0.04	–	–	1.20	0.08	1.40	0.41	0.50	0.84
Sum[a]	100.07	98.3	97.57	99.37	100.11	99.41	99.61	99.54	99.71
Ti	93	92	113	200	270	168	324	1564	482
Ga	58	60	70	70	74	67	67	64	68
Rb	584	437	660	377	12,8	461	352	35	305
Sr	20	18.8	23	296	185	194	202	283	187
Y	10	22	33	322	1055	276	406	1232	210
Zr	728	1486	1403	6624	3607	5538	3363	4228	8348
Nb	905	1196	1401	2056	3249	2117	2168	1628	2233
Cs	1.62	1.54	2.22	0.51	<0.1	0.47	0.70	<0.1	0.66
Ba	1013	855	944	1423	1859	1201	1488	2255	1521
La	11.9	11.3	25	11.8	10.6	9.3	26	31	7.4
Ce	39	65	131	42	43	32	78	99	26
Pr	4.7	9.2	17	5.7	6.0	4.5	9.7	14	3.3
Nd	14	30	53	21	26	16	36	60	13
Sm	2.1	6.1	9.7	8.9	16.9	7.4	12.4	27.3	5.7
Eu	0.08	0.11	0.15	0.52	3.03	0.66	1.50	3.24	0.33
Gd	1.4	3.4	5.5	14.2	35	11.7	18.4	55	11.4
Tb	0.22	0.54	0.73	3.91	9.70	3.14	4.71	14.8	2.88
Dy	2.2	5.1	6.5	36.6	102	30.6	43.7	137	29.1
Ho	0.6	1.2	1.4	8.2	25.7	7.3	10.1	31.0	7.6
Er	2.8	5.7	6.3	34	131	32	45	119	36
Tm	0.65	1.27	1.35	6.36	33.5	6.63	10.20	21.1	7.55
Yb	6.3	11.6	12.1	54	321	58	94	166	67
Lu	0.97	1.78	1.98	7.36	43.7	7.86	12.75	21.41	10.03

(continued)

Table 1 (continued)

Oxide	Faction								
	I	I	I	II	II	II	III	III	III
Hf	35	87	64	356	211	317	257	240	312
Ta	117	172	195	280	396	287	345	274	157
Pb	100	137	150	202	218	352	477	180	106
Th	31	52	34	402	2102	593	1832	500	254
U	23.7	51.3	57.1	17.6	85.3	23.1	20.7	19.8	43.0

Note 1–Arfvedsonite granites (faction I), 2–Leucogranites (faction II), 3–Albites (faction III)
[a] Amount converted to fluorine

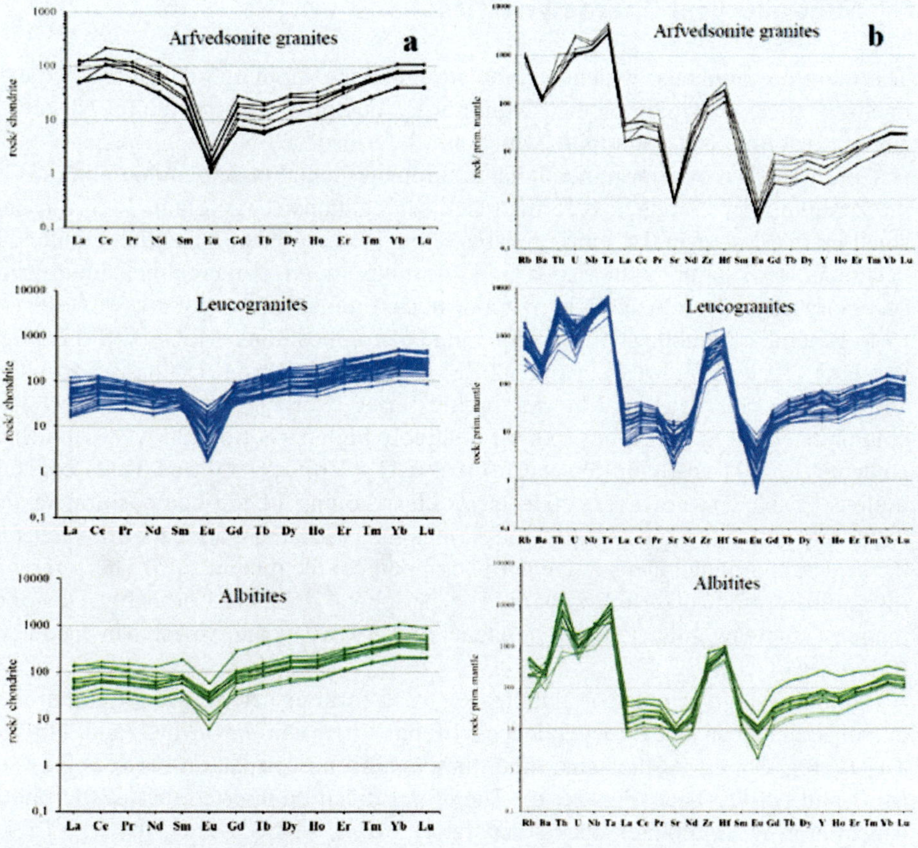

Fig. 2 Spidergrams of rare-earth elements (**a**) and trace element spidergrams (**b**) in the granite of the massif

The distribution of trace elements in the spidergrams is also similar to that of the rocks of all three facies varieties (see Fig. 2b). The spidergrams are characterized by the negative anomalies of Ba, Sr, U, LREE, Ti, and the positive anomalies of Th, Nb, Ta, Pb, Zr, and Hf. HREE and Y patterns reveal a positive slope.

Linear dependences of rock compositions and similar HFSE and REE distribution patterns [8] indicate the subsequent enrichment of the rocks with the corresponding elements from the first to the third facies and point to similar magma sources for all granites. We believe that all granites of the complex could originate from alkali-rich silicate melt.

5 Mineralogical Charactericstics

The main ore minerals, which accumulate tantalum, niobium and other trace elements in rocks and ores of the Zashikhinsky complex, are columbite, Nb-rutile, zircon, xenotime and gagarinite, which are described below.

Columbite is the major mineral that accumulates tantalum and niobium in ores of the Zashikhinsky complex. It forms black segregations (large grains 2–5 mm and small grains−less than 0.5 mm), in cases, elongated crystals. In alkaline granites, it occurs in association with major rock forming (quartz, microcline, albite) and accessory (arfvedsonite, aegirine, zircon, micas) minerals.

In general, columbite shows a wide range of compositions: MnO (0.38–12.06 wt %), FeO (7.19–19.71 wt%), Ta_2O_5 (1.85–18.94 wt%), Nb_2O_5 (58.88–75.46 wt%) and forms a complete Fe–Mn isomorphic series evolving from columbite-Fe to columbite-Mn (Fig. 3a). They exhibit relatively high TiO_2 (to 3.26 wt%) and SnO contents (to 0.41wt%) and lower Ce_2O_3, Nd_2O_3, Yb_2O_3, UO_2 and ThO_2 concentrations (Table 2). However, there is no clear zoning of the composition in the columbite grains from the studied rocks: in some cases, there are dark gray areas in the center, and lighter areas in the peripheral parts of the minerals [8]. The zoning is insignificant, with unclear contours. The peripheral zones are probably later formations, showing higher Ta_2O_5 contents (10–19 wt%) and lower MnO content (0.5–7 wt%).

The plots of paired correlations of rock-forming elements in columbites demonstrate clear linear dependences (trends) between FeO/MnO and Nb_2O_5/Ta_2O_5 (Fig. 3b, c). At the same time, they exhibit no correlation between Ta_2O_5–MnO and Nb_2O_5–FeO (Fig. 2d, e). Therefore, it can be inferred that isomorphous substitution in columbites occurs separately $Fe_2+ \leftrightarrow Mn_2+$ and $Nb_5+ \leftrightarrow Ta_5+$. There is also no correlation between the ratios Nb_2O_5/Ta_2O_5 and FeO/MnO, MnO/Ta_2O_5 and FeO/Nb_2O_5 (see Fig. 3f, g). TiO_2 and SnO_2 form a clear trend in columbites from the rocks of the Zashikhinsky deposit (see Fig. 3h).

Zircon, a typical accessory mineral in all facies varieties of rare-metal granites, is the major concentrator of zirconium and hafnium in alkaline rocks. Its contents in the Zashikhinsky rocks reaches as high as 7%. Zircon forms bipyramid crystals up to 1 mm across, frequently with abundant non-transparent inclusions in the cores of

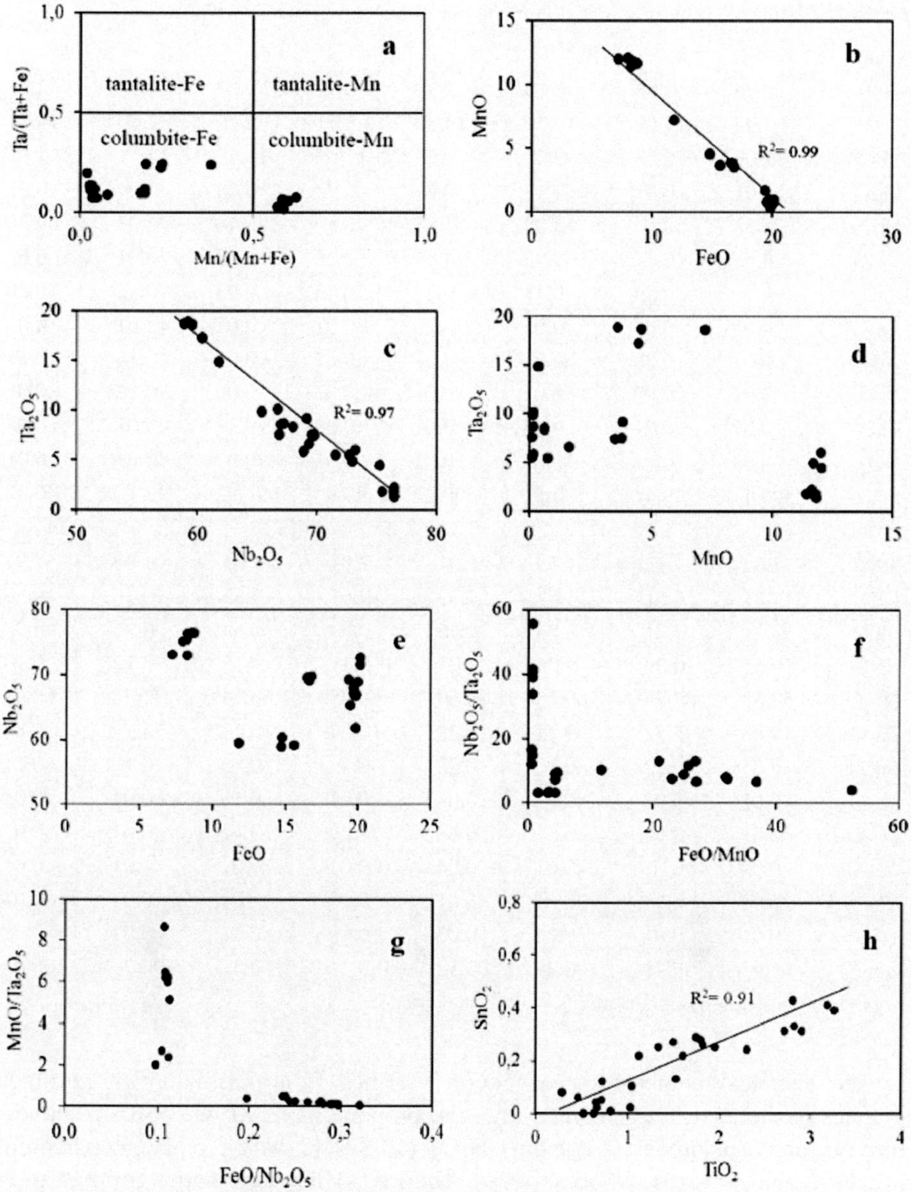

Fig. 3 Variations in the composition of columbites from granites

grains. It occurs in association with mica, cryolite, microcline, quartz and columbite. There are also small, no more than 0.2 mm, transparent and faceted crystals of the mineral.

Table 2 Chemical composition of columbites in rare-metal granites (wt%)

Oxide	1	2	3c	3r	4	5c	5r	6
TiO_2	0.46	1.75	3.19	2.72	0.72	0.67	0.73	3.26
FeO	7.92	14.8	19.71	19.86	8.71	8.2	7.19	19.37
MnO	12.06	4.58	0.62	0.38	11.6	11.42	12	1.64
MgO	<0.01	<0.01	0.17	0.2	0.02	0.05	<0.01	<0.01
Nb_2O_5	75.28	58.9	68.02	61.81	76.5	75.46	73.23	69.36
Ta_2O_5	4.5	18.8	8.32	14.87	2.26	1.85	6.01	6.6
SnO_2	0.06	0.29	0.41	0.31	0.05	0.02	0.12	0.39
Ce_2O_3	0.06	0.03	<0.01	0.06	<0.01	<0.01	<0.01	0.02
Nd_2O_3	<0.01	0.28	<0.01	<0.01	<0.01	<0.01	0.16	0.17
Yb_2O_3	0.01	0.09	<0.01	<0.01	<0.01	0.06	0.26	<0.01
ThO_2	<0.01	<0.01	0.02	<0.01	<0.01	0.12	<0.01	0.04
UO_2	0.03	<0.01	0.01	0.01	<0.01	0.07	<0.01	<0.01
Sum	100.4	99.4	100.5	100.2	99.9	97.92	99.7	100.9
Oxide	7	8	9	10	11	12	13	14
TiO_2	1.34	1.12	1.53	1.61	2.81	1.95	1.05	2.91
FeO	11.81	16.51	16.67	14.79	19.81	19.85	20.16	19.89
MnO	7.21	3.81	3.83	4.46	0.79	0.54	0.96	0.77
MgO	0.03	0.09	0.06	0.07	0.11	0.18	0.12	0.19
Nb_2O_5	59.58	69.76	69.15	60.39	66.85	66.74	72.74	68.91
Ta_2O_5	18.66	7.52	9.21	17.28	7.59	10.13	5.5	5.88
SnO_2	0.25	0.22	0.13	0.22	0.43	0.25	0.1	0.31
Ce_2O_3	<0.01	<0.01	<0.01	<0.01	<0.01	0.15	<0.01	0.08
Nd_2O_3	<0.01	0.68	<0.01	0.68	1.49	0.42	<0.01	0.92
Yb_2O_3	0.09	<0.01	<0.01	0.04	<0.01	<0.01	<0.01	<0.01
ThO_2	<0.01	<0.01	<0.01	<0.01	0.07	0.04	0.04	<0.01
UO_2	<0.01	<0.01	0.06	0.03	0.1	<0.01	<0.01	<0.01
Sum	98.97	99.71	100.79	99.57	100.1	100.3	100.7	100.0

Note c-center of grain, r-grain radius

The chemical composition of zircon varieties from Zashikhinsky rare-metal granites is similar to the theoretic one; the proportions of Zr and Si correspond to formula units in zircon. It contains HfO_2 (1.83–4.42 wt%), replacing zirconium (Table 3) and ZrO_2 (61.39–66.38 wt%). The ZrO_2/HfO_2 ratio varies from 14 to 36. Zircon is characterized by slightly higher Y_2O_3 concentrations (0.24–2.33 wt%). This mineral demonstrates elevated Y_2O_3 (from 0.20 to 0.84wt%), ThO_2 (to 1.61 wt%), UO_2 (to 0.14 wt%) concentrations and lower Nb_2O_5 и Ta_2O_5 (less than 0.1 wt%) levels. The compositional variations from the core towards the rims are insignificant.

Cyrtolite is a metamict type of zircon with a high content of uranium, thorium, hafnium and rare earth elements in variable amounts. The crystal habitus is similar to zircons, but it is different by non-transparent grains and brown color. In thin-sections, it mainly occurs in paragenesis with thorite and fluorite.

Table 3 Chemical composition of zircons in rare-metal granites (wt%)

Oxide	Zircon						Cirtolite	
	1c	1r	2	3	4c	4r	6	7
SiO_2	31.41	31.63	31.69	30.71	31.76	30.99	30.37	30.70
TiO_2	0.01	0.01	0.00	0.00	0.00	0.03	0.00	0.00
Al_2O_3	0.08	0.00	0.00	0.00	0.00	0.02	0.00	0.00
FeO	0.07	0.03	0.05	0.06	0.05	0.03	0.02	0.03
MnO	0.00	0.03	0.00	0.02	0.02	0.00	0.01	0.00
MgO	<0.01	<0.01	<0.01	<0.01	<0.01	<0.01	<0.01	<0.01
CaO	0.00	0.00	0,01	0.07	0.01	0.04	0.01	0.00
Na_2O	<0.01	<0.01	<0.01	<0.01	<0.01	<0.01	<0.01	<0.01
K_2O	<0.01	<0.01	<0.01	<0.01	<0.01	<0.01	<0.01	<0.01
SrO	0.13	0.09	0.07	0.04	0.09	0.07	0.09	0.10
ZrO_2	64.39	64.56	64.33	62.16	63.41	62.55	62.43	61.39
HfO_2	3.11	3.44	3.46	2.73	3.93	4.42	2.74	2.71
Nb_2O_5	<0.01	<0.01	<0.01	<0.01	<0.01	<0.01	<0.01	<0.01
Ta_2O_5	0.00	0.04	0.06	0.00	0.05	0.08	0.00	0.03
P_2O_5	<0.01	<0.01	<0.01	<0.01	<0.01	<0.01	<0.01	<0.01
Y_2O_3	0.54	0.27	0.68	1.30	0.34	1.38	1.80	2.33
La_2O_3	0.02	0.03	0.05	0.00	0.00	0.00	0.04	0.00
Ce_2O_3	0.12	0.08	0.13	0.05	0.00	0.09	0.00	0.16
Nd_2O_3	0.09	0.04	0.00	0.00	0.00	0.00	0.00	0.00
Sm_2O_3	<0.01	<0.01	<0.01	<0.01	<0.01	<0.01	<0.01	<0.01
Eu_2O_3	<0.01	<0.01	<0.01	<0.01	<0.01	<0.01	<0.01	<0.01
Gd_2O_3	0.04	0.00	0.00	0.01	0.00	0.00	0.01	0.05
Dy_2O_3	0.06	0.03	0.01	0.12	0.04	0.08	0.19	0.26
Er_2O_3	0.11	0.09	0.09	0.18	0.06	0.12	0.20	0.19
Yb_2O_3	0.29	0.24	0.28	0.84	0.20	0.20	0.82	0.63
ThO_2	0.00	0.01	0.04	0.89	0.00	0.67	1.12	1.61
UO_2	0.05	0.00	0.00	0.11	0.04	0.00	0.14	0.13
PbO	<0.01	<0.01	<0.01	<0.01	<0.01	<0.01	<0.01	<0.01
F	0.03	0.01	0	0	0.03	0.08	0.08	0.03
Sum	100.65	100.74	101.04	99.39	100.11	100.95	100.17	100.45

Note c-center of grain, r-grain radius

Xenotime (Y) occurs in all granite varieties of the Zashikhinsky deposit as small yellowish-greenish crystals with a perfect cleavage. Concentrations of Y_2O_3 (41.31–43.42 wt%), and P_2O_5 (36.03–38.94 wt%) correspond to the formula composition of xenotime.

Xenotime shows high concentrations of lanthanides of the yttrium subgroup (Gd–Lu), the total content of HREE is up to 17.76 wt% (Table 4). It contains insignificant amount of LREE (Ce–Eu subgroup) (<1%) (see Fig. 4). In addition, the mineral contains ThO_2 (0.05–1.52 wt%), UO_2 (0.02–0.24 wt%), PbO (0.17–0.31 wt%).

The Nb-rutile concentrate tantalum, niobium, titanium. He form isometric faceted crystals with strong diamond luster. Mineral occur in all facies varieties of the Zashikhinsky rare-metal granites. In terms of the chemical composition, he are an intermediate variety between ilmenorutile and struverite (Table 4) at similar contents of Nb and Ta amounting in total over 25 wt%. In addition, mineral contain up to 1.5 wt% Sn.

Gagarinite is very rare fluoride of alkaline granites, which accumulates a significant share of yttrium and rare earth elements. At present, it occurs in several rare-metal granites: Zashikhinsky (Irkutsk Region) [7], Katugin (Transbaikalia) [10], and Ulug-Tanzek (East Tyva) [17], Verkhnee Espe, Kazakhstan [18], Pitinga, Brazil [19], Strange Lake, Canada [20]. Gagarinite in a number of thin-sections is found in accessory amounts and is confined to intergranular spaces in rock-forming minerals and, as a rule, forms an impregnation of grains with elongated and irregular shape. Like in carbonates, it has a large positive relief and strong birefringence. In the granites of the complex, this mineral is crystallized in association with cryolite, fluorite, columbite, and mica. The content in the rock is not more than 0.01%.

From the theoretical composition, the Zashikhinsky granites are different by slightly lower content of CaO (8.65–11.62 wt%), higher Na_2O (10.41–10.64 wt%) and Y_2O_3 (40.87–44.00 wt%) levels (Table 4). In addition to significant concentrations of yttrium, the mineral demonstrates the prevalence of medium and heavy rare earth elements: gadolinium (1.49–1.70 wt%), dysprosium (2.98–4.05 wt%), erbium (1.45–2.65 wt%), ytterbium (0.52–1.64 wt%). The increased concentrations of Gd, Dy, Er, Yb are expressed by positive anomalies in the spectra of rare earth elements (see Fig. 5).

Therefore, the Zashikhinsky rare-metal granites are marked by high alkali contents ($Na_2O + K_2O$ up to 11.59 wt%), higher concentrations of incompatible elements, lower contents of Ca, Mg, Al, P, Ba, Sr as well as the occurrence of alkaline dark-color minerals (amphibole, pyroxene, micas). A unique rare-metal Ta-Nb-REE mineralization in these rocks is related to a prolonged differentiation of melts. Substantial enrichment of alkaline rocks in incompatible elements (up to ore concentrations), fluorine, volatiles, higher Rb/St values (to 34) and agpaitic index (Ka up to 1) point to a higher differentiation degree of primary alkaline-granitic magmas, specialized for Ta, Nb, Zr, REE.

One of the key problems in modern petrology and geochemistry is revealing the genesis of peralkaline granites and the associated rare-metal mineralization. Numerous studies of alkaline granites provided strong evidence for the magmatic genesis of these granites and rare-metal mineralization [2, 7, 8, 10, 16, 19, 21]. Moreover, the studies of ongonites, which are subvolcanic analogues of rare-metal granites [22] also supported the magmatic genesis of the rare-metal granites. The thermobarogeochemical studies of minerals revealed melt inclusions in ongonites and rare-metal granites, which are also indicative of the magmatic genesis of rocks and ores of complexes under study [23–25].

Table 4 Chemical composition minerals accumulating rare elements (wt%)

Oxide	Xenotime				Nb-rutile		Gagarinite	
	1	2	3	4	5	6	7	8
SiO_2	0.11	0.16	0.14	0.06	–	–	0.04	0.02
TiO_2	0.03	0.00	0.00	0.00	61.90	64.77	0.00	0.01
Al_2O_3	<0.01	<0.01	<0.01	<0.01	0.05	0.08	0.00	0.00
FeO	<0.01	<0.01	<0.01	<0.01	7.73	7.92	0.19	0.10
MnO	<0.01	<0.01	<0.01	<0.01	0.01	0.01	0.00	0.00
MgO	<0.01	<0.01	<0.01	<0.01	0.01	0.00	0.00	0.00
ZnO	<0.01	<0.01	<0.01	<0.01	–	–	–	–
CaO	0.03	0.07	0.03	0.03	–	–	8.65	11.62
Na_2O	<0.01	<0.01	<0.01	<0.01	–	–	10.64	10.41
K_2O	<0.01	<0.01	<0.01	<0.01	–	–	0.12	0.13
BaO	0.01	0.03	0.00	0.03	0.06	0.01	0.00	0.00
SrO	0.00	0.00	0.00	0.06	–	–	0.00	0.03
PbO	0.27	0.28	0.23	0.21	–	–	0.29	0.36
P_2O_5	36.68	36.70	37.09	37.15	–	–	–	–
SnO_2	0.00	0.01	0.00	0.02	1.44	1.22	–	–
Nb_2O_5	0.02	0.08	0.02	0.00	15.01	17.74	–	–
Ta_2O_5	<0.01	<0.01	<0.01	<0.01	13.54	8.41	–	–
Y_2O_3	43.07	41.77	41.47	43.18	–	–	40.87	44.00
La_2O_3	0.00	0.00	0.04	0.00	–	–	0.32	0.23
Ce_2O_3	<0.01	<0.01	<0.01	<0.01	–	–	1.37	0.43
Nd_2O_3	<0.01	<0.01	<0.01	<0.01	–	–	0.90	0.70
Sm_2O_3	0.87	1.72	1.74	1.02	–	–	0.71	0.76
Eu_2O_3	0.25	0.39	0.41	0.22	–	–	–	–
Gd_2O_3	2.15	2.74	2.90	2.51	–	–	1.49	1.70
Dy_2O_3	4.31	5.08	5.24	4.74	–	–	4.05	2.98
Ho_2O_3	1.44	1.26	1.83	1.59	–	–	1.04	0.96
Er_2O_3	3.39	3.31	3.39	3.37	–	–	2.65	1.45
Tm_2O_3	0.60	0.60	0.67	0.53	–	–	0.23	0.16
Yb_2O_3	4.39	3.91	3.72	4.31	–	–	1.64	0.52
HfO_2	0.22	0.12	0.16	0.28	–	–	–	–
ThO_2	0.72	0.46	0.40	0.14	–	–	0.10	0.07
UO_2	0.13	0.08	0.10	0.04	–	–	–	–
F	0.30	0.30	0.30	0.30	–	–	36.91	37.79
Sum	98.57	98.57	99.43	99.28	99.93	100.18	112.21	114.43
F ~ O	0.13	0.13	0.13	0.13	–	–	15.54	15.91
Sum[a]	98.82	98.81	99.72	99.63	–	–	96.67	98.52

Note F- ~ O-excess oxygen, which replaces fluorine during calculations, [a] total II is sum without excess oxygen

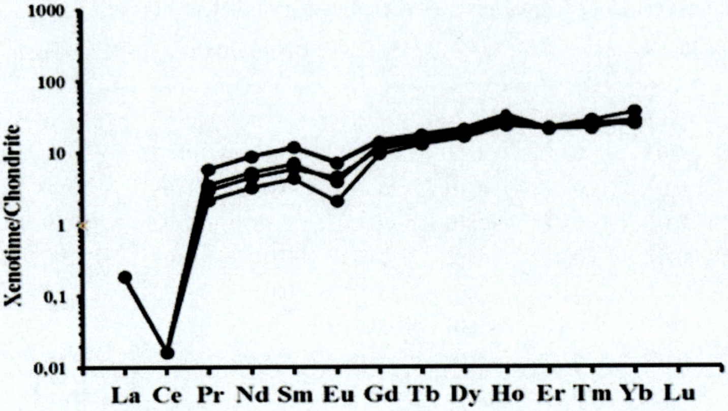

Fig. 4 Spectrum of rare-earth elements in xenotypes normalized by chondrite [15]

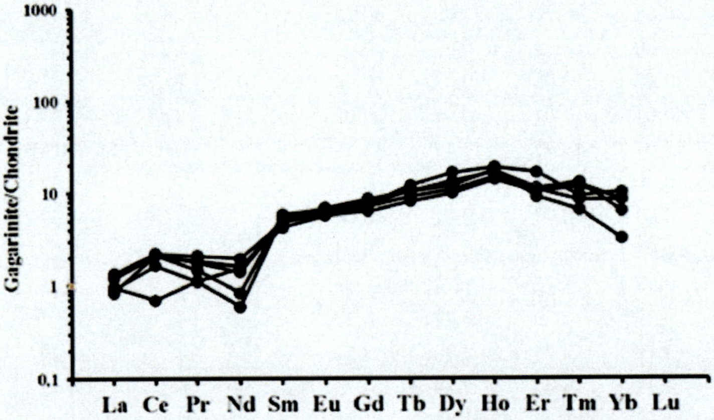

Fig. 5 The spectrum of rare-earth elements in gagarinites normalized by chondrite [15]

In composition, the Zashikhinsky rocks are similar to the rare-metal granites of the Katugin deposit, whose mineralization is localized in two relatively small (3 and 18 sq. km) areas. On the Katugin deposit, the processes of magma differentiation are more profound, that resulted in higher mineralization potential of rocks. Five facies varieties of alkaline granites have been distinguished on the Katugin deposit: I−biotite; II−ribekite-biotite; III−biotite-arfvedsonite; IV−arfvedsonite; V −aegirine-arfvedsonite and aegirine. The final products of the magmatic differentiation are rocks, which contain abundant fluorides (e.g. cryolite and gagarinite). These granites belong to the peralkaline type. They are characterized by very high content of most incompatible elements, high alkali content (up to 11.49 wt% $Na_2O + K_2O$), high agpaitic index (Ka−1.4) and are predominantly sodium

(K_2O/Na_2O = 0.52–2.44). In addition, these granites demonstrate higher Rb, Li, Y, Zr, Hf, Ta, Nb, Th, U, Zn, Ga, REE contents and lower Ca, Mg, Al, P, Ba, Sr levels.

The distribution REE patterns in the alkaline granites of the complex are quite similar, revealing insignificant slopes and a characteristic Eu negative anomaly. The distribution REE spectra in the host unaltered rocks (amphibolites and syenite-porphyries) are almost chondritic without Eu fractionation. The highest concentrations of rare-earth elements are characteristic of pyrochlore deposits and fluid-hydrothermal rocks with cryolite and gagarinite (see Fig. 6). As compared to granites, the near-contact phenites that replaced gneisses and shales have lower concentrations of rare-earth elements, although their distribution patterns are similar to granites, which indicates that the source of the elements were granites themselves.

Spider diagrams of rare-metal granites and contact metasomatites (phenites) differ quite sharply. The latter are characterized by lower concentrations of rare elements with negative and positive anomalies of yttrium and various variations of U, Th, Nb, Ta. The spectra of host rocks have almost horizontal distribution (see Fig. 7).

On the Katugin deposit, the minerals, that accumulate rare elements are quite different. In this deposit, the strategic rare elements are mainly concentrated in zircon, pyrochlore, and gagarinite, indicating different mechanisms of rare elements accumulation.

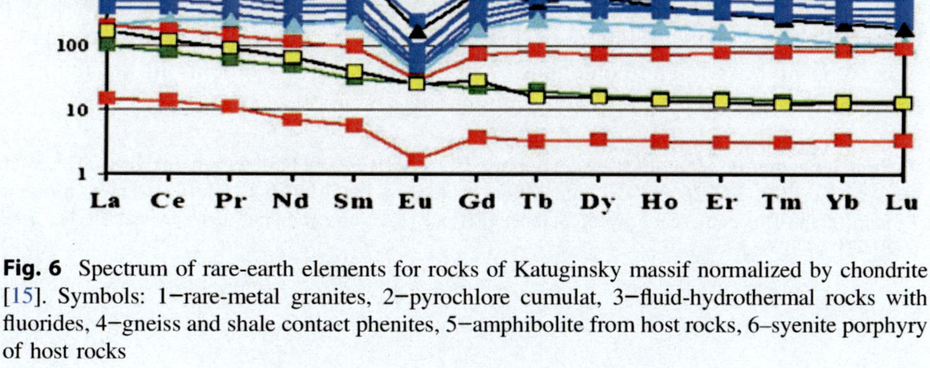

Fig. 6 Spectrum of rare-earth elements for rocks of Katuginsky massif normalized by chondrite [15]. Symbols: 1−rare-metal granites, 2−pyrochlore cumulat, 3−fluid-hydrothermal rocks with fluorides, 4−gneiss and shale contact phenites, 5−amphibolite from host rocks, 6–syenite porphyry of host rocks

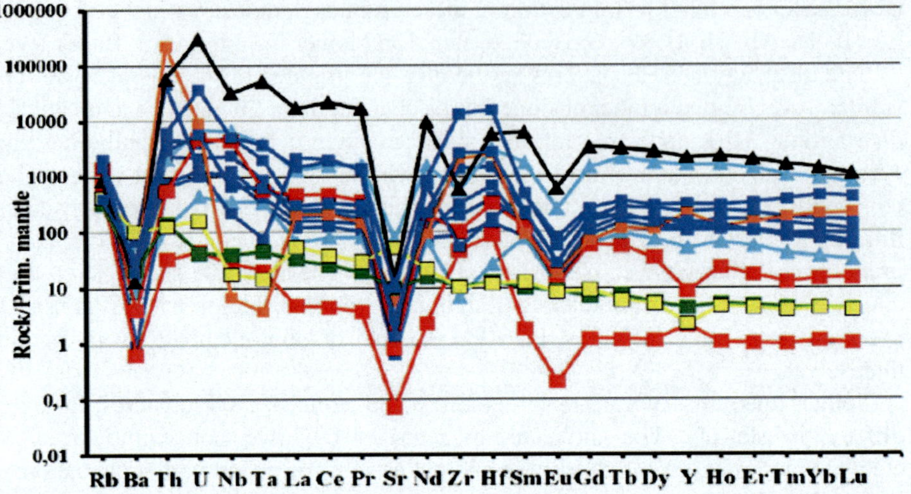

Fig. 7 Spider diagrams of rare elements for rocks of the Katuginsky massif, rationing was performed to the composition of the primitive mantle by [15]. Symbols in Fig. 6

6 Conclusion

The Zashikhinsky deposit is a typical example of alkaline granitic magmatism in the East Sayan rare-metal zone [1, 11]. The rare-metal granites of the deposit are divided into three facies varieties: (I) amphibole-bearing quartz-microcline-albite; (II) leucocratic quartz-albite-microcline; and (III) leucocratic quartz-albite transferring to albitites.

The Zashikhinsky granites show higher alkali contents (up to 12.68 wt% $Na_2O + K_2O$), and agpaitic index of 1.0–1.2 and present sodium type of alkalinity ($K_2O/Na_2O < 1$). In addition, they exhibit elevated Rb, Li, Y, Zr, Hf, Ta, Nb, Th, U, Zn, Ga, REE contents and lower Ca, Mg, Al, P, Ba, Sr levels.

On the Zashikhinsky deposit, the minerals, which accumulate trace and rare earth elements include columbite-tantalite, zircon, xenotime and struverite group minerals. The genetic relation of these minerals is shown and their magmatic genesis is proposed. The contents of rare and rare-earth elements in the minerals from the Zashikhinsky rare-metal granites are given.

Acknowledgements The study was supported by the Russian Foundation for Basic Research (project No. 20-05-00261_a, the integration project of the Irkutsk Scientific Center, Siberian Branch of the Russian Academy of Sciences (Block 1.4), and the state task project IX.129.1.4., RNF -21-17-00015.

References

1. Yarmolyuk, V.V., Lykhin, D.A., Shuriga, T.N., Vorontsov, A.A., Sugorakova, A.M.: Age, composition of rocks and geological setting of the Snezhnoe beryllium deposit: substantiation of the late paleozoic east sayan rare-metal zone. Russia Geol. Ore Deposits **53**(5), 390–400 (2011)
2. Yarmolyuk, V.V., Kuzmin, M.I.: Pozdnepaleozoic and early mesozoic rare metal magmatism of central asia: stages, regions, and formation settings. Geol. Ore Deposits **54**(5), 313–333 (2012)
3. Mashkovtsev, G.A., Bykhovsky, L.Z., Rogozhin, A.A., Temnov, A.V.: Prospects of rational development of complex niobium-tantalum-redkometallic deposits in Russia. Expl. Prot. Mineral Res. **6**, 9–13 (2011)
4. Arkhangelskaya, V.V., Ryabtsev, V.V., Shuriga, T.N.: Geological structure and mineralogy of the Russian tantalum deposits. Mineral Raw Mater. **25**, 155–189 (2012)
5. Archangelskaya, V.V., Shuriga, T.N.: Geological structure, zoning and mineralization of Zashikha tantalum-niobium deposit (in Russian). Domest. Geol. **5**, 7–10 (1997)
6. Kudrin, V.S., Shuriga, T.N.: Specificity of mineral composition of large rare-metal deposits in plutogenic alkaline quartz-albit-microcline metasomatites. In: Large and unique deposits of rare and noble metals, pp. 110–112. State mining institute, Saint-Petersburg (1998)
7. Alymova, N.V.: Metallogenic specialization and ore-bearing capacity of alkaline rare-metal granites of Zashikhinskoye deposit (Irkutsk region). Bull. Siber. Branch Earth Sci. Sect. Russian Acad. Nat. Sci. Geol. Prosp. Expl. ore Deposits **2**(55), 9–20 (2016)
8. Vladykin, N.V., Alymova, N.V., Perfil'iev, V.V.: Geochemical features of rare-metal granites of the Zashikha massif. Eastern Sayan Petrol. **24**(5), 554–568 (2016)
9. Kostitsyn, YuA, Altukhov, E.N.: The khäilama and aryskan massifs of alkali granitoids, easyern sayan: age and formation conditions by Rb-Sr isotopic and geochemical data. Geochem. Int. **42**(3), 195–204 (2004)
10. Gladkochub, D.P., Donskaya, T.V., Sklyarov, E.V., Saveleva, V.B., Mazukabzov, A.M., Bazarova, E.P., Kotov, A.B., Larin, A.M., Salnikova, E.B., Tolmacheva, E.V., Velikoslavinsky, S.D., Kovach, V.P., Zagornaya, N.Y., Vladykin, N.V., Alymova, N.V., Pisarevsky, S.A., Sharygin, V.V., Starikova, A.E., Khromova, E.A.: The unique Katugin rare-metal deposit (southern Siberia): Constraints on age and genesis. Ore Geol. Rev. **91**, 246–263 (2017)
11. Yarmolyuk, V.V., Kozlovsky, A.M., Nikiforov, A.V., Travin, A.V., Lykhin, D.A.: Composition, sources, and mechanisms of origin of rare-metal granitoids in the Late Paleozoic Eastern Sayan zone of alkaline magmatism: a case study of the Ulan Tolgoi massif. Petrology **24**(5), 477–496 (2016)
12. Frost, C.D., Frost, B.R.: On ferroan (A-type) granitoids: their compositional variability and modes of origin. J. Petrol. **52**, 39–55 (2011)
13. Whalen, J.B., Currie, K.L., Chappell, B.W.: A-type granites: geochemical, characteristics, discrimination and petrogenesis. Contrib. Miner. Petrol. **95**, 407–419 (1987)
14. Wedepohl, K.H.: The composition of the continental crust. Geochim. Cosmochim. Acta **59** (7), 1217–1232 (1995)
15. McDonough, W.F., Sun, S.-S.: The composition of the Earth. Ghemical Geol. **120**, 223–253 (1995)
16. Donskaya, T.V., Gladkochub, D.P., Sklyarov, E.V., Kotov, A.B., Larin, A.M., Starikova, A. E., Mazukabzov, A.M., Tolmacheva, E.V., Velikoslavinskiy, S.D.: Origin of the Paleoproterozoic rare-metal granites of the Katuginsky massif. Petrology **26**(1), 52–71 (2018)
17. Grechishchev, O.K., Zhmodik, S.M., Shcherbov, B.L.: Redcometalloye deposit Ulug-Tanzek (Tuva, Russia). Geo, Novosibirsk (2010)
18. Baisalova, A.O.: Features of metasomatic processes of rare-metal manifestations of granite massif Akzhaylyautas and adjacent areas (Ph.D. thesis). Kazakhstan National Research Technical University, Almaty (2018)

19. Bastos, N., Pereira, V.: The world-class Sn, Nb, Ta, F (Y, REE, Li) deposit and the massive cryolite associated with the albite-enriched facies of the Madeira A-type granite, Pitinga mining district, Amazonas State. Brazil The Canadian Mineral. **47**(6), 1329–1357 (2009)
20. Salvi, S., Williams-Jones, A.E.: Orthomagmatic fluid inclusions in the strange lake complex, Quebec/Labrador: implications for Y Zr and ree concentration. Plinius **5**, 189–190 (1991)
21. Kotov, A.B., Larin, A.M., Gladkochub, D.P., Salnikova, E.B., Sklyarov, E.V., Tolmacheva, E.V., Donskaya, T.V., Velikoslavinsky, S.D., Yakovleva, S.Z.: New data on the age of ore formation in the unique Katugin rare-metal deposit (Aldan Shield). Dokl. Earth Sci. **463**(2), 782–786 (2015)
22. Kovalenko, V.I., Kovalenko, N.I.: Ongonites−subvolcanic analogues of rare-metal Li-F granites. Science, Moscow (1976)
23. Naumov, V.B., Kovalenko, V.I., Vladykin, N.V.: Thermometric research of inclusions in topazes from ongonites (in Russian). Reports USSR Acad. Sci. **199**(3), 681–683 (1971)
24. Andreeva, I.A.: Genesis and mechanisms of formation of rare-metal peralkaline granites of the Khaldzan Buregtey massif, Mongolia: evidence from melt inclusions. Petrology **24**(5), 462–476 (2016)
25. Kovalenko, V.I., Yarmolyuk, V.V., Naumov, V.B., Dorofeeva, V.A.: Volatile components (H2O, CO_2, CL, F, AND S) in magmas of intermediate and acid compositions from distinct geodynamic settings: evidence from melt inclusions and chill glasses. Petrology **8**(6), 525–556 (2000)

Mineralogical and Geochemical Features on the Tsagaan-Uul Pantellerite Dike Swarm (Northern Mongolia)

Yu. D. Shcherbakov, A. B. Perepelov, S. S. Tsypukova, E. Dalai-Eredeine, and M. Yu. Puzankov

Abstract

The article shows first results of the large pantelleritic dike swarm study. Dike swarm breaks through the Middle Paleozoic granites and has a length of up to 40 km with the general Northeastern extension. The swarm rocks composition corresponds to pantellerite trachytes (AGP $\sim$ 1.1) and pantellerites (AGP $\sim$ .3). Felsic feldspars and amphiboles (arfvedsonite and cataphorite) as the main phenocrysts, in the microlite paragenesis, represent the mineral composition and the presence of such rare minerals as lorenzenite, Ca-ancylite, catapleiite, monazite, enigmatite eudialite and titanium is noted, such a set of minerals corresponds to typical agpaitic rocks. The distribution of rare elements in the Tsagaan-Uul dike swarm rocks both in character and concentration level corresponds to the petrotype. The key difference is the presence of Nb–Ta anomaly in pantellerite trachytes and pantellerites, which may indicate the involvement of fluid in the formation of the swarm rocks.

Keywords

Alkaline rocks · Pantellerites · Lorenzenite

Yu. D. Shcherbakov (✉) · A. B. Perepelov · S. S. Tsypukova
Vinogradov Institute of Geochemistry SB RAS, Irkutsk 664038, Russia
e-mail: scherb@igc.irk.ru

E. Dalai-Eredeine
Institute of Palaeontology and Geology, Ulaanbaatar 13331, Mongolia

Irkutsk State University, Irkutsk 664003, Russia

M. Yu. Puzankov
Institute of Volcanology and Seismology, FEB RAS Petropavlovsk-Kamchatsky, Kamchatsky 683006, Russia

N. Vladykin (ed.), *Alkaline Rocks, Kimberlites and Carbonatites: Geochemistry and Genesis*, Springer Proceedings in Earth and Environmental Sciences,
https://doi.org/10.1007/978-3-030-69670-2_5

1 Introduction

Interest in the study of alkaline felsic rocks is due to their diverse mineral composition, as well as their unusual origin, which is associated with deep differentiation of basite magma in the transit chambers or crust rocks anatexis. One of the most acidic among alkaline volcanic rocks are pantellerites—rocks with large disseminated feldspars (anorthoclases and sanidines), alkaline amphiboles and pyroxenes. They are most often found in differentiated volcanic series, not rarely having bimodal distribution of chemical compositions with predominance of alkaline and felsic rocks over basite ones. Intracontinental rifts are the most typical structure for such magmatism, for example, the central Kenyan alkaline province [1–3], the Gobi-Tienshan rift zone [4] sometimes-bimodal associations are found in conditions of active continental margins [5].

Recent studies of the Late Paleozoic-Early Mesozoic North-Mongolian rift zone [6] revealed a large Tsagaan-Uul pantellerite dike swarm (named after the Tsagaan-Uul samon in the Hovsgol aimag of Mongolia). Its length is up to 40 km, width up to 15 km, and general stretch of the northeast. It is represented by a complex of numerous sub-parallel dikes with the length of some of them up to 12 km. Dikes break through the massif of Middle Paleozoic granites, and in the southwestern part of the swarm overlap by Neogene basalts, which in some cases streamline the remains of dikes. The whole swarm is confined to a large fault (Fig. 1).

The Tsagaan-Uul dike swarm breeds are not changed, they are massive, dark gray, light gray, often with a bluish tint, medium and fine grained. The most powerful dikes have a zonal structure, the central zone is clearly distinguished, as a rule, a medium light bluish-gray, color of the rock becomes black closer to the quenching part. Some dikes have a quenching part, which contains captured leucocratic material. Small dikes have a light gray color and "infected" with leucocratic material. The complex also includes basaltoid dikes, they are poorly distributed, most often found in the form of broken boulders, however, there are also non-destroyed dikes, which break through the trachitis and pantellerite dikes. In general, the Tsagaan-Uul pantellerite dike swarm, can be attributed to the bimodal association of rocks in the ratio of salic and basite rocks, clearly observed the prevalence of pantellerites and trachitis, in compare with basitic dikes.

2 Analytical Techniques

Analytical studies were carried out at the "Isotope and Geochemical Research" Center of the Institute of Petrochemistry of the Siberian Branch of the Russian Academy of Sciences. The contents of petrogenic oxides in rocks were determined by XRF method using multichannel X-ray spectrometer CPM-25 ("Orelnauchpribor", Russia). The main parameters of surveying, measurements were made using X-ray tube with Rh anode at 30 kV and 40 mA current. The samples were

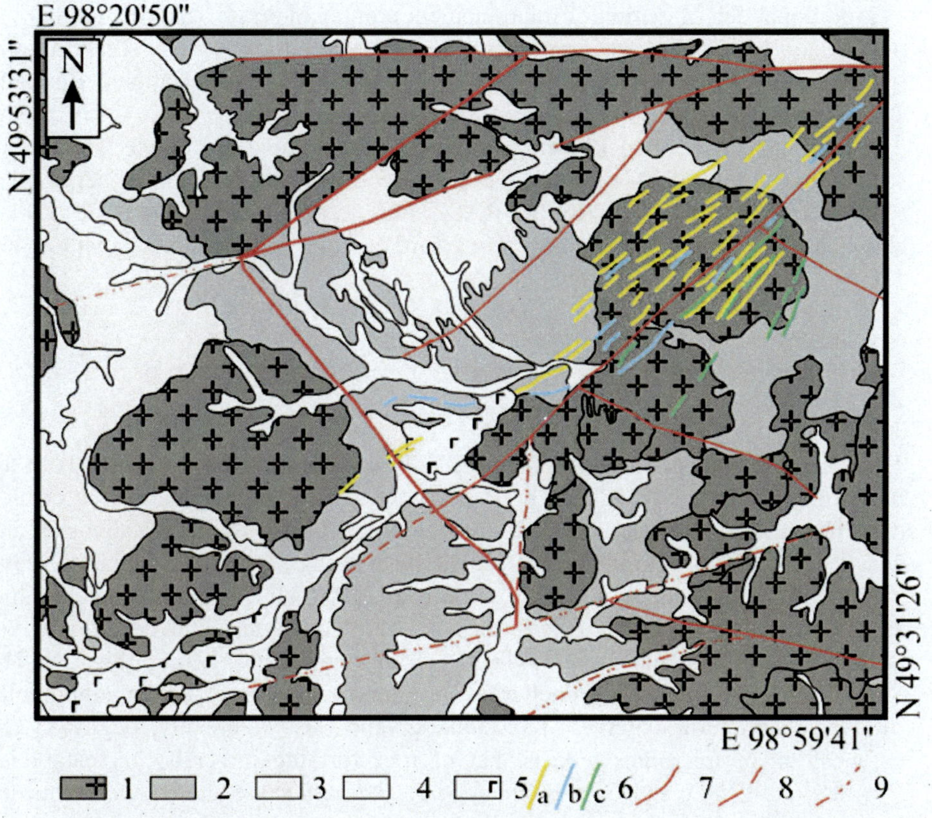

Fig. 1 Geological scheme of Tsagaan-Uul dike swarm location. 1—Pz_2 intrusive complex (granites and granitoids); 2—V-Є deposits (sandstones siltstone, pelit, gravelite); 3—Cambrian system (meta sandstones, limestones, quartzites); 4—Upper Quaternary (modern sediments. Deluvially proluvial rubble); 5—Neogene basalts; 6–dikes of pantellerites (a), pantelleritic trachytes (b), trachybasalts (c); 7—breaking faults; 8—alleged breaking faults; 9—alleged breaking faults under quaternary sediments

homogenized using flux fusion, lithium metaborate (LiBO2) in an induction furnace in carbon fiber crucibles at 1100 °C. Quality control of rock-forming oxides determinations was carried out using standard samples SG-1A (granite, Russia), SG-2 (granite, Russia) and JG-2 (granite, Japan). The concentrations of trace elements in the samples were obtained by mass spectrometry with ionization in inductively coupled plasma (ICP-MS). Measurements were performed on Perkin Elmer NexION 300D magnetic sector mass spectrometers (USA). Correct determination of microelement concentrations and drift of the device were controlled using international standard samples of basalts BHVO-1, BHVO-2 (USGS) and andesites AGV-1, AGV-2 (USGS) in every 5–6 samples.

Additional determination of microelement composition of rocks was carried out on the Agilent 7700 × quadrupole mass spectrometer by Agilent Technologies ("Baikal Center of Nanotechnologies", Technopark at Irkutsk National Research Technical University).

The composition of rock-forming minerals was studied on the electron microscope "LEO 1430VP" (Carl Zeiss, Germany) with the analyzer "Inca Energy 300" (Oxford Instruments Ltd) (GIN SO RAS) at an accelerating voltage of 15–20 kV and a current of ~0.5 nA (probe size <1 μm, the time of spectrum acquisition 40–120 s).

3 Mineralogy

The mineral composition of the Tsagaan-Uul dike swarm rocks corresponds to the rocks of agpaitic type [7] (Table 1). In pantellerite trachytes, the main rockforming minerals are acid feldspar and alkaline amphiboles (harfvedsonite, cataphorite) and quartz. Pantellerite microlites represented by aegirine, enigmatite, elpidite, catapleiite, monazite and apatite in addition to KPS, albite, quartz and alkaline amphiboles. A characteristic feature for pantellerite trachytes is the development of a rare mineral—lorenzenite ($Na_2Ti_2Si_2O_9$) in amphibole grains (Fig. 2). The rock-forming minerals of pantellerites are represented by alkaline amphiboles (harfvedsonite, ferrarichterite), KPS, quartz, and, less frequently, aegirine. The composition of microlites repeats that of rock-forming minerals, but catapleiite ($Na_2ZrSi_3O_9$–$2H_2O$) and Ca-ancylite appear. The basites are altered and composed by amphibole, plagioclase, chlorite, ilmenite, biotite and quartz.

4 Major Oxides

The main rock varieties composing the Tsagaan-Uul pantellerite swarm are pantellerites and pantellerite trachytes (Fig. 3). Distribution of rock compositions of the series is bimodal; at present, no species corresponding to trachandesites and basaltic trachyandesites have been found. The most siliceous dike swarm rocks are defined as pantellerites and pantellerite trachytes according to the classification diagram and are agpaitic (AGP 1.0–1.4). It should be noted that pantellerite trachytes are characterized by an predominance of Na over K (K/Na ~ 1.4), while pantellerites and basalts have approximately the same Na and K content. The graphs show that, in general, for the dike swarm rocks, there is a decrease in the concentration of basic oxides with the growth of silic content (Table 1). A significant gap between the figurative points of basalt compositions with trachytes and pantellerites due to lack of information about intermediate phases. The difference between trachytes and pantellerites is well noticeable in elevated concentrations of Al_2O_3, MgO, CaO, and Na_2O in trachytes (Table 1). The group of pantellerites having large phenocrysts of

Table 1 Major and trace elements concentrations of Tsagaan-Uul dike swarm

Rock	Basites		Trachytes			Pantellerites					
SiO_2	45.30	45.37	60.17	61.92	61.56	72.56	71.94	72.09	72.18	72.49	72.85
TiO_2	2.34	1.89	0.74	0.77	0.71	0.48	0.52	0.53	0.53	0.51	0.53
Al_2O_3	15.37	14.81	14.32	14.68	14.66	9.06	9.50	9.96	9.10	9.08	9.63
Fe_2O_3	12.21	10.90	9.30	9.32	9.32	7.44	7.67	7.85	7.99	8.32	7.75
MnO	0.21	0.17	0.28	0.29	0.28	0.16	0.15	0.16	0.19	0.21	0.19
MgO	4.96	5.39	0.97	0.32	0.36	0.01	0.05	0.06	0.05	0.05	0.06
CaO	8.10	8.24	1.23	1.47	1.13	0.31	0.50	0.43	0.44	0.46	0.39
Na_2O	3.21	2.94	7.22	6.25	6.79	4.30	4.03	4.31	4.51	4.44	4.18
K_2O	1.39	1.51	4.93	5.04	4.94	4.43	4.53	4.58	4.43	4.43	4.55
P_2O_5	0.84	0.92	0.10	0.07	0.06	0.02	0.03	0.02	0.02	0.02	0.04
LOI	4.57	7.02	0.63	0.46	0.48	0.47	0.38	0.36	0.60	0.40	0.31
Total	98.68	99.35	100.03	100.58	100.29	99.48	99.30	100.36	100.05	100.40	100.48
Rb	20.9	24.7	202	194	199	303	260	262	292	286	280
Sr	896	1193	21.2	45.9	9.05	9.05	16.4	6.18	9.57	13.5	7.11
Y	34.3	34.8	123	113	117	129	126	98.5	147	139	129
Zr	232	256	1764	1509	1525	1311	1529	1490	1801	1596	1772
Nb	16.0	17.5	83.1	82.1	84.6	77.0	71.9	68.5	85.2	70.0	79.0
Ba	625	867	17.4	73.4	28.8	15.5	20.5	15.7	21.6	24.7	27.2
La	33.5	46.0	158	148	162	181	165	160	202	153	189
Ce	75.4	96.5	322	302	320	383	329	320	402	305	370
Pr	9.79	12.6	38.0	35.4	38.1	41.5	39.5	39.5	48.6	36.9	45.3
Nd	41.0	49.5	133	121	132	145	138	138	171	132	160
Sm	8.59	9.99	25.6	23.9	25.0	27.1	27.2	26.5	33.7	25.8	31.0
Eu	2.57	2.76	1.72	1.56	1.68	2.12	1.91	1.80	2.16	1.58	2.19
Gd	7.59	8.12	21.2	19.5	21.0	22.1	22.5	21.0	28.7	22.7	25.6
Tb	1.11	1.14	3.47	3.29	3.43	3.58	3.66	3.50	4.62	3.76	4.22
Dy	6.42	6.44	21.0	20.4	21.6	21.9	22.9	21.8	28.2	23.8	25.8
Ho	1.22	1.22	4.39	4.26	4.51	4.59	4.69	4.43	5.72	5.04	5.34
Er	3.31	3.30	12.9	12.5	13.6	13.8	13.7	13.3	16.7	15.1	15.5
Tm	0.46	0.46	1.97	2.01	2.16	2.20	2.14	2.11	2.65	2.31	2.43
Yb	2.80	2.77	13.5	13.5	14.4	14.8	14.4	14.3	17.6	15.5	16.4
Lu	0.42	0.42	2.14	2.09	2.22	2.23	2.25	2.19	2.71	2.46	2.48
Hf	4.48	5.05	31.9	30.5	32.6	25.1	30.6	31.5	37.5	33.7	36.8
Ta	0.88	0.86	4.30	4.30	4.60	4.52	4.31	4.37	5.38	4.34	5.07
Pb	7.02	8.35	47.5	46.6	49.8	36.9	47.5	51.3	63.0	58.5	61.0
Th	1.11	1.55	19.1	18.0	19.2	32.2	26.9	26.7	32.3	24.2	25.5
U	0.44	0.57	5.33	4.25	6.24	9.08	7.66	5.69	8.88	8.87	7.14

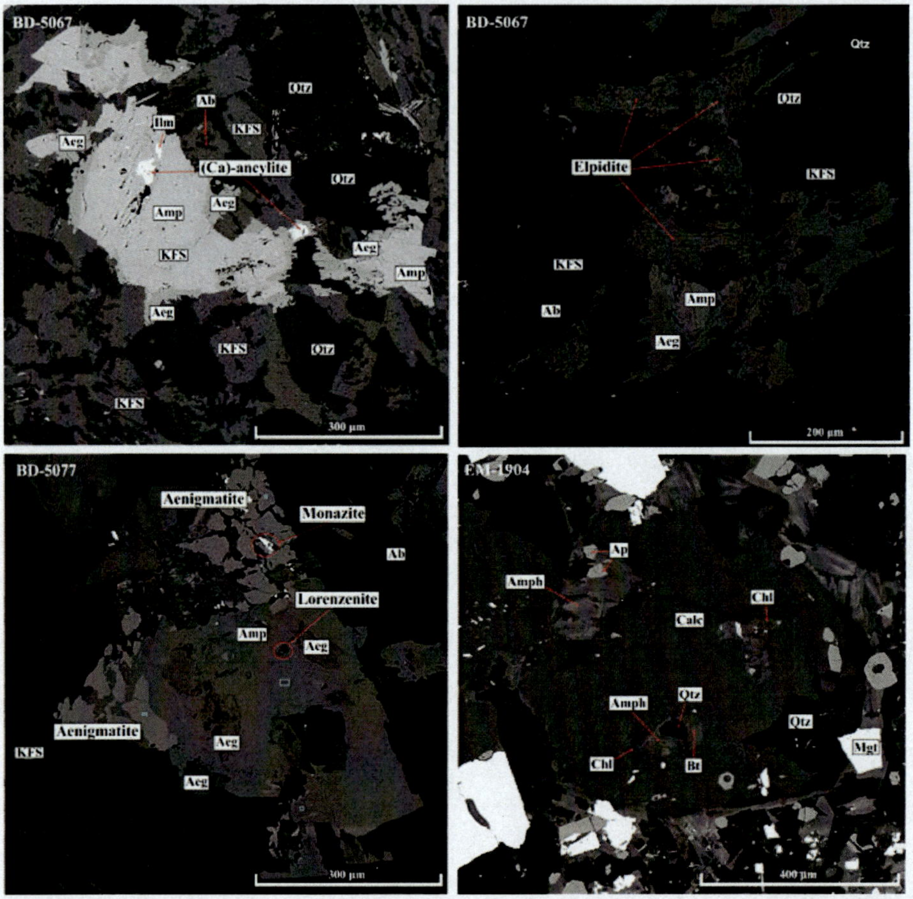

Fig. 2 BSE image of common mineral assemblages of the Tsgaan-Uul dike swarm rocks. Ab–albite, Aeg–aegirine, Amp–amphibole, Ilm–ilmenite, KFS–potassic feldspar, Qtz–quartz, Calc–calcite, Mgt–magnetite, Chl–chlorite, Bt–biotite, Ap–apatite

alkaline feldspar differs from usual pantellerites in increased concentrations of Al_2O_3 and K_2O and lower concentration of FeO. Basalts are not of alkaline type, and also have high losses (up to 7%) and are carbonated.

5 Trace Elements

In comparison with the petrotype, the pantellerite trachytes of the Tsagaan-Uul dike swarm have lower concentrations of Ba, Nb and Ta, but are enriched with Zr and Hf, which may be associated with not typical catapleitic mineralization [8]. In comparison with rocks of Southern Mongolia of the Noen and Toast ranges, relative

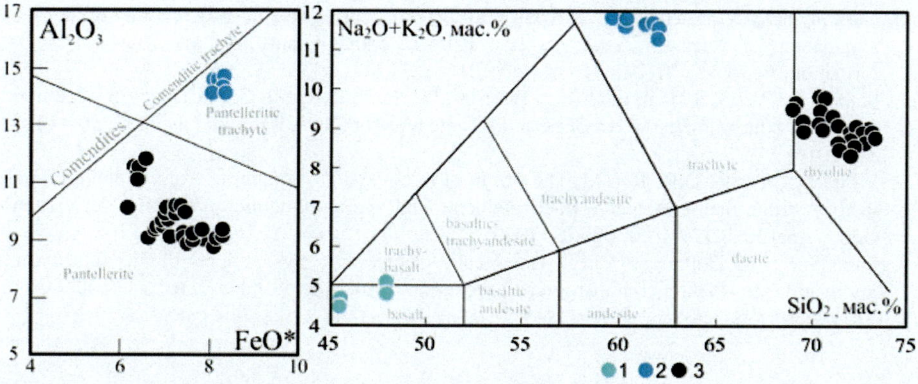

Fig. 3 Diagrams of FeO versus Al_2O_3; SiO_2 versus Na_2O + K_2O for the dike swarm rocks. *Notes* 1–trachybasalts; 2–pantelleritic trachytes; 3–pantellerites

enrichment with rare earth elements Zr and Hf is clearly seen [4]. It is important to note the presence of negative Nb–Ta anomaly, which may be a consequence of the participation of metasomatized mantle in magma formation. In comparison with trachytes of the active continental margin of Kamchatka, enrichment with REE and HFSE elements is also noted [5].

The pantellerites inherit all signs of trachytes of the Tsagaan-Uul dike swarm in comparison with the petrotype, pantellerites of Southern Mongolia, and Comenditis of the active continental margin of Kamchatka; they are also relatively enriched with rare earth elements, Zr and Hf, have Nb–Ta anomaly and are depleted Ba.

Acknowledgments The study was conducted with financial support from RFBR, grants No. 18-35-00,294 mol-a, 18-55-91,049 mg.

References

1. Macdonald, R., Baginski, B., Leat, P.T., White, J.C., Dzierzanowski, P.: Mineral stability in peralkaline silicic rocks: information from trachytes of the Menengai volcano. Kenya. Lithos **125**(1–2), 553–568 (2011)
2. Macdonald, R., Scaillet, B.: The central Kenya peralkaline province: insights into the evolution of peralkaline salic magmas. Lithos **91**(1–4), 59–73 (2006)
3. Scaillet, B.: Experimental constraints on the relationships between Peralkaline Rhyolites of the Kenya Rift Valley. J. Petrol. **44**(10), 1867–1894 (2003)
4. Kozlovsky, A.M., Yarmolyuk, V.V., Kovalenko, V.I., Savatenkov, V.M., Velevitskaya, T.A.: Trachytes, commendites and pantellerites of the Late Paleozoic riftogenic bimodal association of the Noyeon and Toast ranges of Southern Mongolia: peculiarities of differentiation and contamination of Alkaline-Salic melts. Petrology **15**(3), 257–282 (2007)
5. Flerov, G.B., Perepelov, A.B., Puzankov, M. Yu., Koloskov, A.V., Filosofova, N.M., Shcherbakov, Yu. D.: Space-time relations of volcanic associations of different alkalinity of the Belogolovsky Massif (Sredinny Ridge of Kamchatka). Part II. Geochemistry of volcanic rocks and sources of magmas. Volcanol. Seismol. **4**, 3–26 (2016)

6. Yarmolyuk, V.V., Kuzmin, M.I., Kozlovskiy, A.M.: Late Paleozoic-Rannemesozoic intraplate magmatism of Northern Asia: Trappes, Rifts, Batolites-Giants and geodynamics of their formation. Petrology **21**(2), 115–142 (2013)
7. Stabile, P., Webb, S., Knipping, J.L., Behrens, H., Paris, E., Giuli, G.: Viscosity of pantelleritic and alkali-silicate melts: effect of Fe redox state and Na/(Na + K) ratio. Chem. Geol. **442**, 74–82 (2016)
8. White J.C., Parker D.F., Ren M. The origin of trachyte and pantellerite from Pantelleria, Italy: insights from major element, trace element, and thermodynamic modelling. J. Volcanol. Geothermal Res. **179**(1–5), 33–35 (2009)
9. Sun, S.-S., McDonough, W.F.: Chemical and isotopic systematics of oceanic basalts: implications for mantle composition and processes. Geol. Soc. London, Spec. Publ **42**(1), 313–345 (1989)

Indicator Role of Spinel-Olivine Paragenesis in Determining Conditions of Dunites Origin in the Central Cart of the Konder Massif

G. P. Ponomarev, N. V. Vladykin, and I. A. Sotnikova

Abstract

Calculations of conditions for crystallization of spinel-olivine paragenesis from rocks of the core of the Konder alkaline ultrabasic massif were performed. Empirical system of equations for equilibrium calculations of mineral paragenesis was used.

Keywords

Olivine · Spinel · Balance · Konder massif

1 Introduction

The increased and industrial concentrations of platinum group elements in the continental segments of the Earth's crust [1] are mostly confined to ultrabasite associations—folded regions, stages of tectono-magmatic activation of stabilized regions and their associations in the oldest regions of the Earth. Platinoids in

G. P. Ponomarev (✉)
Institute of Volcanology and Seismology of FEB RAS, 9, Piypa Blvd., Petropavlovsk–Kamchatsky 684006, Russia
e-mail: ponomarev@kscnet.ru

N. V. Vladykin · I. A. Sotnikova
Vinogradov Institute of Geochemistry SB RAS, 1A Favorsky str., Irkutsk 664033, Russia
e-mail: vlad@igc.irk.ru

Irkutsk Scientific Center of Siberian Branch of Russian, 134 Lermontov St., Irkutsk 664033, Russia

N. Vladykin (ed.), *Alkaline Rocks, Kimberlites and Carbonatites: Geochemistry and Genesis*, Springer Proceedings in Earth and Environmental Sciences,
https://doi.org/10.1007/978-3-030-69670-2_6

ultrabasites of these associations are, as a rule, spatially within the rock, concentrated together with chromite [2].

Genetically, the massifs of ultrabasites from these associations are rather heterogeneous, which generates a variety of genetic models, which are part of the mineral association of chromite-magnetite-minerals of platinoids.

The genesis of concentric-zonal ultrabasic massifs from the association of the stage of tectonic and magmatic activation of stabilized areas is also a subject of discussion and is represented by a wide range of models. The interest to these massifs is connected with the fact that some of them served as sources of ore component for alluvial deposits formation of platinum group elements. The results of the studies of rocks composing these massifs have led to the publication of a large number of analytical data, both on the rock compositions of these massifs and minerals participating in their structure and inclusions in these minerals. Such detailed analytical studies of the phase composition of rocks make it possible to use, for determining their genesis, the system of equations for the paragenesis of melt-spinel-olivine presented in Ponomarev and Puzankov [3, 4] and the criterion of the threshold content of calcium in olivine [3–6].

2 Phase Composition Data and Calculation Results

Perhaps the most studied among similar massifs in Russia (Konder, Inagli, Chad, etc.) dated to the south-east of the Siberian platform is the Konder complex. Data on the structure of the massif, structural features of dunites, and their mineralogy were borrowed from Prikhodko and Ponomarev [7], Nekrasov et al. [8], Simonov et al. [9], Vasilyev and Gora [10], Ponomarev and Puzankov [3], Pushkarev et al. [2] and presented in generalized form. Common for these magmatic bodies is a concentric zonal structure with a central dunite core. This core, on the example of the Konder Massif, is usually heterogeneous in terms of structure and texture features. Fine-grained dunite varieties are developed mainly in the marginal parts of the core, while its main volume is composed of medium-grained varieties occasionally morphing into coarse-grained ones. The central body of the massif is composed of dunites proper, among which there is also some olivinite. The rocks have mainly hypidiomorphic structure and massive texture. The main rock-forming mineral is magnesian olivine (Fo 85–92). In cases where olivine crystals are included in chromite crystals, the former have minimal glandularity. The studied olivine crystals in the overwhelming majority of cases contain from 0.1 to 0.7 wt% CaO. The share of crystals of low calcium olivine (containing <0.1 wt% of CaO) does not exceed 1–2%. There is no distinct connection between glandularity of olivine crystals and contents of CaO in them. The main accessory mineral in the rock is several generations of spinel, represented mainly by ferrichromite and the less common chromagnetite. The structural relationships of spinels in the rock are distinguished by the interstitial difference, which is quantitatively predominant, and chadacrystals in olivine. Clinopyroxenes, amphiboles, phlogopite, and native minerals such as iron, copper, tin, antimony, and graphite may also be considered to be accessory in dunites. In addition, in some blocks of dunite core there are a limited number of veins and nests composed of sulfides (pyrrhotite, pentlandite, pyrite, chalcopyrite, arsenopyrite). The main ore component of the complex, concentrated in the dunite core, are nests, veins, lenses and slices of ore spinel—mainly ferrichromite. These spinel clusters in the rock contain platinoids, mainly in the form of solid solutions. Secondary changes are serpentinization, affecting up to 25% of the rock, but with the depth decreasing, and absent below ~0.5 km from the day surface.

Of interest is the genesis of a number of dunite core rocks and the mineral phases composing them. Data on the compositions of olivine, spinel and melt inclusions from dunites were collected [7, 9] (Tables 1 and 2).

The following data were used. From Prikhodko and Ponomarev [7], 4 pairs of compositions of olivine spinel association from core dunites and chromite ores representing the most and least magnesian differences of these phases were used; from Simonov et al. [9], compositions of olivines from melt inclusions in chromspinelids and compositions of chromspinelids were used, some of which represent their compositions in close proximity to the largest melt inclusions. 14 olivine spinel paired analyses were collected and used for the studies. These are 4 olivine spinel pairs (1–3 medium dunites and 4 chromite ores) and 10 pairs of different combinations of olivine compositions from melt inclusions (analyses 5, 6) with spinel compositions (analyses 7–11). All these pairs were tested for equilibrium of the assumed olivine and spinel paragenesis in case of their crystallization in a "dry" melt using 5 formulas from Ponomarev and Puzankov [3], for the case of a

Table 1 The compositions of olivine and spinel crystals used for calculations (weight, %)

No.	SiO_2	TiO_2	Al_2O_3	Cr_2O_3	Fe_2O_3	FeO	MnO	MgO	CaO
1	40.39	0.03	0.04	0.02		9.08	0.19	49.07	0.08
		0.55	6.86	51.28	13.01	21.16	0.51	8.01	
2	39.92	0.01	0.05	0.02		8.52	0.18	49.57	0.22
		0.68	5.68	42.6	21.77	21.8	0.51	7.2	
3	39.75	0.03	0.04	0.04		8.5	0.15	50.27	0.29
		0.57	6.75	54.14	10.49	19.17	0.44	9.26	
4	40.01	0.03	0.03	0.01		9.88	0.21	48.04	0.29
		0.77	6.24	52.08	12.83	17.19	0.4	10.42	
5	41.25			1.18		6.67		49.66	1.24
6	40.54			1.36		4.78		52.78	0.53
7		0.65	7.06	50.95	12.24	21.58	0.51	7.65	
8		0.71	7.35	50.74	12.47	20.3	0.49	8.56	
9		0.67	6.85	50.31	13.63	20.96	0.49	8.13	
10		0.72	7.34	51.14	11.87	21.2	0.44	8.04	
11		0.65	6.99	51.62	11.98	19.54	0.47	8.9	

Note Analyses of 1–4—olivine spinel pairs [7], and this table); 5, 6—olivines from melt inclusions in spinels [9], Table 4); 7–11—spinels [9], Tables 2 and 3)

Table 2 Glass compositions from crystallized melt inclusions in spinels, used for calculations (weight, %)

No. п/п	SiO_2	TiO_2	Al_2O_3	Cr_2O_3	FeO	MgO	CaO	Na_2O	K_2O
1	44.57	0.44	14.53	2.31	4.68	16.48	9.29	2.79	4.91
2	46.63		8.75	1.34	12.43	10.29	14.05	1.44	5.06
3	50.46		13.25	1.38	5.12	8.94	10.99	5.55	4.30

fluid-containing melt using 4 formulas (69, 71, 75, 76) from Ponomarev and Puzankov [4]. The criterion of equilibrium or disequilibrium of olivine-spinel pairs was the value of the standard square error. These calculations allow us to state the following. All olivine-spinel pairs are not in equilibrium in the variant of their crystallization in "dry" conditions; olivine-spinel pairs (analyses 1–4, Table 1) can be considered as equilibrium for the variant of fluidized (water) melt in the pressure range ~1–1.5 GPa (Formula 76 allows to take into account the influence of pressure on equilibrium); olivines (analyses 5, 6) from melt inclusions in spinels are not in equilibrium with any spinel composition (analyses 7–11), both for the case of crystallization under "dry" conditions and with participation of fluid component. This indicates their crystallization in the melt inclusion proper and allows to use compositions of melt-olivine pairs for calculation of crystallization conditions. Compositions of olivines from melt inclusions (analyses 5, 6, Table 1) and glasses (analyses 1–3, Table 2) were used for the calculation of crystallization conditions of melt inclusions. 6 pairs of compositions of the olivine melt system were formed, which were used to calculate possible conditions of olivine crystallization. Calculations were performed using 2 groups of formulas created from an array of experimental data with an unsaturated fluid melt (as the most probable one) (Table 3) and created from an array of data with known melt water contents (Table 4). The used formulas are given in Ponomarev and Puzankov [4]. In Formulas 23–29 of this work, the compositions of melt and olivine are used to calculate the required value; in Formulas 31–34 only the melt composition is used.

Under ideal conditions, in case of equilibrium of crystal and melt and equilibrium of the melt itself, all calculated values of temperature, pressure, content of elements in the melt, regardless of the set of values used in this or that formula, should be within the limits of RMS errors. This is the criterion for choosing the values of the calculated values. Taking into account the disparity of the obtained values of the same value and the values of standard errors to each of the formulas given in Ponomarev and Puzankov [4], we can assume that the closest to equilibrium is the system (No. n/p 4, Table 3), consisting of olivine (Table 1, Analysis 6) and melt (Table 2, Analysis 1). The crystallization conditions of the unsaturated melt correspond to the equilibrium crystallization temperature ~1320–1360 ± 40 °C and the pressure ~2 ± 0.3 GPa.

Table 3 Calculation results of the equilibrium crystallization condition of olivine in the melts inclusions from chromium spinel in a variant of unsaturated fluid melt

No.	1	2	3	4	5	6	7	8	9	10
Mod.	t, °C		Atomicquantities, %				t, °C	W_{GPa}	Atomic qty, %	
1	1395	1284	−2.5	−2.5	−2.8	−6.30	1361	1.8	−2.91	−0.9
2	1235	1238	4.51	4.26	0.94	3.41	1228	0.5	3.91	−4.6
3	1288	1437	5.84	5.78	5.38	8.04	1265	0.5	8.30	−6.7
4	1336	1319	0.39	0.42	-4.1	-3.41	1361	2.0	−2.91	−0.1
5	1183	1277	7.41	7.16	-0.2	6.33	1228	0.7	3.91	−3.8
6	1235	1522	8.79	8.70	4.24	10.96	1265	0.6	8.30	−5.9

Note Model 1–3 is a melt-olivine system with composition of olivine 5 (Table 1), 4–6 is a melt-olivine system with composition of olivine 6 (Table 1). n/p 1—equilibrium temperature value (Formula 23); 2—equilibrium temperature value (Formula 24); 3 and 4—difference between calculated (by Formulas 25 and 26) and real Al content in the melt (glass); 5 and 6—difference between calculated (by Formulas 28 and 29) and real Mg content in the glass; 7—equilibrium temperature value obtained only by the melt composition (Formula 31); 8—equilibrium pressure value obtained by the melt composition (Formula 32); 9 and 10—difference between the calculated (by Formulas 33 and 34) and the real Mg content in the melt

Table 4 Calculation results of olivine crystallization conditions in melt inclusions from chromium spinel in the variant with known water content in the melt

	1	2	3	4	5	6	7	8	9	10	11	12	13
	T°C							$P_{\text{ГПа}}$	Атом. кол. %			T°C	$P_{\text{ГПа}}$
1	1375	1367	1205	1302	1328	1273	1289	1.2	0.68	−3.0	−4.4	1243	14.7
2	1287	1279	1114	1212	1238	1144	1160	1.2	4.9	0.8	1.5	1169	13.0
3	1354	1346	1169	1271	1297	1203	1218	1.2	6.8	3.2	5.4	1219	13.2
4	1341	1334	1288	1302	1328	1315	1331	1.8	−0.3	−1.7	−4.4	1244	15.8
5	1259	1251	1196	1216	1242	1187	1203	1.8	3.9	1.8	1.5	1169	14.1
6	1324	1316	1251	1274	1300	1245	1261	1.8	5.8	4.4	5.4	1219	15.7

Note No. n/p: 1—equilibrium temperature value (Formula 47, melt water content 1 wt%); 2—equilibrium temperature value (Formula 47, melt water content 2 wt%); 3—equilibrium temperature value (Formula 48); 4—equilibrium temperature value (Formula 49, pressure—1.4 GPa, 2 wt% H_2O); 5—value of equilibrium temperature (Formula 49, pressure—1.8 GPa, 2 wt% H_2O); 6—value of equilibrium temperature (Formula 50, pressure—1.4 GPa); 7—value of equilibrium temperature (Formula 50, pressure—1.8 GPa,); 8—the value of equilibrium pressure (Formula 51, 2 wt% H_2O); 9—the difference between the calculated (Formula 52, 2 wt% H_2O) and the real content of Al in the melt (glass); 10—the difference between the calculated (Formula 53) and the real content of Mg in the glass; 11—the difference between the melt composition calculated (Formula 56, 2 wt% H_2O) and the real content of Mg in the glass; 12—the value of the equilibrium temperature obtained only by the melt composition (Formula 57, pressure —1.8 GPa, 2 wt% H_2O); 13—the value of the equilibrium pressure obtained by the melt composition (Formula 59, 2 wt% H_2O)

To calculate the conditions of crystallization of the system water-containing melt olivine by the formulas from Ponomarev and Puzankov [4] we used the value of H_2O content in melt 1 and 2 wt%. These values are close to the minimum supposed water content in fluidized island-arc magma [4], as the most studied. At the same time, the numerical values of water content in magma of different geodynamic settings given in the literature differ and have wide enough variations. For calculations, the same 6 variants of compositions of the melt-olivine system were used, namely, compositions of olivines from melt inclusions (analyses 5, 6, Table 1) and analyses of glasses (1–3, Table 2).

The most equilibrium, judging by the differences in calculated values taking into account the standard errors, is model 4 (Table 4), which consists of olivine (Table 1, Analysis 6) and melt (Table 2, Analysis 1). The crystallization conditions closest to real correspond to the melt water content ~1 wt%, equilibrium crystallization temperature ~1300–1340 ± 50 °C and pressure ~1.6–1.8 ± 0.2 GPa.

The compositions of homogenized melt inclusions in spinels (Simonov et al. [9], Table 5, Analyses 2.2, 11.1, 11.3, 9.3, 8.6) were used to calculate the crystallization conditions of the spinel melt system, listed in Table 5 in the descending order of magnesium content in glass. The wide range of compositions of quenching glasses (Table 5) suggests that they are not in equilibrium with sufficiently homogeneous spinel composition (Simonov et al. [9], Table 1, analyses 7–11). The most and least magnesial differences in spinel composition were used (Simonov et al. [9], Table 1, analyses 10, 11). 10 pairs of melt spinel analyses were formed and used for calculations.

All calculated values, taking into account RMS errors, demonstrate the lack of equilibrium in all formed pairs as between spinel compositions and melt as well as between elements in melts. This is indicated by the following results: the obtained numerical pressure values do not make any physical sense—there are significant differences between the calculated and real concentrations of silicon and magnesium in the melt. Of the pairs created, the melt is closest to equilibrium with spinel compositions (Tables 3 and 5), containing ~21 wt% MgO and giving calculated equilibrium temperatures ~1330–1340 + 40 °C with spinel compositions.

The values obtained, as in the case of the calculation for the variant of an undersaturated fluidized melt, clearly demonstrate the lack of both the complete equilibrium between the spinel and the melt compositions and the element content

Table 5 Compositions of heated melt inclusions from spinel crystals (weight. %)

No.	SiO_2	TiO_2	Al_2O_3	Cr_2O_3	FeO	MnO	MgO	CaO	Na_2O	K_2O	P_2O_5
1	39.47	0.18	7.82	1.52	10.57	0.24	29.84	4.21	1.86	2.31	0.38
2	41.07	0.28	11.85	1.30	6.50	0.13	25.68	4.51	3.3	2.72	0.49
3	42.26	0.37	13.87	1.70	6.54	0.1	20.93	5.48	4.22	3.29	0.43
4	45.71	0.22	8.60	1.02	9.63	0.16	12.28	14.87	1.51	3.96	0.49
5	47.87	0.37	11.69	1.20	5.10	0.06	9.20	14.00	2.86	4.53	0.84

in the melt itself (repeatedly exceeding the values of RMS errors). With some caution we can take as probable pressure values—1.6–1.9 ± 0.3 GPa.

The results of calculations of crystallization conditions of spinel-olivine, melt olivine, and melt spinel systems allow to draw the following conclusions: spinel-olivine paragenesis was crystallized from a fluid-containing melt; olivine formed in the melt inclusions in spinel was crystallized at equilibrium temperature ~1300–1340 ± 50 °C and pressure ~1.6–1.8 ± 0.2 GPa; the most equilibrium with this olivine composition is the melt with MgO content ~16–17 wt%; spinel composition is partially equilibrium with the melt with MgO content ~21 wt % and crystallization conditions 1300–1340 ± 40 °C and pressure 1.6–1.9 ± 0.3 GPa. Melt evolution was accompanied by a decrease of MgO content from ~21 to ~16–17 wt%. The obtained numerical values of phase crystallization conditions are not, in the full sense of the word, equilibrium. Thus, for example, the system of melt inclusions-spinel glass by the contents of a number of elements in the melt (see Tables 6 and 7) can be considered suitable for calculations by equations corresponding to equilibrium conditions, with some degree of convention. This is due to the wide range of melt inclusions in the spinel at a sufficiently narrow range of compositions of the spinel itself. One of the variants of explanation can be the influence of the spinel crystal of the inclusions host. According to Simonov et al. [9], the compositions of melt inclusions contain up to ~6 wt% Cr_2O_3, which is not typical for magmatic melts that formed, for example, stratified intrusive massifs. The effect of a spinel including the melt inclusion can be twofold. On the one hand, it may be diffusion enrichment with chromium from the spinel crystal into the melt. On the other hand, these high chromium concentrations in the melt inclusion may be partly imaginary values due to the influence of the substrate in the electron probe

Table 6 Calculation results of the crystallization conditions of the spinel melt system in a variant of an unsaturated fluidized melt

No.	1	2	3	4	5	6	7	8
	T°C			P_{GPA}	At. quant. %			
1	1478	1268	1148	2.15	25.54	−19.88	−27.7	−1.72
2	1404	1178	1374	0.32	26.45	−18.03	−24.9	−2.61
3	1328	1152	1316	−0.91	24.79	−13.6	−20.31	0.19
4	1213	1157	1181	−1.76	8.53	0.6	−9.15	12.83
5	1157	1074	1119	−2.94	10.33	1.92	−7.02	11.18
6	1487	1295	1463	3.00	24.44	−18.66	−26.91	−1.72
7	1413	1204	1389	1.16	25.35	−16.81	−24.4	−2.61
8	1337	1179	1331	−0.07	23.7	−12.37	−19.83	0.19
9	1222	1184	1196	−0.91	7.43	1.82	−8.6	12.83
10	1167	1101	1134	−2.09	9.23	3.14	−6.6	11.18

Note n/a: 1, 2, 3—equilibrium temperature values (Formulas 1–3); 4—equilibrium pressure value (Formula 4); 5—difference between calculated and real content of silicon in the melt (Formula 5); 6, 7—difference between calculated and real content of magnesium in the melt (Formulas 6 and 7); 8—equilibrium distribution of elements in the melt (Formula 8)

Table 7 Results of calculation of crystallization conditions of the spinel melt system in the variant with known water content

No. п/п	1	2	3	4	5	6	7	8	9	10	11
Mod.	T°C	P_{GPA}	Атом. кол. %								T°C
1	1095	1.6	1.8	−0.4	−1.3	0.4	−1.0	16.4	−22.7	−11.4	1628
2	1042	1.7	1.9	−1.4	−3.2	−1.3	−2.9	0.46	−20.4	−13.5	1536
3	1037	1.7	1.9	−1.4	−2.9	−1.5	−2.8	1.28	−17.1	−11.4	1438
4	1294	0.9	0.9	6.1	5.9	5.7	5.5	4.97	−6.7	7.33	1348
5	1237	0.8	0.8	4.7	4.3	4.3	4.0	3.94	−5.51	5.06	1262
6	1093	1.6	1.8	0.1	−1.6	0.2	−1.2	26.8	−21.3	−11.4	1629
7	1042	1.7	1.9	−1.7	−3.4	−1.6	−3.1	2.59	−19.3	−13.5	1536
8	1038	1.7	1.9	−1.7	−3.2	−1.7	−2.9	2.78	−16.2	−11.4	1439
9	1298	0.9	0.9	5.8	5.6	5.4	5.3	5.96	−5.75	7.33	1349
10	1243	0.8	0.8	4.4	4.1	4.1	3.8	4.64	−4.74	5.06	1262

Note No. n/p: 1—equilibrium temperature values (Formula 14, pressure—1.8 GPa, 2 wt% H_2O); 2, 3—calculated equilibrium pressure values (Formula 15, temperature values are taken from Sects. 2 and 3 of Table 6, 2 wt% H_2O); 4, 5—difference between calculated and real aluminum contents in the melt (Formula 16, temperature values are taken from Sects. 2 and 3 of Table 6); 6, 7—difference between calculated and real aluminum contents in the melt (Formula 17, temperature values are taken from Sects. 2 and 3 of Table 6); 8, 9—difference between real and calculated magnesium contents in the melt (Formulas 18 and 19, temperature values are taken from Table 6, Sect. 2, pressure values are taken from this table7, Sect. 3); 10—equilibrium distribution of elements in the melt (Formula 21, 2 wt% H_2O); value of equilibrium temperature by melt composition (Formula 22, pressure value 3 of this table)

analysis. However, in general, the calculated estimates of conditions for the formation of spinel-olivine paragenesis coincide with the results given in Simonov et al. [9]: ultrabasite magma nature; water-containing magmas; melt inclusions homogenization temperatures >1300 °C; and the minimum calculated temperature of olivine crystallization from melt inclusions in the spinel 1230 °C. Ideal calculated equilibrium values of crystallization conditions (calculated values of all values, including homogeneous values calculated by different equations, differ not more than 1–2 values of standard-square error) can be obtained by solving systems of equations with iterations and minimization of discrepancies, taking into account values of standard-square errors for each equation in the variant of program complex.

In case of magmatic genesis of mineral phases (spinel, olivine) composing dunites, one of the variants of formation of this rock can be the accumulation of these phases from the melt [10]. In this variant, the limitations are the absence of overbalancing of the mineral phases, since in many intrusive rocks the calculated temperatures (700–800 °C) correspond to the closing temperatures of the exchange interphase reactions, testifying to the duration of their cooling, as well as baric conditions (~40–50 km) of their formation, which suggests the accumulation of phases, or in the peripheral focus with subsequent squeezing of the cumulate in the form of protrusion, or the accumulation of olivine-spinel crystals in the supply channel. The absence of intercumulus phases and structures is unclear, although it is possible to allow removal of the intercrystalline melt by squeezing it out. Another problem is the size of the core of the massif (~5.5 km) and its shape, because usually the supply channels at different depth erosive sections are represented by dikes or their fields.

Chromitis [2] of the dunite core of Konder are composed of a number of non-equilibrium mineral associations indicating the pneumatolithium-hydrothermal nature of chromospinelids from chromplatinum ores. A good argument for this conclusion is the low calcium composition of olivines from inclusions in chromium spinelids from cement ores. In Pushkarev et al. [2], 32 compositions of these olivine crystals are presented, 31 of which are low calcium varieties. It is not clear whether the loss of calcium from these crystals occurred or whether they were originally of metasomatic nature. It is only clear that their composition is not in equilibrium with the melt and indicates a metasomatic nature. Equilibrium olivine-spinel pairs with the melt (Tables 1 and 4) are also preserved in chromitite, which allows us to believe that the process of formation of platinum-bearing chromites began in the magmatic stage of formation of dunites and continued until the pneumatolite-hydrothermal stage.

According to the data collected in the work [6], for a number of platinum-bearing massifs of the Ural-Alaskan type the low calcium olivine varieties are predominant and the spinel-olivine association is overbalanced at metasomatic transformations of rocks. But in this group there are also massifs, for example, Nizhny Tagil, in which co-existing genetic differences of olivine are calcium-containing crystals. This requires a detailed analysis of the olivine genesis in the massif.

3 Conclusion

The genesis of concentric-zonal ultrabasic massifs was studied on the basis of data on phase compositions of the spinel-olivine-melt (melt inclusions) association from the dunite core of the Konder massif. A system of equations created on the basis of experimental data was used to estimate the equilibrium of phases in this system. Based on their equilibrium it is possible to calculate numerical values of equilibrium temperature, pressure and magnesium content in the melt. In addition, the calcium content of these crystals was used for compositions of olivines formed from the melt and changed during metasomatic transformations. The results obtained suggest that the crystallization of spinel-olivine paragenesis occurred from a fluid-containing high-magnesia melt (MgO from ~21 wt%) at 1300–1340 ± 40 °C and pressure of 1.6–1.9 ± 0.3 GPa. This suggests that the process of cumulation of these phases in the peripheral focus can be realized. The formation of chromitites begins in the magmatic stage and continues in the metasomatic stage of formation of the massif.

The research was done with the financial support of the RFBR, grant project No. 20-05-00261, No. 19-05-00388 as well as the Integration Project of the Research Center of the Russian Academy of Sciences, Block 1.4. and the study was performed by the governmental assignment in terms of Project IX. 129.1.4., RNF-21-17-00015.

References

1. Magmatic rocks. V. 5, 508 p. Moscow, Science (1988)
2. Pushkarev, E.V., Kamenetsky, V.S., Morozova, A.V., Hiller, V.V., Glavatskikh, S.P., Rademann, T.: Ontogeny of ore chromospinelids and composition of inclusions as indicators of pneumatolithium-hydrothermal formation of platinum-bearing chromites of the Konder massif (Aldansky shield). Geol. Ore Deposits **57**(6), 394–423 (2015)
3. Ponomarev, G.P., Puzankov, M.Y.: Distribution of rock-forming elements in the system of the main ultramafic melt spinel, olivine, orthopyroxene, clinopyroxene, plagioclase according to experimental data: geological application. 664 p. M., PROBEL (2012)
4. Ponomarev, G.P., Puzankov, M.Y.: Distribution of rock-forming elements in the system of melt-spinel-olivine with participation of aqueous fluid on experimental data. Vestnik KRAUNTS, Earth Sci. **32**(4), 59–72 (2016) (in Russian)
5. Ponomarev, G.P.: Calcium content in olivine crystals grown from experimental melts. Lithosphere, Part 1, 4, 66–79 (2014a)
6. Ponomarev, G.P.: The content of calcium in natural olivine crystals as an indicator of their genesis. Lithosphere, Part 2, 5, 57–70 (2014b)
7. Prikhodko, V.S., Ponomarev, G.P.: Compositions of rock-forming minerals of the Kondera massif dunites. Pac. Geol. **2**, 59–69 (1990)
8. Nekrasov, I.Y., Lennikov, A.M., Oktyabrsky, R.A., Zalishchak, B.L., Sapin, V.I.: Petrology and Platinum-bearing ring Alkali-Ultra Basic Complexes, 381 p. Science, Moscow (1984)
9. Simonov, V.A., Prikhodko, V.S., Koviazin, S.V.: Conditions of the platinum-bearing ultrabasic massifs formation in the south-east of the Siberian platform. Petrology **19**(6), 79–598 (2011) (in Russian)
10. Vasiliev, Y.R., Gora, M.P.: The nature of dunites and olivinites in the alkaline ultrasonic intrusive complexes of the Siberian platform. NA reports V. **442**(3), 356–359 (2012) (in Russian)

Chemical Compositional Variation of Clinopyroxenes of Alkaline-Ultrabasic and Alkaline Rocks of the Inagli Massif

T. A. Radomskaya, N. V. Vladykin, and E. V. Kaneva

Abstract

The paper presents data on the chemical composition of pyroxenes from ultrabasic and alkaline rocks of Inagli massif (Aldan Shield). It is found that the composition of pyroxenes of a series of chromium-bearing diopside—diopside—aegirine changes from early to late differentiates in accordance with the evolution of the magmatic melt from which the rocks crystallized as a result of fractional crystallization. It is shown that crystallization differentiation was the main mechanism of dunite formation in the Inagli massif. Separation of the early crystallized phase of dunites led to the accumulation of a more alkaline residual melt from which alkaline-basic and alkaline rocks of the Inagli massif were formed as a result of fraction crystallization. Part of the residual interstitial melt was pressed into cooling cracks or formed in situ as small veins and nests of chromium-bearing diopside. The main series of bodies of chromium-bearing diopside pyroxenites, in some places enriched with phlogopite, was formed by the introduction of pyroxenite residual melt on large cracks formed in dunites after their crystallization. The magmatic scheme of sequential crystallization of rocks of the massif is proposed.

T. A. Radomskaya (✉) · N. V. Vladykin · E. V. Kaneva
Vinogradov Institute of Geochemistry, Siberian Branch of Russian Academy of Sciences (IGC SB RAS), 664033 Irkutsk, Russia
e-mail: taniaojigova@mail.ru

T. A. Radomskaya · E. V. Kaneva
Irkutsk National Research Technical University (INRTU), 664074 Irkutsk, Russia

N. V. Vladykin
Irkutsk Scientific Center, Siberian Branch of the Russian Academy of Sciences (ISC SB RAS), 664033 Irkutsk, Russia

N. Vladykin (ed.), *Alkaline Rocks, Kimberlites and Carbonatites: Geochemistry and Genesis*, Springer Proceedings in Earth and Environmental Sciences,
https://doi.org/10.1007/978-3-030-69670-2_7

Keywords

Clinopyroxenes · Chromium-bearing diopside · Aegirine · Alkaline-ultrabasic rocks

1 Introduction

Inagli massif situated on the Aldan shield is an ore-bearing (Pt, Au) object. Its central part is composed of 70–80% ultrabasic rocks (mostly dunites), and the rest of the massif is represented by K-alkaline rocks located on the periphery. The deposits of chromium-bearing diopside and vermiculite, as well as large alluvial deposits of platinum group elements and, to a lesser extent, gold, are associated with the massif. Pegmatite bodies fill radial cracks in the central part of the dunite nucleus, circular and radial cracks at the periphery of thedunite nucleus and partially in the alkaline gabbroid zone. Pegmatites of Inagli massif are divided into three types by their mineral composition: diopside-orthoclase (type I), amphibole–orthoclase (type II) and microcline-albite-amphibole (arfvedsonite) with aegirine and natrolite (type III) [1]. An unusual type of pegmatites of hackmanite–orthoclase composition was found in dunites of the central part of the Inagli massif [2].

Since the discovering of alkaline-ultrabasic complexes of the Aldan Shield, there have been longstanding disputes. One of the assumptions is that intrusive rocks combined within a single massif are the manifestation of a homodromic series of successively generated phases of a single magmatic hearth [3–8]. Another one is that there are two different-aged complexes (ultrabasic and alkaline) spatially combined in long-lived structures of the central type [9–22].

At the same time, among K-alkaline-ultrabasic complexes there are annular-massifs with a full set of differentiations from ultrabasic, through alkaline-syenite to granite. The Bilibinsky massif situated in the eastern part of the Central Aldanis an example [23–26]. There is no doubt about the homodrome sequence of rock formation in such objects. It should be noted that in alkaline-ultrabasic carbonatite complexes of Na-specialization, rocks of early crystallization are also olivinite and dunites (Kovdor, Guli massifs, etc.). Nevertheless, in this case, nobody doubts the genetic connection of ultrabasic and alkaline rocks in these complexes. The study of the chemical composition of rock-forming minerals makes it possible to track changes in their compositions during crystal growth from the evolving magmatic melt. Clinopyroxenes are found in all intrusive rock varieties of the Inagli massif. Therefore, the study of the composition of this rock-forming mineral in combination with geological and petrochemical data will make it possible to determine the change in its composition during crystallization of the massif rocks and to correlate their comagmaticity.

2 Materials and Methods

The work is based on materials of long-term (1981, 1986–1988, 1991, 2004, 2015 and 2019) field works within the Inagli massif. The work was carried out using the equipment of the Center for Collective Use "Isotopic-geochemical investigations" at the Vinogradov Institute of Geochemistry, Siberian Branch of Russian Academy of Sciences. Petrographic research of rocks of Inagli massif was carried out in transmitted light using an OLYMPUS BX-51 polarizing microscope.

The contents of petrogenic elements in the rocks of the Inagli massif were determined using X-ray fluorescent silicate analysis on a CPM-25 multichannel X-ray spectrometer. K, Na, Li, Rb, Cs were determined by flame photometryusing a DFS-12 atomic emission flame spectrophotometer.

The chemical composition of pyroxenes was obtained by means of a JeolJXA-8200 electron microprobe analyzer, equipped with a high-resolution scanning electron microscope, an energy dispersive (ED) spectrometer with the Si (Li) detector (resolution—133 eV) and five wavelength-dispersive (WD)spectrometers. $K\alpha$ lines of Na, Mg, Al, Si, K, Ca, Ti, Mn, Fe, Cr, V, and $L\alpha$ lines of Ba were used as analytical lines. Conditions of excitation and registration of analytical signals are as follows: acceleration voltage—20 kV, electron beam current—2 nA, beam diameter—10–20 μm depending on the analyzed object, counting time on peak—10 s, the background was measured on both sides of line peak—5 s. The analytical signal was corrected for matrix effects by the ZAF method. The following standards, certified as laboratory reference samples at the Institute of Geology and Mineralogy SB RAS (Novosibirsk) were used: albite—Na, Al, diopside—Mg, Ca, Si, orthoclase—K, synthetic V_2O_5—V, chrome spinel—Cr, pyrope—Fe, barite—Ba. The obtained compositions of minerals were calculated into coefficients in the formula on the basis of four cations. The lack of charge was compensated by the transfer of a part of ferrous iron to ferric. To confirm the correctness of the calculation of ferric and ferrous iron, the results of the recalculated microprobe analysis and the chemical analysis of monomineral fractions of some pyroxenes were compared.

3 Chemical Compositional Variation of Clinopyroxenes in Rocks of the Inagli Massif

Dunites composed of olivine (90–99 vol.%) are the most common rocks of the massif. The secondary chromspinelide (1–2 vol.%), clinopyroxene (0–3 vol.%), mica (0–0.5 vol.%) were found. Secondary minerals in altered dunites are serpentine, magnetite and calcite. Changes in the structure of dunites along the section have been noted. At a depth of more than 800 m, dunites are light green rocks with medium and coarse-grained structure and massive texture. Dunites, enriched with chromspinelide, have a grayish tint. Chromspinelide individuals sometimes form bands in the rock. Individual idiomorphic crystals of chromspinelide as inclusions are located inside olivine grains or are confined to the grain boundary. At a depth of 600 m, the structure of dunites is replaced by a granoblastic one. Olivine grains are larger and often have block extinction. Closer to the earth's surface, the dunites are significantly modified by secondary processes and mainly serpentinized. Clinopyroxeneis found in dunites as xenomorphic with respect to olivine and elongated grains of light green color of various shades. Almost all investigated grains correspond to chromium-bearing diopside by their chemical composition (Table 1, Fig. 1). The color saturation of a mineral directly depends on its Cr_2O_3 concentration. The contents of the main components and impurities vary within the following limits (in wt.%): SiO_2 53.1–55.3; CaO 23–25.8; MgO 16.1–17.9; ΣFe_2O_3 1.3–3.5; Al_2O_3 from trace to 1.3; Cr_2O_3 from 0.1 to 1.6; Na_2O 0.2–1.0; TiO_2 from trace to 0.4.

In comparison with other rocks of the massif, wehrlites are rather rare. They are serpentinized rocks of dark-gray color, composed of olivine and clinopyroxene. Serpentinization of wehrlites is accompanied by the formation of magnetite veins that form the central parts of serpentinite veins. Clinopyroxenes of wehrlites correspond to diopsides by their chemical composition and differ from chromium-bearing diopside of dunites by increased iron content and reduced calcium content (Table 1, Fig. 1). The impurity of Cr_2O_3 (up to 0.28 wt.%) is observed in these chromium-bearing diopsides.

There are two main types of pyroxenites in the Inagli massif: chromium-bearing diopside with phlogopite, forming the "Siberian Emerald" deposit of chromium-bearing diopside, and phlogopite pyroxenite. Chromium-bearing diopside pyroxenites with phlogopite are bright green rocks that form complex veins among dunites. Usually, clinopyroxene prevails over phlogopite, but in some areas, there is a change in mineral ratios up to the formation of anchimonomineral clinopyroxenites. The structure of chromium-bearing diopside pyroxenites varies from fine to giant grained. The color range of chromium-bearing diopside is diverse and varies from pale grass shades to bright saturated emerald and olive colors. For a long time, it was believed that chromium-bearing diopside pyroxenites were metasomatically formed by means of the influence of melts on dunites, from which alkaline pegmatites were crystallized [11, 18]. However, this statement is contradicted by the following facts. Firstly, chromium-bearing diopside nests and veins are often observed in dunites, which lack any minerals other than chromium-bearing diopside and phlogopite. Besides, veins and bodies of alkaline pegmatites repeatedly traverse chromium-bearing diopside pyroxenite (Fig. 2). At their contact with dunite, a metasomatic zone of richteriteis formed, which is replaced by the zone of tetraferriphlogopite (Fig. 3). The study of contacts between dunite and alkaline pegmatites has shown that insignificant in-depth serpentinization of dunites develops most frequently on the exocontact. The serpentine veins are expanded along with the contact, and serpentine fibers inside the veins are oriented perpendicularly. The structure of apodunite serpentinite is looped. The zalbandian part of the pegmatite is composed of fine fibrous fan-shaped aggregates of blue richterite, oriented mainly perpendicular to the contact, and fine scales of tetraferriphlogopite. In a direction to the center of pegmatite the size of scales of

Table 1 Chemical composition of clinopyroxenes from rocks of Inagli massif

Name of rock	Dunite						
Sample code	86/740	86/740	86/740	86/740	88	88	d(27)1
Mineral	Cr–Di	Cr–Di	Cr–Di	Cr–Di	Cr–Di	Cr–Di	Cr–Di
SiO_2	54.68	54.72	54.81	54.55	54.42	55.22	54.66
TiO_2	0.13	0.13	0.15	0.22	0.21	0.11	0.22
Al_2O_3	0.36	0.30	0.42	0.66	0.52	0.30	0.72
Cr_2O_3	0.44	0.61	0.53	0.38	0.45	1.57	0.64
V_2O_3	n.d.	n.d.	n.d.	n.d.	0.03	0.02	0.04
Fe_2O_3	1.24	1.25	1.19	1.19	1.00	1.15	0.00
FeO	0.55	0.77	0.99	1.62	0.55	0.79	1.93
MnO	b.d.l.	0.10	b.d.l.	b.d.l.	0.03	0.02	0.03
MgO	17.58	17.24	17.39	16.98	17.65	17.34	17.46
CaO	25.21	24.72	24.82	24.07	24.71	22.66	23.88
Na_2O	0.30	0.46	0.38	0.61	0.37	1.18	0.42
Total	100.49	100.30	100.68	100.28	99.90	100.34	99.96
Formula based on 4 cations. apfu							
Si	1.97	1.98	1.98	1.98	1.97	1.99	1.98
Ti	0.00	0.00	0.00	0.01	0.01	0.00	0.01
Al	0.02	0.01	0.02	0.03	0.02	0.01	0.03
Cr	0.01	0.02	0.02	0.01	0.01	0.04	0.02
V	0.00	0.00	0.00	0.00	0.00	0.00	0.00
Fe^{3+}	0.03	0.03	0.03	0.03	0.03	0.03	0.00
Fe^{2+}	0.02	0.02	0.03	0.05	0.02	0.02	0.06
Mn	0.00	0.00	0.00	0.00	0.00	0.00	0.00
Mg	0.95	0.93	0.94	0.92	0.95	0.93	0.94
Ca	0.98	0.96	0.96	0.93	0.96	0.88	0.93
Na	0.02	0.03	0.03	0.04	0.03	0.08	0.03

Name of rock	Wehrlites	Chromium-bearing diopside pyroxenites	Phlogopite pyroxenites		Shonkinites	
Sample code	200/7	209/2	207/1	203/4	202/6	202/6
Mineral	Di	Di	Di	Di	Di	Cr–Di
SiO_2	53.68	55.59	54.57	55.42	51.95	52.52
TiO_2	0.30	0.07	0.24	0.12	0.53	0.39
Al_2O_3	1.11	0.20	0.72	0.22	2.43	1.62
Cr_2O_3	0.27	0.22	b.d.l.	0.03	0.22	0.35
V_2O_3	0.03	0.01	0.01	0.03	0.06	0.03
Fe_2O_3	1.05	0.00	0.00	0.00	2.59	1.62
FeO	3.52	1.36	5.18	1.86	4.08	3.83
MnO	0.12	0.03	0.16	0.08	0.15	0.14
MgO	17.31	17.96	15.80	17.91	15.67	16.24
CaO	22.01	24.05	22.80	24.09	21.44	21.97

(continued)

Table 1 (continued)

Name of rock	Wehrlites	Chromium-bearing diopside pyroxenites	Phlogopite pyroxenites		Shonkinites	
Sample code	200/7	209/2	207/1	203/4	202/6	202/6
Na_2O	0.41	0.34	0.38	0.30	0.68	0.48
Total	99.77	99.82	99.85	100.02	99.72	99.15
Formula based on 4 cations. apfu						
Si	1.96	2.01	2.01	2.01	1.91	1.94
Ti	0.01	0.00	0.01	0.00	0.01	0.01
Al	0.05	0.01	0.03	0.01	0.11	0.07
Cr	0.01	0.01	0.00	0.00	0.01	0.01
V	0.00	0.00	0.00	0.00	0.00	0.00
Fe^{3+}	0.03	0.00	0.00	0.00	0.07	0.05
Fe^{2+}	0.11	0.04	0.16	0.06	0.13	0.12
Mn	0.00	0.00	0.00	0.00	0.00	0.00
Mg	0.94	0.97	0.87	0.97	0.86	0.89
Ca	0.86	0.93	0.90	0.93	0.85	0.87
Na	0.03	0.02	0.03	0.02	0.05	0.03

Name of rock	Alkaline syenites					
Sample code	202/4	202/4	201/8	202/15	200/5	200/5a
Mineral	Di	Di	Di	Aeg	Aeg	Aeg
SiO_2	53.01	51.68	53.56	52.45	56.03	55.11
TiO_2	0.34	0.27	0.54	3.44	2.03	2.44
Al_2O_3	1.00	1.39	1.21	0.54	0.59	0.45
Cr_2O_3	b.d.l.	b.d.l.	0.02	0.03	b.d.l.	b.d.l.
V_2O_3	0.07	0.04	0.06	0.31	0.33	0.35
Fe_2O_3	2.56	3.52	1.33	22.28	21.50	22.09
FeO	5.34	7.11	4.59	0.00	0.84	0.06
MnO	0.23	0.46	0.19	0.19	0.03	0.10
MgO	14.31	12.03	15.34	3.72	4.31	4.41
CaO	22.43	22.16	21.91	4.69	5.12	4.47
Na_2O	0.56	1.00	0.96	11.88	11.61	11.75
Total	100.41	99.53	99.66	100.35	99.70	99.99
Formula based on 4 cations. apfu						
Si	1.96	1.95	1.97	1.97	2.04	2.03
Ti	0.01	0.01	0.01	0.10	0.06	0.07
Al	0.04	0.06	0.05	0.02	0.03	0.02
Cr	0.00	0.00	0.00	0.00	0.00	0.00
V	0.00	0.00	0.00	0.01	0.01	0.01
Fe^{3+}	0.05	0.09	0.04	0.63	0.59	0.61
Fe^{2+}	0.21	0.23	0.14	0.00	0.03	0.00
Mn	0.01	0.01	0.01	0.01	0.00	0.00

(continued)

Table 1 (continued)

Name of rock	Alkaline syenites					
Sample code	202/4	202/4	201/8	202/15	200/5	200/5a
Mg	0.79	0.68	0.84	0.21	0.23	0.24
Ca	0.89	0.90	0.86	0.19	0.20	0.18
Na	0.04	0.07	0.07	0.87	0.82	0.84

Name of rock	Pegmatite II		Pegmatite III	Alkaline granite	
Sample code	259	260	14	266	266
Mineral	Cr–Di	Di	Aeg	Di	Di
SiO_2	53.69	55.69	54.43	53.91	54.12
TiO_2	0.27	0.07	4.36	0.16	0.28
Al_2O_3	0.53	0.38	0.44	0.52	0.87
Cr_2O_3	0.57	b.d.l.	0.04	0.24	0.37
V_2O_3	0.04	0.04	0.33	0.03	0.03
Fe_2O_3	1.80	0.00	23.49	0.97	0.00
FeO	0.80	3.59	0.00	1.86	3.92
MnO	0.09	0.18	1.30	0.08	0.07
MgO	17.53	16.88	2.01	17.73	0.04
CaO	24.13	22.59	0.65	23.89	16.70
Na_2O	0.33	0.64	14.18	0.13	23.05
Total	99.80	100.02	100.88	99.50	99.59
Formula based on 4 cations. apfu					
Si	1.96	2.03	2.01	1.97	1.99
Ti	0.01	0.00	0.12	0.00	0.01
Al	0.02	0.02	0.02	0.02	0.04
Cr	0.02	0.00	0.00	0.01	0.01
V	0.00	0.00	0.01	0.00	0.00
Fe^{3+}	0.05	0.00	0.65	0.03	0.00
Fe^{2+}	0.02	0.11	0.00	0.06	0.12
Mn	0.00	0.01	0.04	0.00	0.00
Mg	0.95	0.92	0.11	0.97	0.91
Ca	0.94	0.88	0.03	0.93	0.91
Na	0.02	0.04	1.01	0.01	0.02

Note b.d.l.—below detection limit; n.d.—not determined. Aeg—aegirine, Cr–Di—chromium-bearing diopside, Di—diopside

tetraferriphlogopite increases. The grains of arfvedsonite appear and richterite grains disappear. Thus, the study of thin sections from contacts of dunites and alkaline pegmatites has not revealed chromium-bearing diopside in reaction border. Thirdly, clinopyroxene in chromium-bearing diopside pyroxenites corresponds to chromium-bearing diopside by its chemical composition and does not differ from dunite chromium-bearing diopside (Table 1, Fig. 1). Another confirmation of the magmatic origin of chromium-bearing diopside pyroxenites is the melt inclusions

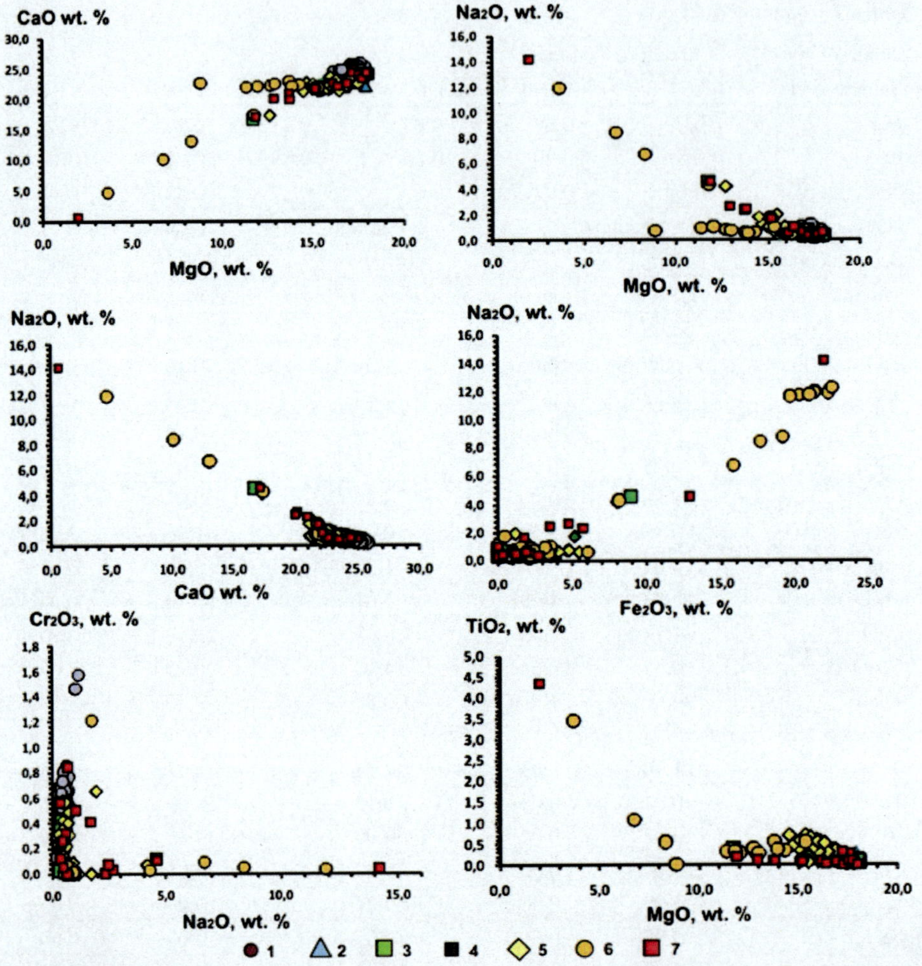

Fig. 1 Diagrams of pair correlation of rock-forming elements in clinopyroxenes from ultrabasic and alkaline rocks. Symbols: 1—dunites, 2—wehrlites, 3—chromium-bearing diopside with phlogopite, 4—phlogopite pyroxenites, 5—olivine melanocratic shonkinites, pseudoleucite and nepheline shonkinites, 6—alkaline melanocratic syenites, alkaline syenites, 7—alkaline pegmatites

found by Naumov and co-authors in the chromium-bearing diopside, which allowed us to conclude about its crystallization from a silicate melt [27].

Phlogopite pyroxenites are fine- and medium-grained black rocks, which form the outcrops between dunites and schonkinites. In terms of chemical composition, phlogopite pyroxenites are enriched with K_2O in comparison with chromium-bearing diopside pyroxenites (Figs. 4 and 5). On the surface, they are very strongly weathered, but in the wells below 20 m, there are quite “fresh” pyroxenites. The mineral composition of phlogopite pyroxenites is richer than that of chromium-bearing diopside pyroxenites. Besides clinopyroxene and mica, grains

Fig. 2 Chromium-bearing diopsides pyroxenite intersecting with alkaline pegmatite

of olivine and apatite are found. In phlogopite pyroxenites, clinopyroxene is found in the form of dark-green prismatic elongated grains with a brownish tint, in thin sections it is clearly colored in pale brown. The microstructure of phlogopite pyroxenite is poikilitic. Pyroxenes contain many inclusions of mica scales and prismatic grains of apatite, less often dusty magnetite. The number of inclusions naturally increases from the central parts to the periphery of the grains. Almost all pyroxene grains have zones formed by plate individuals of titanium magnetite located at some distance from the outer contour of the grains. A mineral can contain up to 3–4 zones (Fig. 6). By chemical composition pyroxenes correspond to diopsides with increased gedenbergitic minal due to higher iron concentration than in chromium-bearing diopsides described above (Table 1, Fig. 1). The impurity of Cr_2O_3 in diopsides is at the level of trace values.

Alkaline and alkaline-ultrabasic rocks of the Inagli massif are represented by melanocratic shonkinites, pseudoleucite and nepheline shonkinites, alkaline melanocratic syenites, alkaline syenites and pegmatites. Melanocratic schonkinites are mainly composed of clinopyroxene (30–40 vol.%), potassium feldspar (20–30 vol.%), olivine (up to 15 vol.%), titanomagnetite (up to 4 vol.%), apatite (up to 2

Fig. 3 Reaction border at the contact of dunite and alkaline pegmatite. Symbols: Arf—arfvedsonite, Fspr—potassium feldspar, Rich—richterite, Tfphl—tetraferriphlogopite

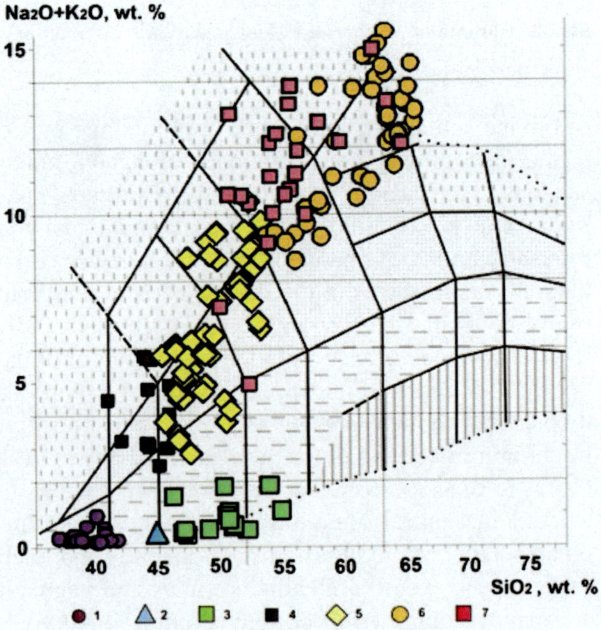

Fig. 4 Position of ultrabasic and alkaline rocks of the Inagli masif in coordinates SiO_2 to $Na_2O + K_2O$. Symbols: 1—dunites, 2—wehrlite, 3—chromium-bearing diopside pyroxenites with phlogopite, 4—phlgopite pyroxenites, 5—olivine melanocratic, pseudoleucite and nepheline schonkinites, 6—alkaline melanocratic syenites, alkaline syenites, 7—alkaline pegmatites

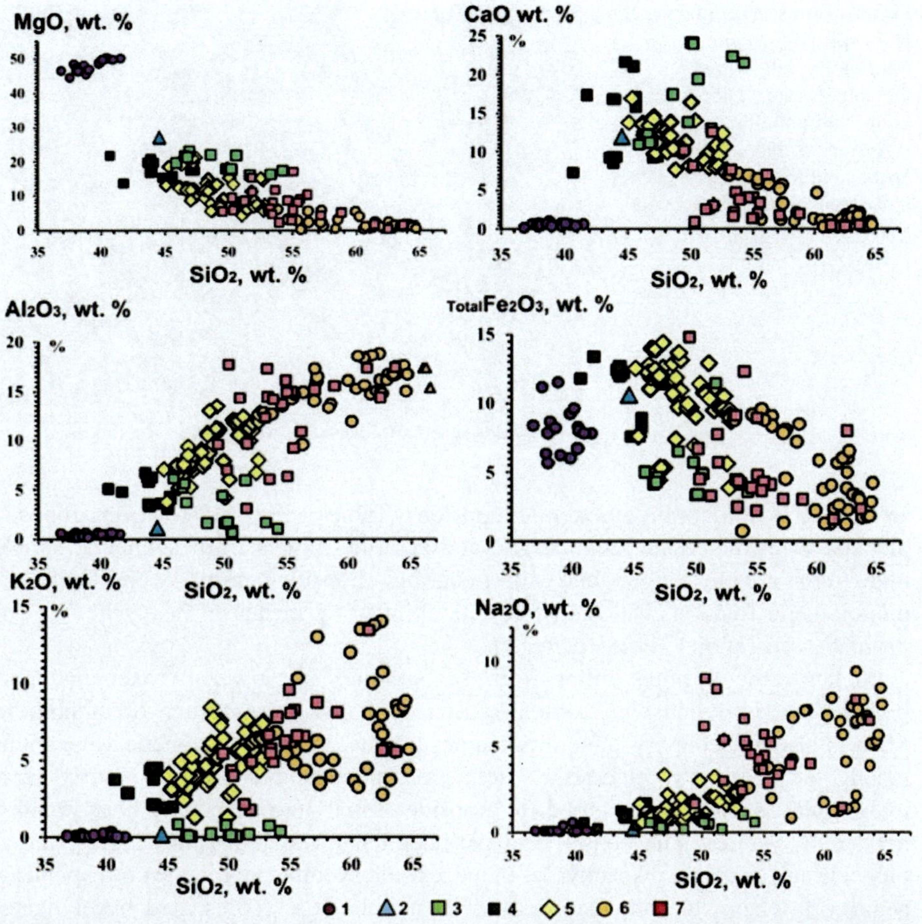

Fig. 5 Position of ultrabasic and alkaline rocks of the Inagli massif on variation diagrams of paired correlation of petrogenic elements. Symbols: 1—dunites, 2—wehrlites, 3—chromium bearing diopside pyroxenites with phlogopite, 4—phlogopite pyroxenites, 5—olivine melanocratic shonkinites, pseudoleucite and nepheline shonkinites, 6—alkaline melanocratic syenites, alkaline syenites, 7—alkaline pegmatites

vol.%), phlogopite (up to 5 vol.%), sometimes nepheline is present. In pseudoleucite schonkinites, characteristic rounded forms were found. They are composed of pseudoleucite, which represents by cryptocrystalline intergrowths of potassium feldspar and kalcilite (structures of decomposition of leucite). In nepheline schonkinites, the content of dark-colored minerals and apatite decreases (clinopyroxene to 15–20 vol.%, olivine to 3 vol.%, phlogopite to 3 vol.%) and leucocratic minerals content increases (potassium feldspar to 50–55 vol.%, nepheline to 10 vol.%). The majority of clinopyroxenes of all varieties of shonkinites by morphological features and chemical composition of the main components

Fig. 6 Thin section image of phlogopite pyroxenite in transmitted light without the compensator. Diopside grain with poikilitic inclusions of phlogopite, apatite and natural zonal plate individuals of titanomagnetite

are similar to phlogopite pyroxenite diopsides (Table 1, Fig. 1). However, grains of diopside with increased hedenbergite and aegirine minals from nepheline shonkinites are noted. Single chromium-bearing diopside grains were found in monomineral fractions, selected from shonkinites by means of heavy liquids and magnetic separation (Table 1, Fig. 1).

Melanocratic syenites differ from schonkinites by lower pyroxene content, higher potassium feldspars content, absence of olivine, presence of amphibole. Mica is also a secondary mineral. Apatite, titanite and titanomagnetie were found among the accessory minerals. In terms of chemical composition, pyroxene in melanocratic syenites corresponds to diopside. Zonal pyroxenes have been found in nepheline syenites. The central and peripheral parts of pyroxenes correspond to diopside and aegirine, respectively. In leucocratic syenites, pyroxene corresponds to aegirine in chemical composition, and most often it is represented by elongated needle-like euhedral and subhedral grains with grassy–green to blue–green pleochroism. Sometimes, in leucocratic syenites, spherical aggregates composed of fine grains of aegirine could be found out together with individual crystals.

As has already been mentioned above, pegmatites of the Inagli massif are divided into three types by their mineral composition [1]: diopside-orthoclase (type I), amphibole-orthoclase (type II), and microcline-albite-amphibole (arfvedsonite) with aegirine and natrolite (type III). By chemical composition, pegmatites correspond to alkaline rocks of the massif. The first type is characterized by the presence of case-shaped crystals of pyroxene, the central part of which is composed of chromium-bearing diopside, and the peripheral part consists of a diopside. In pegmatites of the second type, pyroxene corresponds to diopside, and in pegmatites of the third type, it belongs to diopside-aegirine and aegirine.

It is also worth noting that the alkaline granite dyke was found in the massif. Pyroxene of this dyke by its chemical composition corresponds to the diopside. An impurity of Cr_2O_3 up to 0.37 wt.% was also found in it.

The paired correlation diagrams of rock-forming elements in clinopyroxenes show a direct correlation between Na_2O and total Fe_2O_3 and an increase in their content from ultramafic rocks to alkaline rocks. A direct correlation between CaO and MgO is also noted, but their content decreases from early to late differentiates (Fig. 1).

4 Conclusions

Thus, the given chemical data of pyroxenes from ultrabasic and alkaline rocks of Inagli massif allow to establish their consecutive change in a series of chromium bearing diopside–diopside–aegirine from early differentiates to late ones. Consequently, it is possible to assume that pyroxenes changed according to the evolution of magmatic melt from which rocks crystallized.

The main mechanism of the formation of dunites of Inagli massif is crystallization differentiation. The process of crystallization differentiation and separation of the early crystallized phase of dunites resulted in the accumulation of more alkaline residual melt from which alkaline-basic and alkaline rocks of the massif were formed along the periphery of dunite nucleus as a result of fractional crystallization. Apparently, during crystallization of the dunite nucleus, apart of residual interstitial melt was pressed into cooling cracks or formed in situ as small chromium bearing diopside–phlogopite veins and nests. This mechanism of formation is confirmed by the presence of chromium bearing diopside grains with the composition identical to pyroxenes from chromium bearing diopside bodies in dunites even at great depths. The formation of such veins from the final product of the intergranular melt evolution is known and discussed in detail in classical works [28, 29]. However, the main series of bodies of chromium bearing diopside pyroxenite, which has a depth of 5–10 m and a length of 1–3 km, in some places enriched with phlogopite, was formed by the introduction of pyroxenite residual melt in large cracks formed in dunites after their crystallization.

In the alkaline rocks of the Inagli complex pyroxene, crystallization starts with chromium bearing diopside, though in small amounts, and its composition does not correspond to the chemistry of the alkaline rock from which it crystallizes. The same tendency is typical for the large Bilibinsky massif with a full range of rocks from ultrabasic to granite. In granites of the Bilibinsky massif, diopside-aegirine is crystallized instead of pure aegirine, and mica contains 18 wt% (!) of MgO [23]. The process of chemical differentiation of rock compositions goes faster than the same differentiation of compositions in minerals. The chemical composition of rocks is determined by relations of different minerals with the "lagging" composition of dark-colored minerals [26].

For Inagli massif we can offer the following magmatic scheme of the array.

The magmatic melt of primary magma:

1. Dunites
2. Wehrlites
3. Chromium bearing diopside pyroxenites with phlogopite
4. Phlogopite pyroxenites
5. Olivine melanocratic shonkinites
6. Pseudoleucite and nepheline schonkinites
7. Nepheline syenites
8. Alkaline melanocratic and leucocratic syenites
9. Alkaline pegmatites
10. Alkaline granites

The research was done with the financial support of the Russian Foundation for Basic Research, grant project No. 19-05-00388, as well as the Integration Project of the Research Center of the Russian Academy of Sciences, Block 1.4. and the study was performed by the governmental assignment in terms of Project IX. 129.1.4.

References

1. Efimov, A.F., Kravchenko, S.M., Vlasova E.V.: To mineralogy of alkaline pegmatites in the Inagli massif. In: Proceedings of the Institute of Mineralogy, Geochemistry, and Crystallochemistry of Rare Elements, vol. 16, pp. 141–175 (1963)
2. Radomskaya, T.A., Kaneva, E.V., Shendrik, RYu., Suvorova, L.F., Vladykin, N.V.: Sulfur-bearing sodalite—hackmanite in alkaline pegmatites of the Inagli massif (Aldan shield): crystal and chemical features, photochromism and luminescence. Proc. Russ. Mineral. Soc. CXLIX **2**, 42–54 (2020)
3. Gurovtsch, V.G., Zemlyanukhin, V.N., Emelianenko, V.P., et al.: Geology, Petrology and Orebearing of the Konder Massif. Science, Moscow (1994)
4. Malich, K.N., Efimov, A.A., Badanina, I.Y.: About the age of the Konder dunit massif (Aldan province, Russia): the first U-Pb isotope data. Russ. Acad. Sci. Rep. **446**(3), 308–312 (2012)
5. Nekrasov, I.Y., Lennikov, A.M., Oktyabrsky, R.A., Zalischak, B.L., Sapin, V.I.: Petrology and Platinum-Bearing of the Ring Alkaline-Ultra Basic Complexes. Science, Moscow (1994)
6. Ugryumov, A.N., Kiselev, Y.: About the age of ultramafic rocks of Inagli massif (Aldan shield). Geol. Geophys. **8**, 19–24 (1969)
7. Shukolukov, Y., Yakubovich, O.V., Mochalov, A.G., et al.: New isotope geochronometer for the direct dating of native platinum minerals (^{190}Pt-^{4}Hemethod). Petrology **20**(6), 545–559 (2012)
8. Yakubovich, O. V.: New^{190}Pt-^{4}He method of isotope geochronology for dating platinum minerals. Ph.D. abstract. Saint Petersburg State University, St. Petersburg (2013)
9. Andreev, G.V.: Konder Massif of Ultrabasic and Alkaline Rocks. Science, Novosibirsk (1987)
10. Bogomolov, M.A.: Some peculiarities of the petrology of the central type massifs with the dunite core on the Aldan shield. In: Metasomatism and Other Questions of Physicochemical Petrology, pp. 312–363. Science, Moscow (1967)
11. Glagolev, A.A., Korchagin, A.M., Kharchenkov, A.G.: The Alkaline and Ultramafic Massifs of Arbarastakh and Inagli. Science, Moscow (1974)
12. Elyanov, A.A., Moralev, V.M.: New data on the age of the ultrabasic and alkaline rocks of the Aldan shield. Acad. Sci. Rep. USSR **141**(3), 687–689 (1961)
13. Elyanov, A.A., Moralev, V.M.: About the age of the ultrabasic-alkali rocks of the Aldan and South-Verkhoyansk provinces. In: Proceedings of Universities. Geology and Exploration, vol. 10, pp. 15–23 (1973)

14. Efimov, A.A., Ronkin,Y.L., Malich K.N., Lepikhin G.A.: Sm-Nd and Rb-Sr (ID-TIMS) dating of apatite-phlogopite clinopyroxenites in the dunite core of the Konder massif on the Aldan shield: new data. Acad. Sci. Rep. **445**(5), 564–569 (2012)
15. Ibragimova, E.K., Radkov, A.V., Molchanov, A.V., Shatova, N.V., Shatov, V.V., Lepekhina, E.N., Antonov, A.V., Tolmacheva, E.V., Soloviev, O.L., Terekhov, A.V., Khorokhorina, E.I.: Results of U-Pb (SHRIMP II) dating of zircons from dunites of Inagli massif (Aldan shield) and the problem of genesis of concentric-zonal complexes. Reg. Geol. Metallogeny **62**, 64–78 (2015)
16. Karetnikov, A.S.: To the problem of determination of the Konder massif age. Pac. Geol. **24** (4), 76–83 (2005)
17. Karetnikov, A.S.: Paleomagnetism of ultramafites of the Konder massif and the estimation of its Age. Ph.D. abstract, Khabarovsk (2010)
18. Korchagin, A.M.: Inagli pluton and its minerals. Nedra, Moscow (1996)
19. Orlova, M.P.: Geological structure and genesis of the Konderultramafic massif (Khabarovsk region). Pac. Geol. **10**(1), 80–88 (1991)
20. Orlova, M.P., Avdeeva, O.I., Fedorova, I.V., Yakovleva, L.V.: New data on radiological dating of the Konder massif and the rocks containing it (eastern part of the Aldan shield). Acad. Sci. Rep. USSR **240**(3), 677–680 (1978)
21. Pushkarev, Y.D., Kostostyanov, A.I., Orlova, M.P., Bogomolov, E.S.: Features of Rb-Sr, Sm-Nd, Pb-Pb, Re-Os and K-Ar isotope systems in the Konder massif: mantle substrate enriched with platinoids. Reg. Geol. Metallogeny **16**, 80–91 (2002)
22. Ronkin, Y.L., Efimov, A.A., Lepikhina, G.A., Rodionov, N.V., Maslov, A.V.: U-Pb dating of the “baddeleyite-zircon” system of the platinum-bearing dunite of the Konder massif (Aldan shield): new data. Acad. Sci. Rep. **450**(5), 579–585 (2013)
23. Vladykin, N.V.: Bilibinskiy massif—layered highly differentiated complex of potassium ultrabasic-alkaline rocks. Acad. Sci. Rep. **349**(6), 792–794 (1996)
24. Vladykin, N.V.: Aldan province of K-alkaline rocks and carbonatites: issues of magmatism, genesis and mantle sources. In: Alkaline Magmatism and Problems of Mantle Sources, pp. 18–45, Irkutsk (2001)
25. Vladykin, N.V.: Genesis and crystallization of ultramafic alkaline carbonatite magmas of Siberia: Ore potential, mantle sources, and relationship with plume activity. Russian Geol. Geophys. **57**(5), 698–712 (2016)
26. Vladykin, N.V., Radomskaya, T.A.: K-alkaline magmatic systems: differentiation from ultrabasic rocks to granites. In: Mongush, A.A. (eds.) CONFERENCE 2018, Geology of Magmatism and Metallogeny of the Center of Asia: Ore and Magmatic Systems Sangilene (Alkaline Intrusions, Carbonatite), pp. 15–22, Kyzyl (2018)
27. Naumov, V.B., Kamenetsky, V.S., Thomas, R., Kononkova, N.N., Ryzhenko, B.N.: Inclusions of silicate and sulfate melts in chromium bearing diopside of Inagli deposit (Yakutia, Russia). Geochemistry **6**, 603–614 (2008)
28. Wager, L.R., Brown, G.M.: Layered Igneous Rocks. Oliver & Boyd, Edinburgh, London (1968)
29. Sharkov, E.V.: Formation of the layered intrusions and associated ore bearing. Scientific World, Moscow (2006)
30. Simonov, V.A., Prikhodko, B.C., Koviazin, S.V.: Conditions of the platinum-bearing ultrabasic massifs formation for the South–Eastern Siberian platform. Petrology **19**(6), 579–598 (2011)

Crystal-Chemical Features of Rare and Complex Silicates from Charoite Rocks of the Malyy Murun Volcano-Plutonic Alkaline Complex

E. V. Kaneva, R. Y. Shendrik, N. V. Vladykin, and T. A. Radomskaya

Abstract

The article summarizes the available crystal-chemical data and presents new results of structural and spectroscopic studies of a number of rare and complex silicates found out in the rocks of the Malyy Murun massif—the eastern part of the Murun volcano-plutonic alkaline complex. Murun complex is the world's largest massif of potassium-carbonated alkaline rocks; it is located on the southwestern edge of the Aldan Shield, on the border of Yakutia and Irkutsk region. The work presents descriptions of crystal-chemical and spectroscopic features of tinaksite and tokkoite—silicates with hybrid tetrahedral radical in the crystal structure, agrellite and frankamenite containing tubular silica-oxygen anions, fedorite, and a new mineral in the Murun massif—fluorcarletonite with double tetrahedral layers.

Keywords

Murun massif · Crystal chemistry · Infrared spectroscopy · Tinaksite · Tokkoite · Agrellite · Frankamenite · Fluorcarletonite · Fedorite

E. V. Kaneva (✉) · R. Y. Shendrik · N. V. Vladykin · T. A. Radomskaya
Vinogradov Institute of Geochemistry SB RAS, 1A Favorsky Str., 664033 Irkutsk, Russia
e-mail: kev604@mail.ru

N. V. Vladykin
Irkutsk Scientific Center of Siberian Branch of Russian, 134 Lermontov St., 664033 Irkutsk, Russia

N. Vladykin (ed.), *Alkaline Rocks, Kimberlites and Carbonatites: Geochemistry and Genesis*, Springer Proceedings in Earth and Environmental Sciences,
https://doi.org/10.1007/978-3-030-69670-2_8

1 Introduction

Starting from the 1960s, a number of new minerals with unusual composition and crystal structure were discovered at the Murun massif. The Murun massif's specific geographical location and difficult accessibility do not allow for a comprehensive and systematic study of its mineralization. There are still many open questions related to structural properties and correlations between structural features and chemical compositions of minerals. The choice of a group of rare and complex silicates of the Little Murun alkaline rocks as the object of crystal-chemical study is dictated by the fact that, although the structures of many of these mineral species have already been determined, previous studies were conducted on samples from other deposits and/or did not include complete data based on modern knowledge and technology.

This paper provides an overview of crystal-chemical features of some rare and complex silicates in the light of new data obtained by the authors.

According to the type and complexity of $[SiO_4]$-tetrahedra in the crystal structure, the minerals considered here can be divided into the following groups: (1) silicates with hybrid tetrahedral radical, (2) silicates with tubular tetrahedral radicals, and (3) silicates with double tetrahedral layers.

Infrared (IR) spectroscopy is a rather convenient method to detect some structural features of complex silicates.

2 Silicates with Hybrid Tetrahedral Radical

Tinaksite, $K_2Ca_2NaTi(Si_7O_{18}OH)O$, and **tokkoite**, $K_2Ca_4(Si_7O_{18}OH)(OH,F)$, are found in the charoitites of the Murun massif. Tinaksite often forms radially radiant aggregates or rosette-like clusters. It is associated with potassium feldspar, aegirine, quartz, frankamenite, xonotlite, and a number of rare minerals. The color of the mineral is bright brown to red–orange. The mineral was firstly described in 1965 by Rogov et al. [1]. Von Bissert [2] has been determined, and Rozhdestvenskaya et al. [3] have been refined the crystal structure of tinaksite from the Murun massif. Tinaskite was also discovered in the Rosvumchorr deposit (Khibiny massif), where it is found in pegmatites occurring in rischorrites and urtites [4].

Tokkoite has a light-brown, silver-brown or yellowish color; it crystallizes in the form of radial aggregates in association with charoite, tinaksite, miserite, aegirine, and K-feldspar. The mineral was first described by Lazebnik et al. [5], and the crystal structure was determined in 1989 by Rozhdestvenskaya et al. [6].

A description of the crystal-chemical features of the isostructural tokkoite and tinaksite should be started with the chains of wollastonite ($[Si_3O_9]$) and vlasovite ($[Si_4O_{11}]$) types. Their condensation leads to a special arrangement of tetrahedra along the *c* axis with the formation of four- and eight-membered rings (Fig. 1). Such a radical is called hybrid and has the composition $[Si_7O_{18}OH]^{9-}$. One of the positions of the potassium atoms located in the structural cavities in both minerals is

Fig. 1 Hybrid tetrahedral radical (3) in the structure of tinaksite and tokkoite as a result of condensation of wollastonite (cyan) and vlasovite (yellow) chains

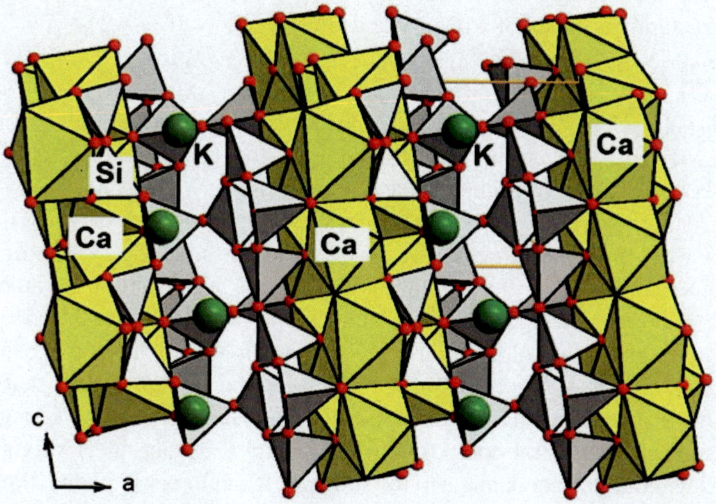

Fig. 2 Tokkoite crystal structure. Si tetrahedra are shown in grey, Ca-octahedra are yellow, and potassium atoms are green spheres

split into several different positions. Ribbons of cation-centered polyhedra are connected through common vertices, forming layers. Silicate radicals and layers of polyhedra combine to form a heterogeneous framework of the structure of minerals (Fig. 2).

Recent detailed studies [7] provided new insights into the crystal-chemistry of tinaksite and tokkoite. A comparison of geometric parameters shows significant similarities between the two structures. Average unit cell parameters for tinaksite (sp. gr. P): are: $a \approx 10.373$, $b \approx 12.176$, $c \approx 7.057$ Å, $\alpha \approx 90.82°$, $\beta \approx 99.22°$, $\gamma \approx 92.80°$, $V \approx 878.5$ Å^3; for tokkoite (P): $a \approx 10.423$, $b \approx 12.477$, $c \approx 7.112$ Å, $\alpha \approx 89.92$, $\beta \approx 99.68°$, $\gamma \approx 92.97°$, $V \approx 910.5$ Å^3 [7].

The derivation of structural formulas for tokkoite and tinaksite is difficult because of the presence of several atoms occupying different cationic positions, atoms with variable charge and coordination number (Fe, Ti), and a small amount of rare-earth elements. The 6- and 7-coordinated atomic positions have a rather diverse chemical composition. The formulas were determined on the basis of Si = 7 atoms per formula unit by combining the results of electron microprobe analysis, Mössbauer data, and crystal structure refinements. Focusing on the octahedral sites, it was assumed that, due to the greater chemical heterogeneity in the tokkoite, Fe, in addition to two oxidation states, can have several structural positions. Thus, it also has a more complex Mössbauer spectrum with respect to tinaksite. The results of the Mossbauer analysis showed that Fe^{2+} is the main state of iron oxidation in both minerals (Fe^{2+} = 60.0(6)% in tokkoite and 87.2(4)% in tinaksite). The spectra indicate the presence of Fe^{2+} and Fe^{3+} in octahedral coordination, while two different local Fe^{2+} environments were found in tokkoite. The Ca atoms in the tokkoite are distributed over all available octahedral sites, while in the tinaksite, Ca atoms occupy only the M3 and M4 octahedra.

The structural formulas of tokkoite and tinaksite, $K_{1.95}(Ca_{0.39}Ti_{0.22}Fe^{2+}_{0.11}Fe^{3+}_{0.10}Mg_{0.17})(Ca_{0.77}Na_{0.23})(Ca_{0.96}Fe^{2+}_{0.04})(Ca_{0.94}Mn_{0.06})(F_{0.72}O_{0.24}(OH)_{0.04})[Si_7O_{18}(OH)]$ and $K_{2.01}(Ti_{0.81}Fe^{2+}_{0.12}Fe^{3+}_{0.02}Mg_{0.05}Mn_{0.01})(Na_{0.86}Ca_{0.14})(Ca_{0.97}Mg_{0.03})(Ca_{0.92}Mn_{0.09})(O_{0.81}OH_{0.19})[Si_7O_{18}(OH)]$, respectively, differ mainly in the composition of the anionic positions M1, M2 and position O20. Based on the ideal tokkoite formula, $K_2Ca_4(Si_7O_{18}OH)(OH,F)$, the substitution reaction $2Ca^{2+}_{(M1+M2)} + (F,OH)^{-}_{(O20)} \leftrightarrow Ti^{4+}_{(M1)} + Na^{+}_{(M2)} + O^{2-}_{(O20)}$ leads to the ideal tinaksite formula, $K_2Ca_2NaTi(Si_7O_{18}OH)O$, as previously was reported by Rozhdestvenskaya and Nikishova [8]. Geometrical features, the difference in the parameters of the unit cells and the distortion parameters of the polyhedra, indicate local stress in the structure, mainly with respect to the M1 position. In particular, the substitution of Ti atoms by Ca atoms in M1 implies the expansion of the polyhedral layer, which affects the distance between the cations and the deformation of the bond angle. It leads to an increase in the unit cell parameters, mainly along the b axis. In conclusion, the chemical composition of M1 and O20 can affect strong geometric distortions without causing damage to the structure. Therefore, M1 and O20 play a major role in crystal chemistry and, possibly, introducing restrictions on the existence of a solid solution between tinaksite and tokkoite [7].

3 Silicates with Tubular Tetrahedral Radicals

Agrellite, $NaCa_2Si_4O_{10}F$, is a rare mineral, first discovered as a rock-forming mineral in the province of Quebec (Canada) in regional metamorphic alkaline rocks [9]. Later, the agrellite was found at the southeastern periphery of the Murun massif, Russia [10]. The following minerals can be found in paragenesis with agrellite: aegirine, K-feldspar, albite, quartz, and alkaline amphibole. Sometimes fluorite, pectolite, miserite, fedorite, eudialyte, wadeite may prevail; calcite, dolomite, sphene, apatite, zircon, narsarsukite, rosenbuschite, strontianite, pyrite, galena, sphalerite, thorosteenstrupine can also be associated with it.

Agrellite is also noted as a rare find on Mount Karnasurt of the Lovozero Massif, Russia [11], and in the pluton Woso, Wisconsin, USA [12]. In addition, the presence of agrellite on the Dara-i-Pioz massif (Tajikistan) was briefly reported in the literature [13, 14].

The symmetry, space group and approximate unit cell parameters were firstly published by Gittins et al. [9], who applied the photographic method to study the Canadian sample. The crystal structure was solved in 1979 using a mineral from Quebec [15]. Rozhdestvenskaya et al. in 1998 published the results of powder X-ray diffraction of strontium agrellite from the Yakutian charoitite (Murun massif) [16], and at about the same time the data of a single-crystal X-ray study were published [17]. Agrellite found on the Murun massif can be described as elongated grayish-white crystals up to 5 cm in size. According to the results of the investigation, it differs from the Murun agrellite described earlier in literature [10, 17]. The formula of the studied mineral, calculated on the basis of 4 silicon atoms per formula unit, is $(Na_{0.99}K_{0.01})(Ca_{1.94}Sr_{0.03}Na_{0.02}(Mn,Fe)_{0.01})[Si_4O_{10}](F_{0.96}OH_{0.04})$ [18]. Thus, a simplified formula of the investigated aggrellite is: $NaCa_2[Si_4O_{10}]F$.

The mineral is triclinic. The unit cell parameters are: $a = 7.7628(1)$, $b = 18.9505(4)$, $c = 6.9847(1)$ Å, $\alpha = 89.774(1)°$, $\beta = 116.581(1)°$, $\gamma = 94.311(1)°$, $V = 915.80(3)$ Å^3 [18].

Silicate tubes consist of two centrosymmetrically connected single silicate chains. Individual chain is formed by four-membered tetrahedral rings similar to the vlasovite chain of the composition $[Si_4O_{11}]$ [19]. These rings through their vertices are combined into chains elongated along the c axis. Two adjacent chains are connected via an oxygen atom, resulting in the formation of a silicate tube of the composition $[Si_8O_{20}]$. Each tube contains two different eight-membered tetrahedral rings, which have four common tetrahedral. Its diameter is determined by a six-membered ring. The eightfold ring joins the calcium polyhedra.

The crystal structure of the Murun agrellite sample studied by the authors [18, 20] corresponds to the model proposed by Ghose and Wan in 1979 [15] for the structure of REE-agrellite.

An EPR signal at room temperature and at 77 K was observed in the agrellite samples and refered to Mn^{2+} ions (I = 5/2). In the crystal structure of agrellite, Mn replaces calcium in a position surrounded by two fluorine ions [18].

Frankamenite, $K_3Na_3Ca_5Si_{12}O_{30}F_3(OH)\cdot H_2O$, is associated with charoite, pectolite, microcline, aegirine, tinaksite, and quartz [21]. The mineral forms flattened, rosette-like, or radially radiant aggregates and may have different coloring: from gray, bluish- and lilacky-grey to greenish and brown colors. In 1992, Nikishova and co-authors presented the results of the study of the mineral, calling it triclinic canasite [22]. The structure of this variety was refined in the space group *P1*. The crystal structure and interatomic distances only slightly differ from similar characteristics of monoclinic canasite from the Khibiny massif. In 1995, triclinic canasite was approved by the Commission on New Minerals, Nomenclature, and Classification as a new mineral frankamenite [23]. It was named after the Russian mineralogist and crystallographer, Professor of St. Petersburg State University V.A. Frank-Kamenetsky. Frankamenite has the following average crystal-chemical formula: $K_{2.93}(Na_{2.93}Ca_{5.05}Mn_{0.07})Si_{12}O_{30}[F_{2.83}(OH)_{1.35}]\cdot 0.79H_2O$ [22].

The parameters of the unit cell of the Murun frankamenite: $a = 10.094(3)$, $b = 12.691(3)$, $c = 7.240(1)$ Å, $\alpha = 90.00(2)^o$, $\beta = 111.02(2)^o$, $\gamma = 110.20(2)^o$, $V = 804.5(7)$ Å^3 (according to [24]).

The crystal structures of Khibiny canasite and Murun frankamenite consist of tubular radical $[Si_{12}O_{30}]^{12-}$ with octagonal section and zigzag walls of octahedra populated with calcium and sodium. In the structure of the canasite Na and Ca cations are usually ordered by different octahedra positions: two positions are occupied by Na^+ ions, the remaining two are filled by Ca^{2+}. In the triclinic structure of frankamenite only one of eight octahedral positions is fully occupied by the Na^+ ion, one position is occupied by Ca^{2+} ion, and the remaining six ones are mixed (Ca, Na)-positions [24]. In contrast to canasite, the H_2O molecule is localized between potassium atoms in the center of the silicate tube of the frankamenite structure [22, 24].

According to Nikishova et al. [22], a high degree of order in monoclinic canasite suggests slower growth of its crystals in the pegmatites of Khibiny. The manifestation of a high degree of ordering of most octahedral positions in frankamenite indicates other conditions of its crystallization.

4 Silicates with Double Tetrahedral Layers

Fedorite, $(Na,K)_{2-3}Ca_4Na_3(Si_{16}O_{38})F_2\cdot 3.5H_2O$ is a rare and complex silicate, firstly discovered in the phenytized rocks adjacent to the alkaline complex of the Turiy Peninsula, Kola Peninsula, Russia [25]. In 1968, the mineral was approved and named after mathematician, crystallographer, and mineralogist Evgraf Stepanovich Fedorov. Subsequently, the second discovery of the mineral in charoite rocks of the Malyy Murun complex was reported [26]. Fedorite has a modular structure [27] and it is part of a group of silicates that consist of modules of tetrahedral and octahedral layers. The structure of the fedorite was solved by Sokolova et al. [28] and later refined by Joswig et al. [29]. However, the remaining uncertainties in the positions of interstitial atoms and the order of cations occupying octahedral and tetrahedral positions were clarified in 2001 by Mitchell and Burns [30].

Crystal-chemical formula of Murun fedorite: $(K_{2.17}Na_{0.12}Ba_{0.01})$ $(Ca_{4.31}Na_{2.61}Mn_{0.07}Fe_{0.01})(Si_{15.98}Al_{0.02})(O_{37.68}OH_{0.32})(F_2OH_{0.40}Cl_{0.03}){\cdot}0.7H_2O$ [31]. The mineral is crystallized in triclinic symmetry (space group *P*). Parameters of the unit cell: $a \approx 9.64$, $b \approx 9.65$, $c \approx 12.62$ Å, $\alpha \approx 102.46°$, $\beta \approx 96.24°$, $\gamma \approx 119.93°$, $V \approx 960$ $Å^3$ (according to [31]).

The crystal structure of the fedorite consists of layers of [8]-coordinated cations associated with a double layer of Si–O-tetrahedra. Tetrahedral layers consist of hexagonal rings but contain tetrahedra of two orientations. The layer of octahedra (*O*-layer) consists of edge-shared (Ca,Na)-octahedra. Tetrahedral nets of adjacent modules are connected through inverted tetrahedra. H_2O molecules, K, and Na atoms are placed between the layers. Fedorite is one of the simplest of the modular structures because it consists only of *O* (octahedra) and *T* (tetrahedra) structural units in a $OT_2\overline{T}_2O$ sequence of stacking. In this scheme, identical sheets linked by a center of symmetry are marked with a "strip" symbol.

In 2020, Kaneva et al. [31] obtained for the first time spectra of EPR, optical absorption, and emission of Mn^{2+} ions for fedorite. Some samples of the mineral containing Mn and Ce have bright raspberry coloring due to the presence of Mn^{4+} ions [31].

Fluorcarletonite, $KNa_4Ca_4Si_8O_{18}(CO_3)_4F{\cdot}H_2O$, a rare mineral belonging to the group of phyllosilicates, is an F-analog of carletonite named after Carleton University [32]. The mineral was discovered in 2019 by Kaneva et al. [33] in charoitites of the Murun massif. Blue fluorcarletonite forms close intergrowths with apophyllite, which indicates their joint crystallization, and it is also associated with pectolite and charoite.

The study of the crystal structure of carletonite was published by Chao in 1972 [34]. At present, it is almost impossible to find samples of carletonite in the Poudrette quarry, Mont Saint-Hilaire, Canada—the only known mineral deposit in the world—because the area where the mineral was found was flooded [35].

The crystal-chemical formula of fluorcarletonite was calculated on the basis of (Si + Al) = 8 atoms per formula unit, while the H_2O content is consistent with the refined occupancies of the positions of water molecules in the crystal structure. The average formula of Murun fluorcarletonite:

$K_{1.04}Na_{3.89}Ca_{3.87}Ti_{0.01}(Si_{7.99}Al_{0.01}O_{18})(CO_3)_{3.86}(F_{0.72}OH_{0.28}){\cdot}1.11H_2O$ [33].

It is noteworthy that Al is almost absent, and Ti concentration is lower in Murun fluorcarletonite compared to carletonite from the Mont Saint-Hilaire (Canada).

The parameters of the fluorocarletonite unit cell (space group *P4/mbm*) refined in TOPAS 4 [36]: $a = 13.219(1)$ Å, $c = 16.707(2)$ Å, $V = 2919.4(6)$ Å3 [33].

The crystal structure of the mineral consists of silicate layers, layers of Na- and Ca-centered polyhedra, as well as K^+ cations occupying cavities inside the eight-membered silicate rings. H_2O molecules are located in the pores of silicate double layers. Finally, the isolated triangles of the CO_3 group are bonded to layers of Na and Ca polyhedra. Silicate layer with composition $[Si_8O_{18}]^{4-}$ in carletonite and fluorcarletonite is the result of condensation of two branched single layers.

5 Discussion and Conclusion

On the infrared spectra of the minerals (Figs. 7, 8 and 9) the Si–O stretching vibrational modes are appeared in the region from 900 to 1200 cm^{-1}: Si–O vibrations with the participation of non-bridging oxygen atoms (~900–1000 cm^{-1}) and bridging Si–O–Si (~1000–1200 cm^{-1}). Si–O–Si bending vibrations, mixed silicon-oxygen rings vibrations and stretching vibrational modes of different cations are characterized by the spectrum region from 500 to 700 cm^{-1}.

Table 1 Band positions (cm^{-1}) in IR spectra of some rare silicates of Murun massif

	Tinaksite (Fig. 7 line 1)	Tokkoite (Fig. 7 line 2)
ν(OH)	3366w, 3469	3374, 3479w
δ(H–O–H)	1640	1634
ν(Si–O–Si)	1111, 1090, 1076, 1025w	1109, 1095, 1047
ν(Si–O)	994sh, 965, 940	995, 940, 955, 922sh
δ(Si–O-Si) and Si–O ring	784w, 727, 699, 654, 644, 635, 543, 508	786w, 733, 698, 656, 637sh, 550, 522
ν(Ti–O)	681	683w
	Agrellite (Fig. 8 line 3)	Frankamenite (Fig. 8 line 4)
ν(OH)	3564w	3559
ν(H–O–H)	3065w, 3470w	3518w, 3609
δ(H–O-H)	1636w	1598, 1701
ν(Si–O–Si)	1141, 1092, 1063, 1041, 1036, 1005	1124, 1091, 1023
ν(Si–O)	963	953
δ(Si–O–Si) and Si–O ring	781, 715w, 689, 654, 612, 590	798, 768, 698, 677, 648, 623w, 516
	Fedorite (Fig. 9 line 5)	Fluorcarletonite (Fig. 9 line 6)
ν(OH)	3562, 3639	3655
ν(H–O–H)	3157, 3433	3564, 3587
δ(H–O–H)	1635	1770
ν(Si–O-Si)	1129, 1041	1197, 1166w, 1148w, 1072sh, 1051
δ(Si–O–Si) and Si–O ring	798, 621, 516	785, 592, 524, 502, 488
ν(C–O)		1396, 1451, 1480, 1527
δ(C–O)		662, 693, 728, 874

Note ν—stretching vibrations, δ—bending vibrations, w—weak band, sh—shoulder

Fig. 3 Crystal structure of agrellite [18]. Si tetrahedra are shown in gray, Ca octahedra are yellow, and sodium atoms are blue spheres

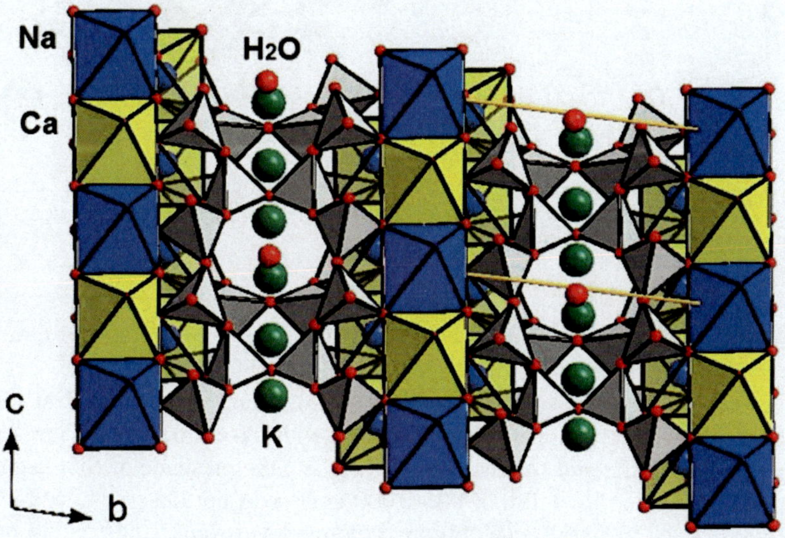

Fig. 4 Crystal structure of frankamenite (by [24]). Si-tetrahedra are shown in gray, Ca-octahedra are yellow, Na-octahedra are blue, potassium atoms are green spheres, oxygen atoms of water molecules are red spheres

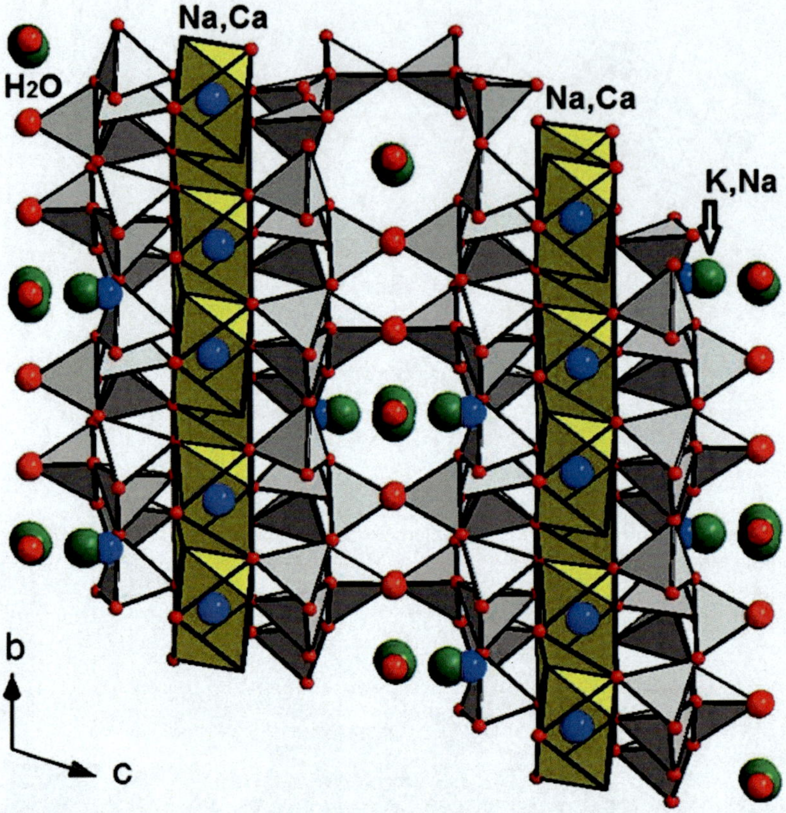

Fig. 5 Crystal structure of fedorite. Si-tetrahedra are shown in gray, Ca,Na mixed octahedra are yellow, sodium atoms are blue spheres, potassium atoms are green spheres, oxygen atoms of water molecules are red spheres

Information on the position of diagnostic bands in the infrared spectra of the silicates under study is given in Table 1.

Tinaksite and tokkoite (Fig. 7) contain a strong and narrow band at 681 and 683 cm^{-1}, respectively, apparently associated with Ti–O stretching vibration. On the spectra of fedorite and fluorcarletonite (Fig. 9), the presence of Si–O stretching vibrational bands in the region of 900–1000 cm^{-1} was not observed. In the region of 660–880 and 1300–1600 cm^{-1} fluorcarletonite has strong bands due to bending and stretching vibrations of CO_3 molecule (Fig. 9).

In the region of O–H stretching vibration (3000–4000 cm^{-1}) the silicates of different groups clearly differ. Fluorcarletonite may have three bands in this area (Fig. 9). Tinaksite and tokkoite (Fig. 7), frankamenite (Fig. 8) have two distinct peaks in this region, while the agrellite (Fig. 8) has a weakly manifested band of O–H vibration and a band of H–O–H. Fedorite (Fig. 9) has a large number of lines.

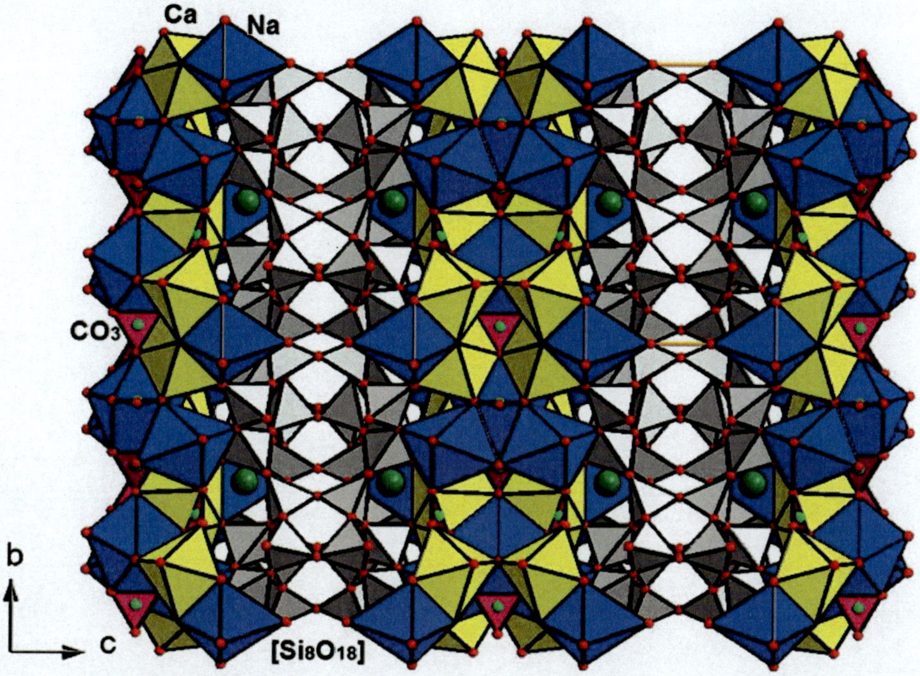

Fig. 6 Crystal structure of fluorcarletonite. Si-tetrahedra are shown in gray, Ca-octahedra are yellow, Na-octahedra are blue, potassium atoms are green spheres, and CO_3 groups are purple triangles

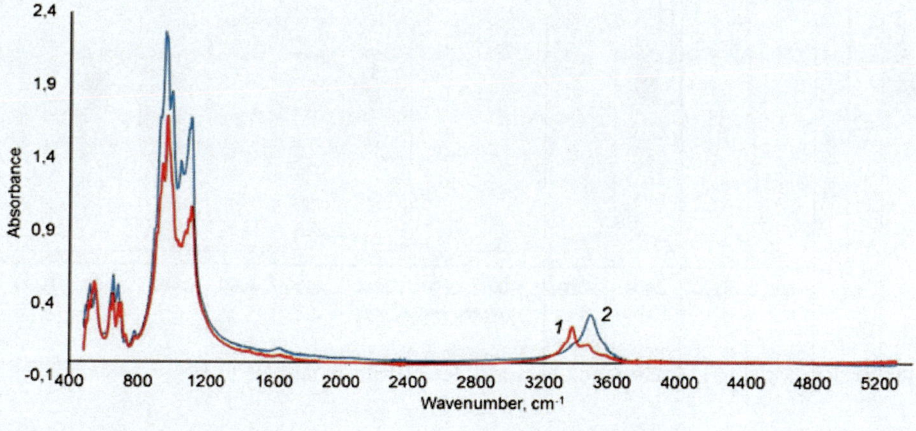

Fig. 7 IR spectra of tinaksite (1, red line) and tokkoite (2, blue line)

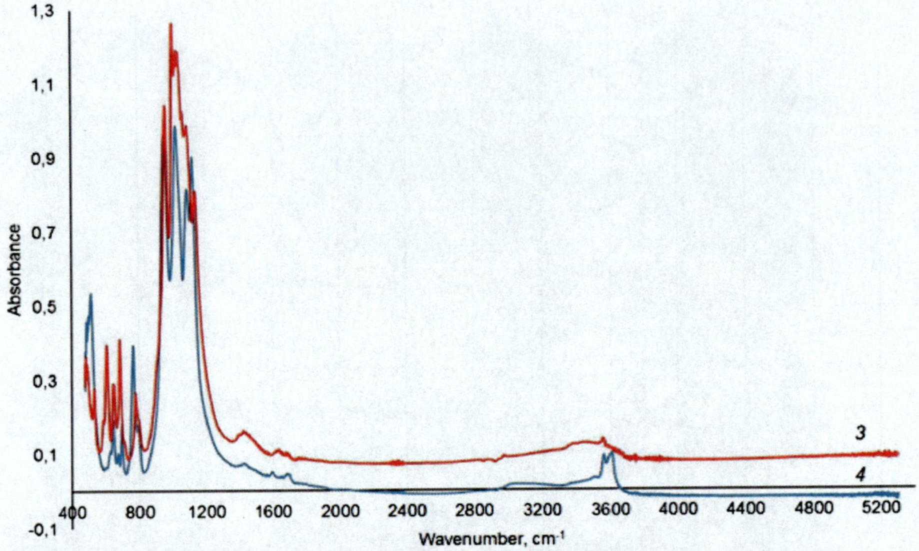

Fig. 8 Infrared spectra of agrellite (3, red line) and frankamenite (4, blue line)

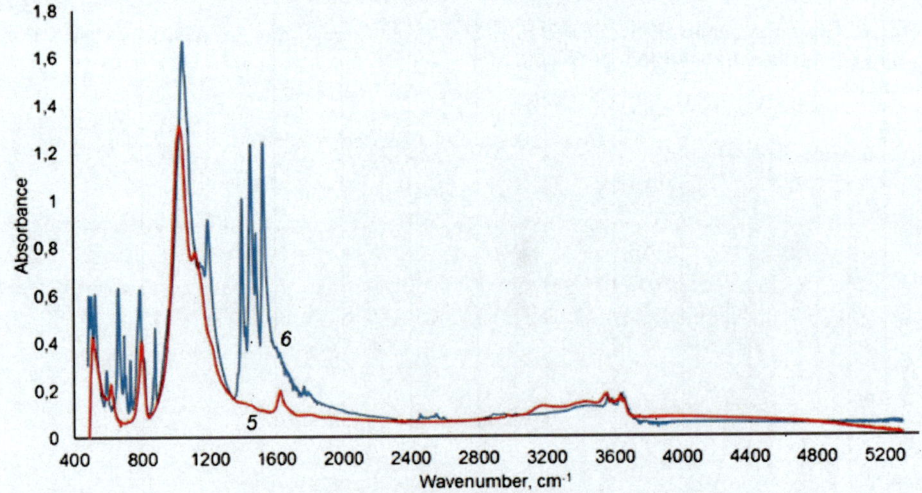

Fig. 9 IR spectra of fedorite (5, red line) and fluorcarletonite (6, blue line)

In the region of 1500–1800 cm^{-1}, all minerals have a band of H–O–H deformation oscillations.

The close intergrowth and germination of minerals crystallizing in the charoitites of the Malyy Murun massif indicate their constitutional relationship. The crystal structures of the alkaline silicates discussed here contain rare-type silicon-oxygen radicals. They are curved ribbons, tubes, layers of silicon-oxygen rings having wide cavities or channels along one or two crystallographic axes. K or Na cations fasten silicon-oxygen chains condensed into tubes or layers. Silicate channels are located between the chains or walls of the polyhedra. From a chemical point of view, it is noted that the minerals of the Murun massif, in comparison with similar minerals from other deposits, contain more F in the positions usually occupied by the OH group and F.

In general, the diversity of structures, chemical compositions and conditions of origin of alkaline minerals of the Murun massif provides for the manifestation of various mechanisms of isomorphic substitutions. Knowledge of the features of the atomic structure, as well as the distinctive properties of the mechanisms of transformation of minerals, is important for understanding the essence of geological processes leading to the formation of such unique deposits.

Acknowledgements The research was done with the financial support of the Integration Project of the Research Center of the Russian Academy of Sciences, Block 1.4. and the study was performed by the governmental assignment in terms of Project IX. 129.1.4.

References

1. Rogov, Y.G., Rogova, V.P., Voronkov, A.A., Moleva, V.A.: Tinaksite, $NaK_2Ca_2TiSi_7O_{19}(OH)$—new mineral. DAN USSR **162**(3), 658–661 (1965). [in Russian]
2. Bissert, G.: Verfeinerung der Struktur von Tinaksit, $Ca_2K_2NaTiO[Si_7O_{18}(OH)]$. Acta Crystal **B36**, 259–263 (1980). https://doi.org/10.1107/S0567740880003068
3. Rozhdestvenskaya, I.V., Nikishova, L.V., Lazebnik, K.A.: Crystal structure features of tinaksite group minerals. Mineralogichesky Zhurnal **13**(4), 3–12 (1991). [in Russian]
4. Kostyleva-Lobuntsova, E.E., Borutsky, B.E., Sokolova, M.N., Shlyukova, Z.V., Dorfman, M. D., Dudkin, O.B., Kozyreva, L.V.: Mineralogy of the Khibiny Massif (Minerals), vol. 2. Science Publishing House, Moscow (1978). [in Russian]
5. Lazebnik, K.A., Nikishova, L.V., Lazebnik, Yu.D.: Tokkoite—a new mineral of charoitites. Mineralogichesky Zhurnal **8**(3), 85–89 (1986). [in Russian]
6. Rozhdestvenskaya, I.V., Nikishova, L.V., Lazebnik, Yu.D., Lazebnik, K.A.: The crystal structure of tokkoite and its relation to the structure of tinaksite. Z Kristallogr **189**, 195–204 (1989). https://doi.org/10.1524/zkri.1989.189.3-4.195
7. Lacalamita, M., Mesto, E., Kaneva, E., Scordari, F., Pedrazzi, G., Vladykin, N., Schingaro, E.: Structure refinement and crystal chemistry of tokkoite and tinaksite from the Murun massif (Russia). Min. Mag. **81**(2), 251–272 (2017). https://doi.org/10.1180/minmag.2016.080.094
8. Rozhdestvenskaya, I.V., Nikishova, L.V.: Crystal-chemical features of alkaline calcium silicates from charoitites. Kristallografiya **47**(4), 602–611 (2002). [in Russian]
9. Gittins, J., Bown, M.G., Sturman, D.: Agrellite, a new rock-forming mineral in regionally metamorphosed agpaitic alkalic rocks. Can. Min. **14**, 120–126 (1976)
10. Konev AA, Vorobiev EI, Sapozhnikov AN, Paradina LF, Lapides IL, Ushchapovskaya ZF (1987)Sr-containing agrellite from Murun alkaline massif. Mineralogichesky Zhurnal **9** (3):73–79. [in Russian]
11. Pakhomovsky, Ya.A., Ivanyuk, G.Yu., Yakovenchuk, V.N.: Loparite-(Ce) in rocks of Lovozero stratified complex of Karnasurt and Kedikvyrpakhk mountains. Zapiski Ross Mineral Obshc **143**(1), 68–87 (2014). [in Russian]
12. Falster, A.: Minerals of the pegmatite bodies in the Wausau Pluton. Rocks and Minerals **62** (3), 188–193 (1987). https://doi.org/10.1080/00357529.1987.11762652

13. Semenov, E.I., Dusmatov, V.D.: Agrellite—the first discovery in the USSR. Miner. of Tajikistan (Dushanbe) **8**, 3–5 (1989). [in Russian]
14. Faiziev, A.R.: Comparative mineralogical characteristics of the Dunkeldyk and Darai-Piez alkaline massifs (Tajikistan). Zapiski Ross Mineral Obshc **145**, 20–29 (2016). [in Russian]
15. Ghose, S., Wan, C.N.: Agrellite, Na(Ca, RE)$_2$Si$_4$O$_{10}$F: a layer structure with silicate tubes. Amer. Min. **64**(5-6), 563–572 (1979)
16. Rozhdestvenskaya, I.V., Nikishova, L.V., Lazebnik, K.A.: Diagnostics of agrellitepolytypes by X-ray powder diffraction data. Apiski Ross Mineral Obshc **127**(1), 89–94 (1998)
17. Rozhdestvenskaya, I.V., Nikishova, L.V.: Crystal structure of strontium agreliite Na(Ca, Sr$)_2$S$_i$4$_{O1}$0F from charoitites of Yakutia. Polytypy of Agrellite. Kristallographiya **43**(4), 637–645 (1998)
18. Kaneva, E., Shendrik, R., Mesto, E., Bogdanov, A., Vladykin, N.: Spectroscopy and crystal chemical properties of NaCa$_2$[Si$_4$O$_{10}$]F natural agrellite with tubular structure. Chem. Phys. Lett. **738**, 136868 (2020). https://doi.org/10.1016/j.cplett.2019.136868
19. Kaneva, E.V., Vladykin, N.V., Mesto, E., Lacalamita, M., Scordari, F., Schingaro, E.: Refinement of the crystal structure of vlasovite from Burpala massif (Russia). Crystallogr. Rep. **63**, 1092–1098 (2018). https://doi.org/10.1134/S1063774518070106
20. Kaneva, E., Bogdanov, A., Shendrik, R.: Structural and vibrational properties of agrellite. Sci. Rep. **10**, 15569 (2020). https://doi.org/10.1038/s41598-020-72631-1
21. Evdokimov, M.D., Regir, E.P.: Canasite in charoitites of Murun alkaline complex ("Lilac stone" deposit). Zapiski Ross Mineral Obshc **1**, 104–118 (1994). [in Russian]
22. Nikishova, L.V., Lazebnik, K.A., Rozhdestvenskaya, I.V., Emelianova, N.N., Lazebnik, Y. D.: Triclinic canasite from charoitites of Yakutia. Mineralogichesky Zhurnal **14**(1), 71–77 (1992). [in Russian]
23. Nikishova, L.V., Lazebnik, K.A., Rozhdestvenskaya, I.V., Emelianova, N.N., Lazebnik, Y. D.: Frankamenite K$_3$Na$_3$Ca$_5$(Si$_{12}$O$_{30}$)F$_3$(OH)·nH$_2$O—a new mineral. Triclinic analogue of canasite from charoitites. Zapiski Ross Mineral Obshc **125**(2), 106–108 (1996). [in Russian]
24. Rozhdestvenskaya, I.V., Nikishova, L.V., Lazebnik, K.A.: The crystal structure of frankamenite. Min. Mag. **60**, 897–905 (1996). https://doi.org/10.1180/minmag.1996.060.403.05
25. Kuharenko, A.A., Orlova, M.P., Bulakh, A.G., Bagdasarov, E.A., Rimskaya-Korsakova, O. M., Nefedov, E.I., Ilyinsky, G.A., Sergeev, A.S., Abakumova, N.B.: Caledonian complex of ultrabasic, alkaline rocks and carbonatites of the Kola Peninsula and North Karelia. Subsoil, Leningrad (1965). [in Russian]
26. Konev, A.A., Vorobiev, E.I., Lazebnik, K.A.: Mineralogy of Murun massif. Siberian Branch of Russian Academy of Sciences, Novosibirsk (1996). [in Russian]
27. Ferraris, G.: Polysomathism as a tool for correlating properties and structure. In: Merlino, S. (ed.) Modular Aspects of Minerals. European Mineralogical Union, Notes in Mineralogy, vol. 1, pp. 275–295. Springer, Berlin (1997). https://doi.org/10.1180/EMU-notes.1
28. Sokolova, G.V., Kashaev, A.A., Dritz, V.A., Ilyukhin, V.V.: Crystal structure of fedorite. Kristallographiya **28**, 170–172 (1983). [in Russian]
29. Joswig, W., Drits, V.A., Sokolova, G.V.: Refinement of the structure of fedorite. Sov. Phys. Crystallogr. **33**, 763–765 (1988)
30. Mitchell, R.H., Burns, P.C.: The structure of fedorite: a re-appraisal. Can. Min. **39**, 769–777 (2001). https://doi.org/10.2113/gscanmin.39.3.769
31. Kaneva, E.V., Shendrik, R.Yu., Radomskaya, T.A., Suvorova, L.F.: Fedorite from Murun alkaline complex (Russia): spectroscopy and crystal chemical features. Minerals **10**(8), 702 (2020). https://doi.org/10.3390/min10080702
32. Chao, G.: Carletonite, KNa$_4$Ca$_4$Si$_8$O$_{18}$(CO$_3$)$_4$(F,OH)·H$_2$O, a new mineral from Mount St. Hilaire, Quebec. Amer. Min. **56**, 855–1865 (1971)
33. Kaneva, E., Radomskaya, T., Suvorova, L., Sterkhova, I., Mitichkin, M.: Crystal chemistry of fluorcarletonite, a new mineral from the Murun alkaline complex (Russia). Eur. J. Mineral **32**, 137–146 (2020). https://doi.org/10.5194/ejm-32-137-2020

34. Chao, G.Y.: The crystal structure of carletonite, $KNa_4Ca_4Si_8O_{18}(CO_3)_4(F, OH)\cdot{}_H2O$, a double-sheet silicate. Amer. Min. **57**, 765–778 (1972)
35. Pohwat, P.W., Cook, R.B.: Carletonite, Mont Saint-Hilaire, La Vallèe-du-Richelieu RCM, Montèrègie, Quèbec, Canada. Rocks and Minerals **91**, 48–51 (2016). https://doi.org/10.1080/00357529.2016.1099133
36. Bruker, A.X.S.: Topas V4: general profile and structure analysis software for powder diffraction data. Bruker AXS Inc., Karlsruhe, Germany (2008)

Ca-Concentrations in Olivines from Ultrabasic Rocks—Genetic Findings

G. P. Ponomarev, N. V. Vladykin, and I. A. Sotnikova

Abstract

We present the results of genetic systematization of olivine compositions formed both in experimental conditions and its natural variations from the rocks of the main ultrabasic compositions of different depth facies and formed in different geodynamic conditions. The criterion for genetic differences is the features of calcium content in olivine crystals.

Keywords

Calcium · Olivine · Equilibrium

1 Introduction

Minerals of olivine groups, being practically constant participants of mineral associations in rocks of ultrabasic and basic composition, are widely used indicators of their genesis. For genetic constructions, two types of tools based on the compositions of olivine crystals are used—discriminating diagrams and equations,

G. P. Ponomarev (✉)
Institute of Volcanology and Seismology of FEB RAS, 9, Piypa Blvd., 684006 Petropavlovsk-Kamchatsky, Russia
e-mail: ponomarev@kscnet.ru

N. V. Vladykin · I. A. Sotnikova
Vinogradov Institute of Geochemistry SB RAS, 1A, Favorsky St., 664033 Irkutsk, Russia
e-mail: vlad@igc.irk.ru

Irkutsk Scientific Center of Siberian Branch of Russian, 134 Lermontov St., 664033 Irkutsk, Russia

N. Vladykin (ed.), *Alkaline Rocks, Kimberlites and Carbonatites: Geochemistry and Genesis*, Springer Proceedings in Earth and Environmental Sciences,
https://doi.org/10.1007/978-3-030-69670-2_9

binding compositions of olivine and other phases from this association with the physical and chemical conditions of their formation. The most important genetic criteria for the content of small elements are, as a rule, the generalizations of a large number of data on their content in the natural variations of olivine and the discriminative diagrams created on this basis. Contents of elements forming the olivine lattice are mainly significant genetic criteria in the form of equations for calculating temperatures, pressures, oxygen fugacity, equilibrium with the melt and other phases in the rock, melting degree, etc.

For olivine, thanks to the development of precise local analysis methods, both groups of elements are in demand. Among the small elements for genetic structures, Ni, Co, Mn, Cr, Ti, Ca are usually used. Probably, one of the most important genetic indicators for olivine is the threshold content of calcium oxide (0.1 wt.%) in its composition. This criterion was identified and shown in Simkin and Smith [1]. Its existence was explained by the authors by the influence of crystallization depth—pressure. Since then, no due attention has been paid to the ratios of calcium-containing and low calcium (less than 0.1 wt.% CaO) olivine variations in the results of study and experiment, which often did not correspond to the conclusions [1]. In Ponomarev and Puzankov [2], Ponomarev [3, 4], Ponomarev and Puzankov [5], based on the generalization of a large number of data on experimental studies of melt systems involving olivine in a wide range of conditions as well as on the results of systematization of its natural variations, the error of this conclusion was shown. However, a number of facts (in terms of calcium content in the olivines) have accumulated among the experimental data as well as among the natural differences of the olivines, both confirming and amplifying, the version of explanation proposed by us and not corresponding to it.

The purpose of this work is to demonstrate the collected facts and consider possible options for their explanation.

2 Genetic Criterion—Threshold Content of CaO in Olivine

Thanks to the development of microanalysis methods, it became possible to locally determine the content of elements in the mineral phase up to 3–4 decimal places. This was the impetus for the creation of geothermometers using the content of elements—impurities in the olivine. On the basis of the interphase distribution of calcium from experimental data several geothermometers were proposed, e.g. geothermometer [6]. In Gavrilenko et al. [7], using precise calculations of calcium content, the dependence of H_2O content in the melt on the value of K” ol/m for CaO was proposed. The verification of the K”–H_2O bond on the inter-phase distribution of calcium on 3 arrays of experimental data (310, 333 and 115 analysis points) used in Ponomarev and Puzankov [5] showed the following. The dependence of the value of K” ol/m for Ca on the magnesium content in the melt reflects the curvilinear dependence of the bond between magnesium and calcium content in the melt in equilibrium with olivine, which is due to the low bond of calcium content in the olivine on its content in the melt (the average value of the correlation coefficient for 3 arrays is 0.37) and varies from sample to sample. Dependence of the value of K” ol/m for Ca on the composition of olivine, melt composition, P–T conditions showed rather high negative dependence of Ca only on the magnesium content in the olivine (average value of correlation coefficient for 3 arrays −0.62) and rather low (−0.23) negative correlation with water content in the melt. Dependence of magnesium content in olivine is controlled by Fe^{2+} content very sensitive to oxygen fugacity, Mg content in melt and variable K” $_{Fe/Mg}$ ol/m. These bonds show the ambiguity of the proposed dependence of the K” ol/m value for CaO on the melt water content.

Probably, one of the most informative criteria for the genesis of olivine, which is included in the rocks of ultrabasic basic composition, is the threshold content of calcium (0.1 wt.% CaO) in its crystals. In Ponomarev and Puzankov [2], Ponomarev [3, 4], Ponomarev and Puzankov [5], on the basis of analysis of a large number of experimental data, both on crystallization of melts of the basic-ultramafic composition and melting of rocks of similar compositions in a wide range of conditions, it was shown that crystals of equilibrium with the basic-ultramafic olivine melt contain at least 0.1 wt.% of CaO in its composition, at contents more than 3 wt.% of CaO in the melt. This conclusion was obtained from the analysis of arrays of experimental data created on the basis of the database of experimental data on phase equilibria of erupted rocks (INFOREX system) [8] and the author's own collections. These samples are as follows: “dry” conditions, atmospheric pressure (~100 GPa)–1033 experiments; dry conditions, pressure up to 17.5 GPa–756 experiments; experiments with “unsaturated” fluid melt at pressures up to 3.5 GPa–310 points of analysis; experiments with “saturated” fluid melt at pressures up to 12.5 GPa–330 points of analysis. An array of experiments with known melt water contents and pressures up to 6 GPa–130 points of analysis (author's study). In addition to these data, the experimental results that were not included in the above samples collected in Ponomarev [Tables 2, 3]–313 analysis points and [5]–40 analysis points were also included. These are mostly experiments in melting peridotite, kimberlite, peridotite-carbonate systems, eclogite-carbonate with maximum pressures up to 26 GPa.

However, a number of exceptions to this pattern were also found among the experimental results, amounting to ~2–7% of cases in the samples where olivines are represented by either calcium-free or low calcium differences. In the first case, the calcium content was not analyzed in olivine crystals, and this practice, unfortunately, has been preserved to date. This may also include analytical errors during the period of low calcium determination accuracy in olivine. In the second case, there may be several options for the presence of low calcium differences in olivine. First, these are solidus conditions of its crystallization, or close to sub-solidus + fluid, when the system under study practically does not contain a melt cementing the mineral phases. Examples are the data [9] (33 crystals out of 36 are represented by low calcium ones), [10], where in 3 out of 4 crystals olivine is represented by low calcium variation and the data [11] (29 crystals out of 30 are represented by low calcium crystals). This group of experiments probably includes the results [12] on the partial melting of peridotite (1.9 GPa) with participation of H_2O–KCl (out of 7 crystals only 2 are represented by calcium-containing crystals, which, according to the authors, is related to the capture of clinopyroxene inclusions during analysis). Melt quenching products in these experiments contained less than 3 wt. % CaO in 5 out of 6 of these analyses, which, in our opinion, causes the formation of low calcium variations. A series of experiments presented in Pirard and Hermann [13], where all the analyses (14 points) are represented by low calcium variations, is a good illustration of the influence of low CaO contents in the melt on its content in olivine and of the formation of low calcium olivine in solidus region. The existence of local equilibria in the experimental system under study can lead to the appearance of 2 olivine variations. Thus, in the experiment (without melting) with fluid [14] olivine (near tablet $CaCO_3$) is represented by calcium-containing variations, and in the matrix by calcium-free variations. These experimental data with the formation of olivine in the solidus region suggest that not achieving the melting point in the experiment and the presence of fluid (water) are the causes of the formation of low calcium olivines. The experimental results of recent years have revealed several other influencing elements, along with calcium, in the melt on the peculiarities of calcium entering the olivine crystals. In the melts formed during the melting of peridotite-carbonate, peridotite-phlogopite-containing, potassium-rich kimberlite-like systems with the participation of fluids, low calcium variations are formed along with calcium-containing olivine crystals [15–16]. This sample of 4 works is represented by 40 analyses of olivine crystals, among which the share of low calcium differences varies from 30 to 100%. These melts are characterized by CaO contents greater than 4 wt.%, but by wide variations in SiO2 contents (~6–46 wt.%) and sufficiently high contents of Σ alkalis ~6–30 wt.% with potassium predominance (kamafugitic differences). In case of moderate concentrations of alkalis in the melt (1–8 wt.%) [17] all presented analyses (10 crystals) refer to calcium-containing variations. One of the variants of formation of low-calcium olivines in these experiments can be the effect of high alkali concentrations in the melt. A special feature of the experimental results with the presence of low-calcium olivines is the data from Tenner et al. [18]. In this system peridotite-fluid (water) with a small amount of locally developed fluidized glass, all 10 cited analyses of olivine crystals are represented by low calcium variations in association with low and high calcium pyroxenes and garnet. Although the contents of CaO in the melt are in the range 11–16 wt. %, SiO_2 35–40 wt. %, Σ alkalis are less than 2 wt. %. This discrepancy may probably be due to several reasons. Because of the limited development of melting (glass), probably, there was a sub-consolidous recrystallization of the starting material with the participation of water to form a low calcium olivine. In experiments [11] on dehydration of serpentinite under milder conditions (2 GPa, 550–850 °C) in comparison with conditions of experiments [18] (10–13 GPa, 1350–1450 °C) in solidus conditions similar mineral association was formed: olivine-low calcium pyroxene-high calcium pyroxene-granate. Olivine in this experimental series is also represented by low calcium variation (29 out of 30 given analyses), but garnet is represented by more calcium variation and in mineral association there is also clinochlorine. Besides, a possible cause of formation of low calcium olivines in experiments [18] may be the non-equilibrium of olivine with melt, as the duration of experiments was equal to 24 h, whereas according to Ariskin and Barmina [8] the equilibrium of olivine melt occurs at duration of experiments more than 48 h.

The main condition for experimental studies is to achieve the equilibrium of compositions of coexisting phases. It allows to assert that crystals of calcium-containing olivine are in equilibrium with melts of basic-ultramafic compositions containing at least 3 wt. % of CaO and normal and moderate alkalinity. All the above-mentioned experimental data collected, together with exceptions, allow us to state that the pressure to 29 GPa does not affect the calcium content of olivine crystals in equilibrium with the melts of the ultramafic-main composition. This disproves the thesis [1] about the influence of the pressure on the calcium content in olivine crystals and necessitates the search for other reasons for the presence of low calcium differences in rocks of known primary magmatic genesis. It should be noted that the number of experimental works on the interaction of peridotite with carbonatite melts under mantle pressure with the formation of calcium containing olivines is constantly growing. An example is the recently published work [19] on the interaction of peridotite with carbonatite melt under conditions of 3.1 and 6.5 Gpa, where the 10 compositions of olivine contain from 0.4 and above weight % CaO. Moreover, the K_2O content in the melt is high enough and lies in the range of ~10–15 wt.%, which should contribute to the formation of low calcium olivine variations. It is possible that the influence of potassium in the melt is compensated by the high content of calcium in the melt (~20 wt. % CaO). Summarizing all collected experimental data, it can be assumed that the presence of low calcium variations of olivine crystals under experimental conditions is due to the composition of the melt: the content of CaO less than 3–4 wt. %; low SiO_2 ~ less than 10 wt. %; high content of Σ alkalis ~10–30 wt. %, with potassium predominance. The boundaries of the influence of SiO_2 content and the sum of alkalis in the melt need to be clarified using more experimental data. It is possible that this will be an integral index taking into account the temperature values and the CaO contents in the melt.

This revised conclusion, based on experimental data, was used to consider the possible genesis of some rocks of the basic-ultramafic compositions with an emphasis on alkaline differences. Since we still find arguments on the basis of the conclusion from Simkin and Smith [1] in published papers.

3 Volcanic Rocks

Features of CaO contents in olivines from lava, komatiites, meimechitis, rocks of camafugitic series, carbonatites and kimberlites were considered.

3.1 Lavae of the Basic Composition

Lavae of the basic composition, formed in different geodynamic conditions mainly in terrestrial conditions in a wide age range contain, as a rule, mainly calcium-containing crystals of olivine. As the degree of metamorphic transformations increases, the share of low calcium olivines increases. In Quaternary lava the fraction of low calcium olivines is small and does not exceed a few percent. For the main lava of the modern second layer of the oceanic crust, represented by pillow lava, calcium-containing olivines are typical. The only exception so far is the discovery of young (6 million years) Shoshoneites in the Japanese trough [20]. The above analysis of olivine phenocrystals (6 crystals) is represented by low calcium variations with the edges of calcium-containing olivine; microlites in the main mass are represented by calcium-containing olivines. In the lavae of the ophiolitic complexes, the preservation of calcium-bearing olivines depends on the facial level of metasomatic transformations of the rock and the degree of its change. But on the whole, we can confidently state that low calcium olivine variations in pillow-lavae of the ophiolitic complexes are becoming dominant. These data, taking into account the generalization from experimental data, allow to assume the removal of calcium from olivine crystals under metamorphic transformations and the non-equilibrium with the melt of low calcium olivines present in quaternary basalts.

3.2 Comathies and Meimecrites

Comatiite lava, which is part of the Pre-Cambrian greenstone belts, is usually modified in the range from green shale to epidote amphibolite facies to match the rocks of the frame. Quantitative ratios of high calcium and low calcium olivines vary in a wide range: there are cases of both the presence of individual crystals of low calcium olivines and complete absence of calcium containing olivines with the appearance of a newly formed metasomatic without calcium olivine. In Cretaceous komatiites of Gorgona Island, practically not affected by secondary changes, olivine crystals are poorly affected by secondary transformations—crack serpentinization, and are represented by calcium-containing variations. According to the above data (188 analyses) [21], these crystals have a sufficiently homogeneous composition in terms of CaO content ~0.3–0.35 wt.% without any distinct differences between the nuclei and edges of phenocrysts and microlites.

Many studies of the gross and phase compositions of these rocks are devoted to the Meimechite genesis. Spatially, they are confined to the alkaline-ultramafic Gulinsky complex. The compositions of olivine crystals given in the publications are represented by calcium-containing differences. As an example, we use data from Sobolev et al. [22]. Meimechites and dunites of Gulinsky plutonium have definitions of absolute age, according to data from Sobolev et al. [22] in the range of 231–245 million years. Crystals of olivine in Meymeychite are quite fresh and are affected by serpentinization through cracks. They are represented by calcium-containing variations (20 analyses) of CaO content in the range of ~0.3–0.5 wt.%. These vulcanites were formed from ultramafic alkaline magma, while plutonium dunites may be olivine cumulates from this magma [22], the derivatives of which may also be carbonatite melts [23].

3.3 Kamafugite Series and Carbonatite Vulcanites

Camafugitic volcanism is rather limited, mainly in the zones of large structural tectonic disturbances. In almost all cases, its manifestations are spatially and genetically closely related to carbonatites. As an example, we use data on the contents of CaO in olivines from ugandite (9 crystals) [24]; mafurites (7 crystals) and ugandite (7 crystals) [25]; and melilitites (25 crystals) (Panina et al. 1989, 26]), calciolithitic and melilithitic (7 crystals) [27]; carbonatite (5 crystals) [28]; calcio-carbonatite (8 crystals, 5 monticellite, 3 olivine) [29]. All these rocks are of quaternary age. Of the 68 above analyses of olivine crystals, only 5 crystals are represented by a lowcalcium difference. In most cases, an increase in the calcium content in the marginal parts of the crystals was found. The content of the sum of alkalis in the rocks of Kamafugite series of used data lies in the range of ~3–9 wt.% with potassium predominance. Higher values of the sum of alkalis are fixed in heated melt inclusions, where the contents can reach up to 15–20 wt.%.

In calciocarbonatite the sum of alkalis (by compositions of melt inclusions) can reach 36 wt.%; in silicate melt inclusions ~9–16 wt.%, sodium prevails over potassium in both groups of compositions.

3.4 Kimberlites

The upper age limit for kimberlite eruptions was considered ~22–32 million years. But kimberlite lava and quaternary age Igwisi Hills volcanoes were found, Tanzania [30]. For almost all kimberlite lava, regardless of age (including quaternary lava) and location, low calcium olivine phenocrystals are typical. Among the microlithic phase of olivine, the calcium-containing differences are quite common. But there are cases when almost all of the phenocrysts and olivine microlites studied are represented by a low calcium variation. An example is the data on kimberlites of the Udachnaya Vostochnaya tube [31], where from ~270 cited phenocrysts and microlites analyses, including the edge parts of microlites, single (less than 10 analyses) are represented by calcium-containing difference. Numerous published data on the compositions of olivine crystals included in the diamond crystals are also represented almost exclusively by the calcium-containing variation. Examples, along with previously collected data [4], include olivine inclusions in microdiamonds of the Yubileinaya (Yakutia) tube [32]. There are also isolated finds of calcium-containing olivines from the diamond crystals (see Ponomarev [4]). Olivines formed in melt polymineral inclusions from phenocrystalline minerals in kimberlites are represented by the calcium-containing variation. Examples include data on olivines from phenocrysts of the Udachnaya-Vostochnaya tube olivine (Yakutia) [33] and olivine crystals from phenocrysts of clinopyroxene and garnet from kimberlites Lac De Grass (Canada) [34].

4 Plutonic Rocks

Previously collected data on hypabissal bodies and large intrusive massifs and xenoliths allowed making the following generalizations for this group of rocks. Similar to effusive variations with increasing age and degree of metamorphic transformations (facial level and volume fraction) of rocks, the share of low calcium olivines in basic and ultrabasic full-crystalline rocks grows and can reach 100 vol. Although there are examples of preservation of calcium-containing variations of olivine in massifs of pre-Cambrian age. In the intrusive bodies of the basic composition (the third layer) of the modern oceanic crust, low calcium variations in the ratio of 3:1 prevail significantly. In the tertiary stratified complex of the island of O. S. and in the Caspian Sea. Along with separate fragments of the complex completely devoid of calcium-containing olivines, there are also fragments of the complex, in which the ratio of low to calcium-containing variations of olivine is ~1. Whereas in the Pre-Cambrian layered massifs of calcium-containing olivine crystals are practically absent. Olivines of homogeneous xenoliths are represented by calcium-containing variations. Olivines of ultramafic xenoliths, for the majority of which mantle nature is assumed, in the overwhelming majority of cases are represented by low calcium variations of crystals. Newly formed crystals of olivines from melting zones in these xenoliths are represented by calcium-containing variations.

Below are some sample data for a number of massifs, which allow to supplement previously obtained generalizations.

There is a distinct difference in the share of low calcium olivines in the complementary complexes of plutonium rocks and in the share of derivatives of this magma—volcanic rocks (although this difference may not be widespread). An example is the data [35] on the Miocene ruined volcanic structure on Fuerteventura Island (Canary Islands), in the structure of which a stratified intrusive complex (peripheral focus?) and its comagmatic lava take part. The olivine crystals are practically unaffected by secondary changes, although sometimes they develop yiddingsite. Among olivine crystals from the intrusive complex, the share of low calcium variations is ~42% (150/364); in lava, the share of low calcium olivines among phenocrysts is ~27% (22/83). Xenolites of the gabbro with Fuerteventura Island [36] are represented by fresh rock variations and contain in the overwhelming majority low calcium differences (8 out of 9 crystal analyses given).

The differences in quantitative ratios of low and calcium-containing variations of olivine in pillow-lavah (modern oceanic crust), where olivine phenocrysts are represented in the overwhelming majority of calcium-containing differences, and in gabbroids of the 3rd layer of modern oceanic crust are stable enough and these differences appear already in rocks with the age of ~1 million years, probably earlier. Data on gabbroids from the Atlantis U1309D well [37] and dredged from the Vityaz fracture zone of the Northern Central Indian COX [38] are used as examples. Gabbroids of the Atlantis massif are ~1–1.3 million years old, olivine crystals are affected by serpentinization, and all the given analyses of 39 crystals are represented by low calcium differences. Gabbroids from the "Vityaz" zone are over 0.4 million years old and contain an equal amount of low and calcium-containing olivines (8/8). However, the overwhelming prevalence of low calcium variations of olivine gabbroids crystals in the modern oceanic crust is typical. Examples are the following data (in the form of the ratio of the number of low calcium to calcium-containing variations): CAC crest zone [22]–8/1; CAC crest zone Kane [39]–8/0; Parece Vela Basin (Philippine Sea) [40]–28/1.

Homogeneous inclusions of gabbroids in volcanites of the main composition can probably be an indicator of the preservation of olivine composition. The majority of such inclusions described in the literature contain calcium-containing crystals of olivine. Xenolites from lava of Antilles, East Pacific Rise, Island Island, etc. are not exceptions. For example, xenoliths gabbro from picrite pillow-lava (under ice eruption) of Iceland island. Iceland [41], the given ranges of contents of CaO in olivine crystals from 7 samples lie in the interval of 0.34–0.45 wt.%.

Intrusive variations in carbonatite rocks are quite often a part of alkaline and ultrabasic complexes. Olivine crystals in these rocks can be represented by low and calcium-containing variations, with the former dominating. In modern carbonatite lava (Oldonie Lengai volcano), finds of xenoliths represented by forsterite phlogopite ioliths with minimal amount of glass have been described, which indicates final crystallization in a near-surface intermediate chamber [42]. Of the given 5 analyses of 2 crystals of CaO content more than 0.1 wt.%, the sum of alkalis in heated melt inclusions ~15–19 wt.% with prevalence of sodium.

5 Data Discussion

The collected experimental data allow us to make a number of generalizations and specify their limits. From melts of the basic ultramafic compositions with restrictions on their compositions (content of CaO more than 3–4 wt. %, content of the sum of alkalis less than 10 wt. %, content of SiO_2 more than 10 wt. %) regardless of the pressure value up to values 26 GPa. olivine crystals are formed, containing in their composition more than 0.1 wt. % CaO. The specified boundary values of the element contents are probably mutually dependent and need to be clarified. In experimental conditions with insignificant melting, or its absence (sub-solidus and solidus areas of existence of the given material at carrying out of experiment) with participation of fluid the formed crystals of olivine are presented by the low calcium variations containing less than 0.1 wt.% of CaO.

These generalizations coincide well with the data on the content of calcium in olivine crystals from basic and ultrabasic volcanites (comatiites and meimechitis). Calcium-containing olivines may sometimes be preserved in comatiites, which are usually pre-Cambrian in age and modified at the level of green shale facies according to the frame [2, 4]. In Cretaceous Komatiites, o. Gorgon and Triassic Meimechetes of the Siberian Platform (see above), olivines are represented only by calcium-containing variations. The presence in the rocks of some portion of low Calcium crystals in the rocks that are very weakly affected (both facially and quantitatively) by secondary changes (Neogene Quaternary age) indicates that they are in disequilibrium with the melt that formed this rock and, as a consequence, xenogenic. Newly formed olivine crystals at metasomatic transformations of rocks are mostly represented by low calcium variations. Calcium-containing olivine crystals are represented by single analyses, which is typical for rocks of intrusive appearance, for example, those included in the ophiolite complexes. However, there are single examples in the literature of metasomatites finds from the ophiolitic complexes with a sufficiently high content of calcium-containing variations. Thus, in peridotites and dunites of the Jurassic officiolite complex of the Central Pontid (Turkey), which were subjected to different degrees of serpentinization, [43] the share of calcium-containing olivines (11 analyses from 72 given) is $\sim$15%.

Cases of finds of natural calcium-containing olivines are usually described in contact metasomatites (scarns, corneas). The following data may serve as examples. Scars: areas of contact between South California batolite (USA) and dolomites, where olivines are represented by low calcium forsterite and monticellite [44]; the Alta rod (Utah, USA) contact zone with dolomites—forsterites contain 0.04–0.11 wt. % CaO [45]; contact zone of magmatic chamber melt with carbonates, Colli Albani volcanic region (Italy)—forsterite in endoscarnets contains 0.44–1.01 wt. % CaO, exoscarnets forsterite contains 0.09–0.31 wt. % CaO [46]. An example of the presence of calcium-containing olivines in corneas is the data [47], where faialite, in corneas from the contact zone of the Khibiny alkaline massif with metavulcanites, contains 0.12–0.42 11 wt. % CaO. It is interesting that in the given examples on contact metasomatites the members of olivine group are represented by the extreme members of this group—faialite and forsterite. Although there are experimental data indicating that calcium-containing olivines may be formed in the solidus region. Thus, the publication [46] presents experimental results of studying peridotite + water system in solidus region (at pressures of 2.5 and 6.5 GPa and temperatures of 680–800 °C). Fourteen analyses are presented, of which only four are presented as low calcium variations, which in the light of the generalizations obtained does not find an explanation.

The data collected suggest that the calcium content of the olivines of volcanoes and their coherent intrusive variations (peripheral foci, etc.) will differ. The most real influencing factors on the mobility of calcium in olivine crystals are the fluid and temperature, providing these differences at the stage of autometasomatosis. Probably, the wide development of low calcium olivine variations in the tertiary stratified massif of Ram Island can be partially explained by autometasomatosis [4]. In the gabbroids of the modern oceanic crust, both at the stage of autometasomatosis during melt solidification and in the process of further metasomatic transformations, practically all olivine crystals are affected by these processes. This may be due to large volumes of aqueous fluid (various forms of sea water) involved in this process. The existence of homogeneous inclusions with calcium-containing olivine suggests that the phenomenon is not widespread. It can be assumed that, like calcium in autometasomatosis, other chemical elements (for example, rare and scattered) and in other crystalline phases become mobile. It can be assumed that due to autometamorphism, the effusive and intrusive variations in comagmatic rocks in some formation complexes will differ geochemically. This is probably indicated by the differences for these rock classes in the family positions on the TAS-diagrams. Although the scales (spatial and quantitative) of this phenomenon are not clear. In contact-metasomatic processes, the behavior of calcium in olivine crystals requires study.

In the vulcanites of Kamafugite series and carbonatite, the alkali content often does not exceed ~10 wt.% and olivine crystals are mostly represented by calcium containing variations. The presence of high-magnesia low calcium olivines, on the example of Italian venn cite lava, is explained by oxidation of primary less magnesia variations [48]. In Ponomarev [4] it was shown that the appearance of such phases is due to secondary transformations of the primary olivine. In heated melt inclusions from kamafugite series volcanites and carbonatites, the sum of alkalis can reach 20 wt.% or more, which should prevent the formation of primary calcium-containing olivines. This discrepancy in the content of alkalis in the primary melts of the rocks of this series and the content of calcium in the olivines from them requires more detailed experimental study of the mutual influence of silicon, calcium and alkalis in the melt, in particular the separate potassium and sodium on the calcium content in the olivine crystals. As in calcium-carbonatite of Kerimasi volcano, despite the content of the sum of alkalis up to 36 wt. % with the prevalence of sodium olivine is represented by the calcium-containing variation. This probably indicates the difference in the influence of sodium and potassium in the melt on the incorporation of calcium from the melt into the lattice of the olivine and, possibly, on their structural position in the melt.

The most significant is the discrepancy between the conclusion based on experimental data on the growth of calcium-containing olivine crystals from the magmatic melts of the main ultramafic composition, regardless of pressure, and the compositions of olivine crystals from kimberlites, as well as those included in the diamonds from this rock. Almost all of these crystals, except for the peripheral edges on the olivine phenocrysts, and the microlite phase, which is also often zonal in calcium, are low calcium, regardless of age and rock variation. The revealed limitations on the compositions of melts (calcium, silicon, and sum of alkalis) from which low calcium variations in olivine crystals can be formed do not apply to natural kimberlite melts, judging by data on reconstructed compositions of primary kimberlite melts given in Sokol and Kruk [9]. In studies on the crystallization of diamond [49], the growth of diamond inoculation occurs in a kimberlite melt in equilibrium (probably, there were local equilibriums in the experiments) with olivine, which in 9 cases out of 14 presented analyses is represented by calcium-containing difference. In Sokol and Kruk [9], Sokol et al. [38], the conditions for the generation of kimberlite magmas and their role in the growth and dissolution of diamond crystals are considered, and in a paper summarizing a wide range of data [50], the mantle carbonatite concept is proposed, according to which the main diamond-forming medium is melts/solutions. In experiments involving the melt phase and diamonds [51, 38], olivine crystals are represented by both calcium-bearing and low-calcium variations, 2015, in 2 cases out of 6 crystals studied, this is a low calcium variation, which can be explained, according to our conclusions, by the potassium content in the co-existing melt ~10–11 wt. % K_2O and, consequently, they fall into the selected region of the composition exceptions. In general, we can conclude that in experiments involving the paragenesis of the melt diamond-olivine, the latter is represented mainly by the calcium-containing difference, while the olivines of the phenocrysts in kimberlites and included in the diamonds are represented by the low calcium difference, and the finds of calcium-containing variations among them are isolated incidents.

As a result, the discrepancy between the compositions of olivine crystals in kimberlite and included in the diamonds in this rock with the experimental results suggest that the compositions of natural variations in kimberlite olivine are in disequilibrium with the melts, with the possible exception of calcium-bearing edges, and diamonds are formed in a solid-phase medium. The most probable parent medium under these conditions for the growth of diamond crystals is the fluid phase, which agrees with the model of fluid metasomatic mantle substance. The experiments show the reality of diamond crystallization in the fluid [49, 52] under mantle P–T conditions. In the experiments from [53] under conditions of corresponding upper mantle, diamond crystals were formed in a mixture of silicate melt, aqueous fluid and carbonates. Moreover, the compositions of olivine crystals (4 analyses) are represented by calcium-free forsterite in wide variations of glass compositions. The absence of calcium in olivine crystals can be explained by the presence of aqueous fluid in the system. The possibility of growth of diamond single crystals under sufficiently "soft" conditions—at atmospheric pressure and temperature below 100 °C in a graphite solution in a solvent from a mixture of acids in water—has also been proved by experiments. Probably, micron and finer diamond crystals present in the main mass of kimberlites could be formed from the fluid in a similar way.

In diamond crystals, along with solid-phase monomeral inclusions, there are also polyphase melt inclusions with calcium-containing olivine, probably corresponding to kimberlite melts, which carry diamond crystals to the surface and form a calcium-containing olivine border around phenocrystals. It is not clear whether this border is of diffusion nature, or whether it is a new phase growth on low calcium nuclei. The duration of formation of such a border is limited to several years or months, probably regardless of the method of formation, and is quite constant regardless of the geological age of kimberlites. This also indicates that these olivine and diamond crystals are not melted. There is a growing number of publications on the studied fluid mantle inclusions far from the compositions of kimberlite melts (see review [4, 5]). These data are also arguments in favor of a fluid mother environment for diamond growth. Judging by the number of data on the compositions of olivines included in the fluid phase, the vast majority of the diamond crystals studied in kimberlites were formed. Single finds of calcium-containing olivines in the diamonds are known [4], and these exceptions point to the rarity of diamond growth from melt under natural conditions.

In Ponomarev [4], Ponomarev and Puzankov [5], based on the fluid genesis of diamond under natural conditions, the possibility of using solid phase inclusions in diamond crystals (olivine, garnet, pyroxene, spinel, perovskite, etc.) as substrates for artificial heteroepitaxial growth of single crystalline diamond films by CVD-technologies was stated. The possible existence of natural diamond accretions on included olivine crystals is shown in Neusser et al. [54]. But there is also a group of phases, both included in the diamond crystals and forming diamond crystals [55]. These phases are developed practically everywhere on the diamond crystals. Their formation ranges from mantle films (there are multiphase films buried inside the diamond crystals [55] to crystals formed in terrestrial conditions on diamond crystals from pores in the lavae of the Tolbachinsky Cracks eruption. 50th Anniversary of the Institute of Physics of the FEB RAS—TTI-50 (Kamchatka, 2012–13) [21]. According to Makeev and Kriulina [55], growth and dissolution of diamond crystals occurs through metal membranes (films) surrounding a diamond crystal. Leaving this supposed growth mechanism aside, it can be argued that the existence of diamond crystals buried in the body of these films is indisputable evidence of the reality of heteroepitaxial growth and growth of single crystalline diamond at the phases composing the films. These phases can be used as substrates for artificial growth of monocrystalline diamond films using CVD and other technologies. Of interest among them, at the first stage of research is a variety of metal films. Many of the phases included in these films have a face-centred cubic lattice with a constant lattice close to the diamond lattice ~0.357 nm. Consistency of lattice parameters for diamond heteroepitaxial growth significantly affects the perfection of the growing crystal due to low mechanical stresses in the transition zone. The phases composing the films are pure metals, their alloys and intermetallic materials. Systematization of data on these films (50 kinds) is given in Makeev and Kriulina [55]. Data on films on diamonds from TTI-50 lava fall into the number of these species, but the set of elements and phases is wider and as the research the number of mineral species forming films and increasing mineral phases is expanding. According to Makeev and Kriulina [55], films mainly contain the following elements—Au, Pd, Cu, Zn, Sn, Fe, Ni, Cr, W, both in the form of monoelement phases and forming bronze and brass alloys, etc. Under experimental conditions [49], newly formed crystals of diamond 30–80 μm were found growing on platinum (in the cold part of the platinum ampoule). Platinum has a lattice period (cubic syngonia) of ~0.392 nm. But there are several other elements with high melting temperatures (more than 1600 °K) of zirconium (cubic singonium, lattice period ~0.361 nm), dysprosium and erbium (hexagonal singonium, lattice period ~0.356 nm), whose lattice parameters practically coincide with the period of the diamond lattice (0.357 nm). The study of films from all these elements and different variants of their compounds as substrates for heteroepitaxial growth, with the participation of diamond seed on their surface, will probably provide an opportunity to finally obtain cheap diamond monocrystalline films. Giving them semiconductor properties will allow replacing in some cases such materials as Si, Ge, GaAs, GaN, and be the basis for creating a new generation of field transistors, radiation detectors, quantum computers, etc. Such monocrystalline diamond films can also be used to create two- and three-dimensional photonic crystals, which are in demand for optical processors.

Olivine crystals from ultrabasic xenoliths carried by alkaline basalts and kimberlites are represented, with rare exception, by low calcium variations. The existing very small number of exceptions, where olivine is represented by calcium-containing variation can be satisfactorily interpreted as cumulata from magmatic systems. Assuming mantle nature of xenoliths and taking into account experimental data according to which olivine is the main phase in the mantle, we can say that in the upper mantle up to the depths of 200–300 km. olivine is represented by low calcium variation. Based on the conclusion that the olivine crystals in equilibrium with the melt of the main ultramafic composition are represented, regardless of pressure, by the calcium-containing variation, we can argue that the melting in the mantle in a wide pressure range must be non-equilibrium. This is well proved by calcium-containing crystals of olivine from melt veins in xenolites. Besides, it also gives the right to assert that the coefficient K" ol/m for the Fe/Mg ratio is not an unequivocal criterion of equilibrium. Equivalent and necessary one more criterion is also boundary calcium content in olivine and it is necessary to use both these criteria together.

An interesting problem is the role of water in removing calcium from the olivine lattice. On the basis of collected data on the composition of olivines from lava and intrusive bodies of the oceanic crust, the complex of rocks of Fuerteventura Island, and stratified intrusive complexes, it can be assumed that the main role in the removal of calcium from the olivine is played by the presence of fluid, its share (water/rock) and temperature. The removal process begins during the cooling period of the melt, capturing and solidifying the solidus stage. The removal of calcium from olivine crystals is the beginning of metasomatic transformations of the rock. And, which may be equivalent, based on structural features of olivine, calcium, as a rule, is not included in the grid of newly formed metasomatic olivines, although there are exceptions. These processes are probably controlled, and the structural position of calcium in the lattice of olivine. These peculiarities of the mobility of calcium in olivine crystals and its connection with the fluid is an actual problem. It is also possible that the levels of calcium and water in the olivine in mantle conditions are interrelated.

On the example of xenolite in the lava of carbonatite volcano Aldoinho Lengai with the contents of the sum of alkalis in the melt inclusions 15–19 wt. % it can be assumed that in the cases of both effusive and intrusive rock variations there is a discrepancy in the experimental data between the content of alkalis in the melt and the content of calcium in the olivine, which requires further study. In particular, it is necessary to identify the reasons for different effects of sodium and potassium on the ability of calcium to enter the olivine lattice.

The ideas about genesis of ultrabasic mantle xenoliths olivines, being in overwhelming majority of cases low calcium variations, are represented by wide spectrum of opinions. But, even if we assume their restite genesis, in this case they should have been initially calcium-containing variations, but a long stay in sub-solibus region (in the presence of fluid?) should have led to losses of calcium.

6 Conclusions

1. The systematization of experimental data allowed to identify factors influencing the CaO content in olivine crystals grown from magmatic melts and formed in the solidus region with fluid participation.
2. At this stage of research they are the contents of CaO, K_2O, SiO_2(?), H_2O in the melt. In metasomatic processes, water and temperature promote the growth of low calcium olivines and, possibly, the removal of calcium from the olivine crystals. The existence of a number of exceptions indicates a number of more factors that have not been identified.
3. In most rocks of ultramafic-main composition, which underwent metasomatic processing of olivine is represented by low calcium variations. This can be most clearly seen in the examples of modern oceanic crust and ophiolitic complexes. Calcium-containing crystals of olivine may be formed in contact-metasomatic rocks, which contradicts the supposed scheme. But this phenomenon and its scale needs further study.
4. To reveal the equilibrium of compositions in the system of magmatic melt olivine it is necessary to use two criteria—coefficient K" ol/m for Fe/Mg ratio and content of CaO in olivine.
5. The use of these findings with respect to olivines from the rocks of the main ultrabasic composition made it possible to understand their genesis in some cases. In particular, the disequilibrium of the main phase of mantle substance (olivine) up to the diamond phase of depth with melts of the main ultrabasic compositions was revealed, and this suggests that in most cases, kimberlite diamonds are of metasomatic nature.
6. A number of substances have been identified that can serve as a basis for heteroepitaxial growth of single crystalline diamond films using CVD technology, aqueous solutions and other technologies, which will significantly reduce the cost of this product.

Acknowledgements The research was done with the financial support of the Russian Foundation for Basic Research, grant project No. 20-05-00261, No 19-05-00388 as well as the Integration Project of the Research Center of the Russian Academy of Sciences, Block 1.4. and the study was performed by the governmental assignment in terms of Project IX. 129.1.4., RNF -21-17-00015.

References

1. Simkin, T., Smith, L.V.: Minor-element distribution in olivine. J. Geol. **78**(3), 304–325 (1970)
2. Ponomarev, G.P., Puzankov, M.Yu.: Distribution of Rock-Forming Elements in the System of Basic-Ultramafic Melt Spinel, Olivine, Orthopyroxene, Clinopyroxene, Plagioclase According to Experimental Data: Geological Application, pp. 664. Probel, Moscow (2012)
3. Ponomarev, G.P.: Calcium content in olivine crystals grown from experimental melts. Part 1: calcium in olivine crystals grown from experimental melts. Lithosphere **4**, 66–79 (2014)
4. Ponomarev, G.P.: The content of calcium in natural olivine crystals as an indicator of their genesis. Part 2: calcium content in natural olivine crystals as an indicator of their genesis. Lithosphere **5**, 57–70 (2014)
5. Ponomarev, G.P., Puzankov, MYu.: Distribution of rock-forming elements in the system of melt-spinel-olivine with participation of aqueous fluid according to experimental data. Kraunts Bulletin. Land Sci. **32**(4), 59–72 (2016)
6. Sharygin, I.S., Shatskiy, A., Litasov, K.D., Golovin, A.V., Ohtani, E., Pokhilenko, N.P.: Interaction of peridotite with Ca-rich carbonatite melt at 3.1 and 6.5 GPa: implication for merwinite formation in upper mantle, and for the metasomatic origin of sublithospheric diamonds with Ca-rich suite of inclusions. Contrib. Mineral. Petrol. **173**, 22 (2018)
7. Gavrilenko, M., Herzberg, C., Vidito, C., Carr, M.G., Tenner, T., Ozerov, A.: A calcium-in-olivine geohygrometer and its application to subduction zone magmatism. J. Petrol. **57**(9), 1811–1832 (2016)
8. Ariskin, A.A., Barmina, G.S.: Modeling of Phase Equilibria during Basalt Magmas Crystallization, p. 363. Science, Moscow (2000)
9. Niida, K., Green, D.H.: Stability and chemical composition of pargasitic amphibole in MORB pyrolite under upper mantle conditions. Contrib. Mineral. Petrol. **135**, 18–40 (1999)
10. Gorbachev, N. C., Kostyuk, A.V., Shapovalov, Yu.B.: Experimental study of peridotite-H_2O system at R 3.8–4 GPa, T 1000–14000S: critical ratios and vertical zoning of the upper mantle. Dokl. AN **461**(4), 442–446 (2015)
11. Merkulova, M., Minoz, M., Vidal, O.: Brunet: role of iron content on serpentinite degydratation depth in subduction zones: experiments and thermodynamic modeling. Lithos **264**, 441–452 (2016)
12. Safonov, O.G., Butvina, V.G.: Interaction of model peridotite with the fluid H_2O–KCl: an experiment at 1.9 GPa and its application to the processes of upper mantle metasomatosis. Petrology **21**(6), 654–672 (2013)
13. Pirard, C., Hermann, J.: Experimentally determined stability of alkali amphibole in metasomatised dunite at sub-arc pressure. Contrib. Mineral. Petrol. **169**, 1 (2015)
14. Chepurnov, A.I., Sonin, V.M., Tychkov, N.S., Kulakov, IYu.: Experimental evaluation of the reality of the leaking (migration) of volatile components (H_2O + CO_2) in rocks of mantle wedge. Dokl. AN **464**(1), 100–104 (2015)
15. Condamine, P., Medard, E.: Experimental melting of phlogopite-bering mantle at 1 Gpa: implication for potassic magmatism. Earth Planet. Sci. Lett. **397**, 80–92 (2014)
16. Sokol, A.G., Kruk, A.N., Chebotarev, D.A., Palyanov, Y.N.: Carbonate melt-peridotite interaction at 5.5–7.0 Gpa: implications for metasomatism in lithospheric mantle. Lithos **248** (251), 66–79 (2016)

17. Dasgypta, R., Hirschmann, M.M.: A modified iterative sandwich method for determination of near-solidus partial melt compositions. II. Application to determination of near-solidus melt compositions of carbonated peridotite. Contrib. Mineral. Petrol. **154**, 647–661 (2007)
18. Tenner, J.T., Hirschmann, M.M., Withers, A.C., Ardia, P.: H_2O storage capacity of olivine and low-Ca pyroxene from 10 to 13 GPa: consequences for dehydration melting above the transition zone. Contrib. Mineral. Petrol. **163**, 297–316 (2012)
19. Shejwalkar, A., Coogan, L.A.: Experimental calibration of the roles of temperature and composition in Ca-in-olivine geothermometer at 0.1MPa. Lithos **177**, 54–60 (2013)
20. Hirano, N., Ymamoto, J., Kagi, H., Ishii, T.: Young, olivine xenocryst-bearing alkali-basalt from the ocean-ward slope of the Japan trench. Contrib. Mineral. Petrol. **148**(1), 47–54 (2004)
21. Silaev, V.I., Karpov, G.A., Rakin, V.I., Anikin, L.P., Vasiliev, V.A., Filippov, V.N., Petrovsky, V.A.: Diamonds in products of the cracked Tolbachi Eruption 2012–2013. Kamchatka. Vestnik of Perm University (Geology) **1**(26), 6–27 (2015)
22. Silantyev, S.A., Kostitsyn, Yu.A., Cherkashin, D.V., Dick, G.J.B., Kelemen, P.B., Kononkova, N.N., Kornienko, M.N.: Magmatic and metamorphic evolution of the oceanic crust of the western flank of the crest zone of the CAH at 15° 44' N: results of the study of the core of wells 1275B and 1275D (209th flight "Joides Resoljushn"). Petrology **16**(4), 376-400 (2008)
23. Ryabchikov, I.D., Kogarko, L.N.: Deep differentiation of Alkali-Ultra-Base Magmas: generation of carbonatite melts. Geochemistry **9**, 771–779 (2016)
24. Gurenko, A.A., Sobolev, A.V., Kononkova, N.N.: New data on petrology of the East African Rift Ugandites based on the results of the study of magmatic inclusions in minerals. Dokl. AN **305**(6), 1458–1463 (1989)
25. Muravyeva, N.S., Senin, V.G.: Carbonate-silicate equilibriums in the high-magnesia ultrakaline volcanoes of Toro-Ankole (East African Rift Zone). Geochemistry **9**, 937–957 (2009)
26. Panina, L.I., Stopa, F., Usoltseva, L.M.: Genesis of the melt inclusions in minerals (Kamafugite Province of Umbria, Central Italy). Petrology **11**(4), 405–424 (2003)
27. Nikolaeva, A.T.: Petrology of the melite-containing rocks of the volcanoes Kupaello and Colle Fabri (Central Italy). Cand. Geo-Min. Sciences. IGM CO RAS, p. 148, Novosibirsk (2014)
28. Stoppa, F., Lipini, L.: Mineralogy and petrology of the Polino monticellite calciocarbonatite (Central Italy). Mineral. Petrol. **49**, 213–231 (1993)
29. Guzmics, T., Mitchell, R.H., Szabo, C., Berkezi, M., Milke, R., Abart, R.: Carbonatite melt inclusions in coexisting magnetite, apatite and monticellite in Kerimaci calciocarbonate, Tanzania: melt evolution and petrogenesis. Contrib. Mineral. Petrol. **161**, 177–196 (2011)
30. Willcox, A., Buismann, I., Sparks, R.S.J., Brown, R.J., Mania, S., Schumacher, J.C., Tuffen, H.: Petrology, geochemistry and low-temperature alteration of lavas and pyroclastic rocks of the kimberlitic Ingwisi Hills volcanoes. Tanzania. Chem. Geol. **405**, 82–101 (2015)
31. Kamenetsky, V.S., Kamenetsky, M.B., Sobolev, A.V., Golovin, A.V., Demouchy, S., Faure, K., Sharygin, V.V., Kuzmin, D.V.: Olivine in the Udachnaya-East kimberlite (Yakutia, Russia): types, compositions and origins. J. Petrol. **49**(4), 823–839 (2008)
32. Sokol, N., Khokhryakov, A.G., Palyanova, A.F., Borzdov, A.G., Sobolev, N.V.: Diamond and graphite crystallization in SON-fluid at PT-parameters of natural diamond formation. Dokl. AN **375**(3), 384–388 (2000)
33. Golovin, A.V., Sharygin, V.V., Pokhilenko, N.P.: Melt inclusions in olivine phenocrysts from unchanged kimberlites of the Udachnaya-Vostochnaya tube (Yakutia): some aspects of kimberlite magmas evolution at late stages of crystallization. Petrology **15**(2), 178–195 (2007)
34. Bussweiler, Y., Stone, R.S., Pearson, D.G., Luth, R.W., Stachel, T., Kjarsgaard, B.A., Menzies, A.: The evolution of calcite-bearing kimberlites by melt-rock reaction: evidence from polymineralic inclusions within clinopyroxene and garnet megacrysts from Lac de Gras kimberlites. Canada. Contrib. Mineral. Petrol. **171**(7), 1–25 (2016)

35. Tornare, E., Pilet, S., Bussy, S.: Magma differentiation in vertical conduits revealed by the complementary study of plutonic and volcanic rocks from Fuerteventura (Canary islands). J. Petrol. **57**(11–12), 2221–2250 (2016)
36. Neumann, E.-R., Vannucci, R., Tiepolo, M.: N-MORB crust beneath Fuerteventura in the easternmost part of the Canary islands: evidence from gabbroic xenoliths. Contrib. Mineral. Petrol. **150**, 156–173 (2005)
37. Suhr, G., Hellebeand, E., Jonhnson, K., Brunelli, D.: Stacked gaabbro units and intervening mantle: a detailed look at a section of IODP Leg 305, Hole U1309D. Geochem. Geophys. Geosyst. **9**(10), 1–31 (2008)
38. Ray, D., Misra, S., Banerjee, R., Weis, D.: Geochemical investigation of gabbro from slow-spreding Northern Central Indian Ocean Ridge, Indian Ocean. Geol. Mag. **148**(3), 404–422 (2011)
39. Fumagalli, P., Poli, S.: Experimentally determined phase relations in hydrous peridotites to 6.5 GPa and their consequences on the dynamics of subduction zones. J. Petrol. **46**(3), 555–578 (2005)
40. Sanfilippo, A., Dick, H.J.B., Ohara, Y.: Melt-rock reaction in the mantle: mantle troctolites from the Parece Vela ancient back-arc spreding center. J. Petrol. **54**(5), 861–885 (2013)
41. Gurenko, A.A., Sobolev, A.V.: Crust-primitive magma interaction beneath neovolcanic rift zone of Iceland recorded in gabbro xenoliths from Midfell, SW Island. Contrib. Mineral. Petrol. **151**, 495–520 (2006)
42. Sekisova, V.S., Sharygin, V.V., Zaitsev, A.N., Strekopytov, S.S.: Liquidation phenomena during crystallization of Forsterite-Flogopite Jolites of Oldoino Lengai, Tanzania: based on the study of melt inclusions in minerals. Geol. Geophys. **56**(12), 2173–2197 (2015)
43. Celik, O.F., Marsoli, A., Marschik, R., Chiaradia, M., Mathur, R.: Geochemical, mineralogical and Re-Os isotopic constraints on the origin of Tethyan oceanic mantle and crustal rocks from the central Pontides, Northern Turkey. Mineral. Petrol. **112**, 25–44 (2018)
44. Alexandrov, S.M.: Metasomatic changes of carbonate rocks in quarries of Riverside, California, USA. Geochemistry **7**, 751–766 (2011)
45. Cook, S.J., Bowman, J.R.: Mineralogical evidence for fluid-rock interaction accompanying prograde contact metamorphism of siliceous dolomites: Alta stock aureole, Utah, USA. J. Petrol. **41**(6), 739–757 (2000)
46. Di Rocco, T., Freda, C., Gaeta, M., Mollo, S., Dallai, L.: Magma chambers emplaced in carbonate substrate: petrogenesis of skarn and cumulate rocks and implications for CO_2 degassing in volcanic areas. J. Petrol. **53**(11), 2307–2332 (2012)
47. Chaschin, V.V.: Mineral paragenesis and conditions of corneal formation in the exocontact zone of Khibiny alkaline plutonium (Kola Peninsula, Russia). Geochemistry **1**, 19–37 (2007)
48. Plechov, P.Yu., Nekrylov, N.A., Shcherbakov, V.D., Tikhonova, M.S.: Extremely magnesian olivines from the Venetian lava Pian di Chelle (Italy). Dokl. AN **474**(3), 331–335 (2017)
49. Palyanov, N., Sokol, A.G., Khokhryakov, A.F., Kruk, A.N.: Conditions of diamond crystallization in kimberlite melt according to experimental data. Geol. Geophys. **56**(1), 254–272 (2015)
50. Litvin, Yu.A., Spivak, A.V., Kuzyura, A.V.: Basics of the mantle-carbonatite concept of diamond genesis. Geochemistry **10**, 873–892 (2016)
51. Bobrov, A., Litvin, Yu.: Peridotite-eclogite-carbonatite systems at 7.0–8.5 GPa: concentration barrier of diamond nucleation and synthesis of its silicate and carbonatite inclusions. Geol. Geophys. **50**(12), 1571–1587 (2009)
52. Chertkova, N., Yamachita, S., Ito, E., Shimojuku, A.: High-pressure synthesis and application of a 13 °C diamond pressure sensor for experiments in a hydrothermal diamond anvil cell. Mineral. Mag. **78**(7), 1677–1685 (2014)
53. Bureau, H., Frost, D.J., Bolfan-Casanova, N., Leroy, C., Esteve, I., Cordier, P.: Diamond growth in mantle fluids. Lithos **265**, 4–15 (2016)

54. Neusser, R.D., Shertle, G.-P., Logvinova, A.M., Sobolev, N.V.: Study of inclusions of olivine in Siberian diamonds by the method of diffraction of backscattered electrons: signs of syngetic growth? Geol. Geophys. **56**(1–2), 416–425 (2015)
55. Makeev, A.B., Kriulina, G.Yu.: Metal films on the surface and in the volume of diamond crystals of the Arkhangelsk and Yakut diamondiferous provinces. Rec. RMO **1**, 101–114 (2012)
56. Condamine, P., Medard, E., Devidal, J.-L.: Experimental melting of flogopite-peridotite in the garnet stability field. Contrib. Mineral. Petrol. **171**, 95 (2016)
57. Sokol, A.G., Kruk, A.N., Palyanov, Y.N., Sobolev, N.V.: Stability of phlogopite in ultrapotassic kimberlite-like systems at 5.5–7.5 GPa. Contrib. Mineral. Petrol. **172**, 21 (2017)

Composition and Mechanisms of Formation of Comendite Melts of the Early Mesozoic Bimodal Association of Sant, Central Mongolia

I. A. Andreeva, S. E. Borisovsky, and V. V. Yarmolyuk

Abstract

Based on the investigation of melt inclusions in minerals using electronic and ion microprobe analysis, we estimated the composition and evolution of the melts involved in the formation of comendites of the bimodal association of Sant's volcanites (Central Mongolia), the mechanisms of their formation were revealed. Primary melt inclusions and crystalline inclusions coexisting with them were studied in quartz in three samples of comendites selected in different parts of the volcanic section. Sanidin, zircon, and chevkinite were determined among crystalline inclusions. It was found that the crystallization of phenocrysts of comendites was carried out in the temperature range of 880–960 °C. It was determined that the glasses of homogeneous melt inclusions correspond to the compositions of trachydacites and rhyolites. They are also characterized by high concentrations of Zr, Nb, Rb, Y, Th and rare earth elements. At the same time, significant differences in the composition of melt inclusion glasses in the concentrations of Li and volatile components (H_2O and F) were revealed. Some of the melts are enriched in these components, whereas other are depleted in them. Analysis of composition of glasses of homogenized melt inclusions in phenocrysts of alkaline salic rocks of bimodal association of Sant allowed to reveal magmatic processes responsible for the formation of comendite melts and leading to the accumulation of rare and rare earth elements in the latter. The dominant role among them is played by the process of magma crystallization differentiation accompanied by the process of liquid immiscibility with participation of fluoride melts.

I. A. Andreeva (✉) · S. E. Borisovsky · V. V. Yarmolyuk
Institute of Geology of Ore Deposits, Petrography, Mineralogy and Geochemistry RAS, 119017 Moscow, Russia
e-mail: andreeva@igem.ru

N. Vladykin (ed.), *Alkaline Rocks, Kimberlites and Carbonatites: Geochemistry and Genesis*, Springer Proceedings in Earth and Environmental Sciences,
https://doi.org/10.1007/978-3-030-69670-2_10

Keywords

Melt inclusions · Crystallization differentiation · Liquid immiscibility

1 Introduction

The problem of the formation of alkaline granites with rare-metal mineralization is one of the priority problems of modern petrology and geochemistry. The solution of this problem is connected with a whole complex of researches aimed at revealing the composition and sources of magma of rare-metal rocks, as well as natural processes that lead to concentration of ore elements. At present, there are various views on the role of processes that determine the accumulation of ore components in alkaline granite systems. One of them assumes the concentration of rare elements directly during the evolution of magmatic melts [1, 2] and others—due to the attraction of an additional source in the form of crust matter, such as partial melting of alkaline basite in the lower parts of the continental crust [3]. The verification of these models can be carried out on the example of magmatic series differentiated by composition, the rocks of which, in one case, should demonstrate the accumulation of rare elements with a growing degree of differentiation, in another case, should be indifferent with respect to the distribution of such elements. The most effective for these purposes are bimodal associations uniting the longest in the evolutionary sense series of magmatic rocks: from basalts to comendites and pantellerites, and ending with alkaline and rare-metal granites.

In the proposed work the first data on the formation conditions and composition of magmas for the comendites of bimodal Sant association in the Central Mongolia are presented, and also the magmatic processes responsible for accumulation of rare elements in them are considered.

2 Short Geological Background and Compositional Characteristics of the Rocks

Within the Mongolia-Transbaikalia zonal magmatic province developed volcanic and plutonic rocks of increased alkalinity, including trachybasalts, comendites, alkaline and lithium-fluorine granites. These rocks form the volcanic-plutonic complexes that are part of the West Transbaikalia, Kharkhorin, and North Gobi rift zones (Fig. 1). Their outcrops are associated with separate volcanic fields, the area of which does not exceed first hundreds of km^2 [4]. The studied volcanic bimodal association is developed in the area of Sant's somon of Kharkhorin rift zone. It is a volcanic field 20 × 12 km in size located directly to the east from the Somon. Rocks of bimodal series are basaltoids, trachydacites, comendites and pantellerites;

volcanic thickness is over 2400 m. The complex also includes alkaline porphyric granites. The U-Pb zircon age of the massif is 206 ± 1 million years. In terms of their mineral and chemical composition, the alkali-salic rocks correspond to acid rocks of the alkaline potassium-sodium series with an agpaite index (Ka) close to or above 1, with an increased content of zirconium, rubidium, and REEs. According to MacDonald classification (1974), which divides alkaline rhyolites into comtndites and pantellerites by the ratio of Al_2O_3 and total iron, compositions of most studied rocks are located in the field of comendites and only a small number of compositions located into the field of pantellerites.

Inclusions of mineral-forming media were studied in phenocrysts of three samples of comendites (SNT-2/13, SNT-2/9, SNT-2/8), selected in different parts of the section of volcanic sequence. Their chemical compositions are given in Table 1. In petrographic terms, the comendites are porphyric rocks containing up to 20% of phenocrysts of potassium-sodium feldspar, quartz, and aegirine. The micro-felsic graundmass is composed of the smallest microlites of potassium-sodium feldspar, quartz, alkaline amphibol, aegirine, Ti-magnetite and ilmenite. The accessory minerals are zircon and monazite.

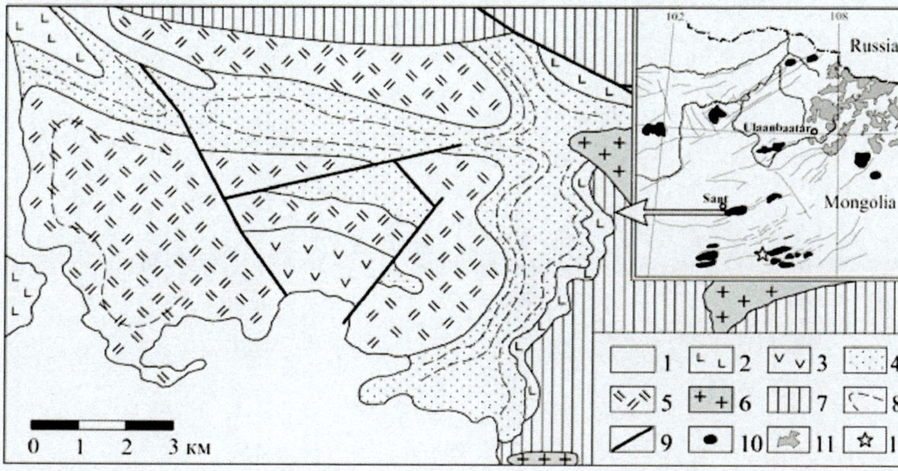

Fig. 1 Scheme of the structure of bimodal Early Mesozoic basalt- comendire association of Sant [14]. 1—Quaternary deposits, 2–8—bimodal association: 2—basalts, 3—trachytes, 4—trachyriolites and comendites of lava flows and their tuffs, 5—comendite extrusions, 6—alkaline granites, 7—Pre-Mesozoic basement, 8—structural lines, 9—faults. The inset shows the location of the bimodal association of Sant in the structure of the Early Mesozoic magmatic province of Mongolia: 10—basalt–comendite associations and alkaline granites, 11—granites of the Khentei batholith granite, 12—bimodal association Dzarta-Khuduk

Table 1 Concentration of the main (wt%), volatile (wt%), and trace (ppm) elements in rocks and melt inclusions in quartz from comendites of the Sant bimodal association

	1	2	3	4	5	6	7	8	9	10	11	12	13
SiO_2	74.2	75.18	73.47	77.95	70.46	70.37	70.24	68.17	69.80	69.49	71.56	72.66	71.68
TiO_2	0.22	0.24	0.31	0.22	0.13	0.19	0.14	0.21	0.19	0.14	0.17	0.22	0.16
Al_2O_3	11.66	11.97	12.80	10.15	11.98	11.75	12.96	12.86	12.88	13.49	11.47	11.97	11.92
FeO	2.84	2.24	2.42	2.66	2.29	2.34	2.29	3.62	2.96	2.67	3.96	3.55	3.79
MnO	0.02	0.02	0.05	0.02	0.00	0.00	0.03	0.05	0.00	0.02	0.05	0.00	0.01
MgO	0.05	0.06	0.10	0.10	0.00	0.00	0.00	0.00	0.01	0.01	0.00	0.00	0.01
CaO	0.34	0.05	0.11	0.12	0.15	0.17	0.15	0.17	0.18	0.16	0.16	0.20	0.17
Na_2O	3.48	4.02	4.13	2.82	5.01	4.93	5.44	6.25	6.37	5.51	5.15	4.79	4.70
K_2O	5.02	4.98	5.19	4.6	4.86	4.95	5.22	5.51	5.23	5.61	5.90	5.89	5.93
P_2O_5	0.03	0.02	0.02	0.02	0.01	0.00	0.00	0.01	0.00	0.00	0.02	0.00	0.00
ppm	1.24	0.70	0.87	0.83	–	–	–	–	–	–	–	–	–
F	–	–	–	–	1.05	1.00	1.09	0.76	0.80	0.76	0.05	0.11	0.13
Cl	–	–	–	–	0.16	0.16	0.17	0.26	0.22	0.20	0.20	0.18	0.17
SO_3	–	–	–	–	0.01	0.03	0.04	0.00	0.00	0.00	0.01	0.03	0.00
Sum	98.37	99.48	99.47	99.49	96.11	95.89	97.77	97.86	98.65	98.06	98.69	99.61	98.66
H_2O	–	–	–	–	2.22	2.16	2.44	0.76	0.80	–	0.34	0.20	0.17
Ka	0.96	1.00	0.97	0.95	1.17	1.25	1.13	1.14	1.25	1.12	1.3	1.19	1.19
Li	15.9	–	–	–	353	320	705	510	406	402	68.7	47.0	53.5
Be	5.7	–	–	–	18.6	17.0	19.2	30.6	24.6	17.9	28.0	24.0	23.1
B	–	–	–	–	36.0	37.6	41.7	56.9	51.7	33.2	44.2	39.6	39.1
V	2.1	4.4	5.4	5.6	6.7	5.8	6.2	8.6	8.7	8.5	7.6	6.9	5.6
Rb	151	152	138	137	321	356	309	525	431	445	426	415	448
Sr	20.8	7.9	18.8	15.2	0.5	0.5	0.6	1.7	1.1	0.8	0.6	0.8	1.0

(continued)

Table 1 (continued)

	1	2	3	4	5	6	7	8	9	10	11	12	13
Y	77.2	72.4	26.4	45.6	172	167	177	348	291	230	266	251	258
Zr	678	726	755	539	1777	1827	1878	3201	2691	2274	2125	2057	2074
Nb	39.4	37.9	37.7	28.7	104	107	119	187	171	144	128	122	132
Ba	73.6	49.0	118	97.9	1.2	1.1	1.6	15.5	2.5	2.8	1.7	2.4	63.2
La	76.8	36.3	33.5	61.6	61.7	61.1	59.7	120	96.6	72.4	89.4	82.9	86.7
Ce	141	84.0	73.8	126	159	163	159	306	233	187	228	237	233
Nd	78.9	42.6	31.4	61.0	79.3	78.5	77.5	160	129	101	120	111	117
Sm	15.9	9.39	5.35	11.7	22.7	22.6	22.1	47.8	35.9	31.0	36.4	35.6	33.9
Eu	0.5	0.39	0.47	0.55	0.3	0.2	0.4	0.5	0.1	0.5	0.5	0.5	0.6
Gd	14.8	9.93	4.77	10.2	32.0	28.3	27.3	52.3	38.6	39.6	41.5	38.7	47.4
Dy	14.0	12.1	4.83	8.47	29.3	29.4	29.0	59.3	46.0	42.1	44.2	42.2	45.0
Er	8.7	7.8	3.47	5.4	17.5	20.7	16.0	43.7	35.3	28.7	32.4	31.9	27.8
Yb	7.6	6.95	4.04	5.21	16.2	20.9	16.6	42.8	37.8	26.7	33.7	30.9	25.2
Hf	24.2	15.2	15.4	10.0	44.2	47.2	48.0	86.6	67.8	61.1	55.1	53,3	54.5
Ta	2.2	2.46	2.23	1.9	8.7	9.3	9.3	16.9	13.1	11.6	12.1	10.8	10.8
Pb	28.2	13.1	16.7	22	40.1	55.3	34.3	108	75.7	72.1	92.0	77.2	68.1
Th	12.8	12.8	11.3	9.27	32.9	39.1	38.6	69.2	62.7	50.1	46.6	44.4	45.2
U	3.7	3.18	1.81	2.08	10.9	12.1	12.9	22.2	20.5	15.4	13.9	14.3	13.7
ΣTR	358	209	162	290	418	425	408	832	652	529	626	611	617

FeO is total Fe. 1–4—comendite samples: 1—SNT-1/11, 2—SNT-/13, 3—SNT-2/9, 4—SNT-8. 5–13–glasses of homogenized melt inclusions: 5–7—in quartz, sample SNT-2/13, 8–10—in quartz, sample SNT-2/9, 11-13—in quartz, sample SNT-2/8. Ka–agpaite index. "–" not determined

3 Analytical Methods

The contents of major elements in the rocks of the Sant bimodal association were determined by classical methods for silicate analysis at the chemical laboratory of the Institute of Geology of Ore Deposits, Mineralogy, Petrography, and Geochemistry, Russian Academy of Sciences. Trace and rare earth elements were analyzed in all the rocks by ICP MS techniques. The analyses were determined at the Institute of Precambrian Geology and Geochronology, Russian Academy of Sciences (St. Petersburg) using a PlasmaQuad 3 inductively coupled plasma mass spectrometer (VG Elemental, Great Britain).

Inclusions in minerals were first examined under an optical microscope in 0.3 mm—thick polished sections. Heating experiments with melt inclusions were carried out using electric furnaces and a Linkam TS 1500 microscopic heating stage designed for high temperature experiments. The error of temperature measurement in the electric furnaces was estimated as ±10 °C. In order to achieve melt–host mineral equilibrium, the samples were kept in the furnace at a desired temperature for 30–120 min.

The chemical composition of crystalline inclusions, mineral phases within melt inclusions, as well as residual and homogeneous glasses of melt inclusions were studied on the JXA-8200 electron microprobe (Japan). The contents of major elements, F, Cl, and S were analyzed under the following conditions: accelerating voltage 20 kV, beam current 10 nA for glasses of melt inclusions and accelerating voltage 20 kV, beam current 20 nA for the mineral phases. The diameter of the beam for glasses of melt inclusions was 6, 8, 12, and 16 μ, for daughter mineral phases in melt inclusions and crystalline inclusions −1 μ. The concentrations of trace elements and H_2O in glasses of melt inclusions were determined by the secondary ion mass spectrometry (SIMS) using Cameca IMS-3F ion microprobes (PTIAS, Yaroslavl (analyst S. G. Simakin) by the method described by Sobolev [5].

4 Crystalline and Melt Inclusions in Minerals

4.1 Investigations of Crystalline and Melt Inclusions in Minerals

Primary melt and coexisting crystalline inclusions were found in quartz from the comendites. Among crystalline inclusions in the quartz of comendites are sanidin, zircon and diortosilicate of rare earth elements—chevkinite. Zircon is characterized by increased concentrations of HfO_2—up to 0.7 wt%. In addition to petrogenic components, 19.8 wt% Ce_2O_3, 10.5 wt% La_2O, 7.2 wt% Sm_2O_3, 2.2 wt% Pr_2O_3, 0.7 wt% Sm_2O_3, 0.46 wt% ZrO_2, 0.26 wt% Y_2O_3, and 0.5 wt% F have been determined in the composition of chevkinite. It is known that Chevkinite, as an accessory mineral, is distributed mainly in syenites and alkaline granites and is much rarer in volcanic rocks [6, 7]. Crystalline inclusions of chevkinite of similar composition were previously determined in quartz quartz of bimodal association Dzarta-Khuduk (Central Mongolia) [8].

Primary melt inclusions are located azonally in the host mineral and have the shape of negative crystal and sizes—from 30 to 90 microns (Fig. 2). In terms of phase composition, two groups are distinguished among them: (1) two-phase inclusions containing glass and gas bubble (sample SNT-2/8) and (2) multiphase inclusions consisting of glass, gas bubble and daughter minerals (sample SNT-2/13 and SNT-2/9). Inclusions of the second type are mica, fluorite and villiaumite as daughter phases. The mica chemical composition corresponds to polylithionite. It contains 59.7 wt% SiO_2, 11.2 wt% Al_2O_3, 5 wt% FeO, 11.3 wt% K_2O, 10 wt% F.

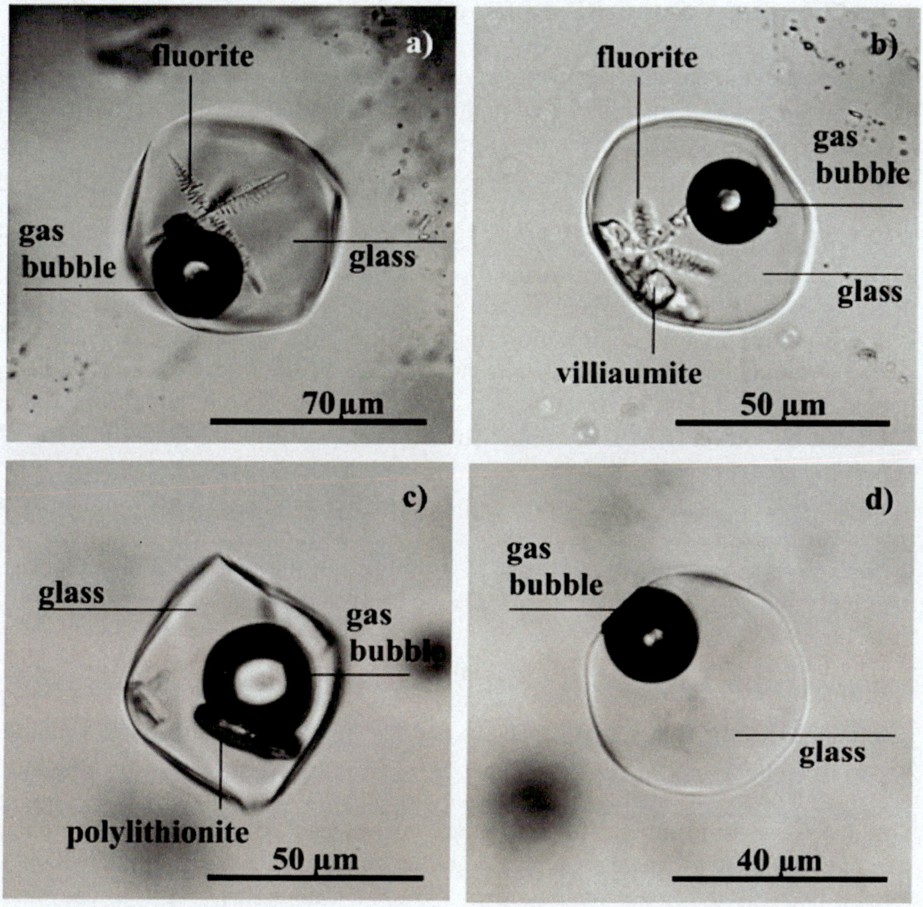

Fig. 2 Melt inclusions in quartz of comtndites at room temperature: **a**, **b** from sample SNT-2/13, **c** from sample SNT-2/9, **d** from sample SNT-2/8. Transmitted light, without analyzer

The deficit in the sum of components including conversion to fluorine is 5.7 wt%, which allows to assume significant Li_2O contents in mica.

Glasses of unheated melt inclusions of all studied samples are characterized by high concentrations of Na_2O and K_2O, in total reaching 10–13 wt%. SiO_2 concentrations vary in the range 66–73 wt%, Al_2O_3—11.5–14.6 wt%. The concentrations in these glasses are different F. For example, for glasses of melt inclusions from SNT-2/13 and SNT-2/9, they are 0.76–1.05 wt%, and for glasses of melt inclusions from SNT-2/8, they does not exceed 0.13 wt%.

Thermometric experiments with melt inclusions in quartz of commandites (samples SNT-2/13 and SNT-2/9) showed that their homogenization was carried out in the temperature range of 880–960 °C. Complete homogenization of melt inclusions from sample SNT-2/8 was not achieved.

4.2 Chemical Composition of Melt Inclusions

Microprobe studies of glasses of homogenized melt inclusions allowed to estimate the composition of the melt involved in the formation of comendites. By the ratio of SiO_2 and (Na_2O + K_2O) glasses of homogenized melt inclusions in quartz, all the studied samples correspond to the compositions of trachidacites and rhyolites and contain (in wt%): SiO_2–68–73, Al_2O_3—11.7–13.5, FeO—2.3–4.8, Na_2O + K_2O—10–13. The agpaite index varies in the range 1–1.3. The analysis of trace element composition in glasses of melt inclusions in quartz of all studied samples showed that they have an identical trace element patterns (Fig. 3) and are enriched relative to the primitive mantle composition by most rare and rare-earth elements, with some predominance of light REE over heavy (La_N/Yb_N = 1.8–2.6). Zr concentrations range from 1780 to 3200 ppm, Rb—from 309 to 525 ppm, Nb—from 104 to 187 ppm, Y—from 167 to 348 ppm, Th—from 33 to 70 ppm, Ce—from 159 to 306 ppm. Contents of REE in the sum are from 408 to 830 ppm. At the same time, the glasses of melt inclusions are sharply depleted of Ba, Sr and Eu, which is probably a result of early crystallization of feldspars from the melt. Comparison of the bulk composition of rocks (Table 1) with those of the glasses of homogenized melt inclusion shows that the latter are characterized by increased concentrations of most rare and rare earth elements.

The behavior of Li and volatile components (F and H_2O) in the melts studied should be considered separately. The maximum concentrations of Li have been determined in the glasses of melt inclusions from the comendites (samples SNT-2/13 and SNT-2/9), where they are 500–700 ppm. At the same time, the level of Li content in the melt inclusions in quartz of the commandites from sample SNT-2/8 is significantly lower and does not exceed 47–69 ppm. Insignificant concentrations of Li are also typical for the bulk rocks (16–28 ppm), which is clearly fixed as a negative anomaly in the spider diagrams (Fig. 3). The same tendency can be observed for F and H_2O. In the glasses of melt inclusions in the quartz of comendites from samples SNT-2/13 and SNT-2/9 the concentrations of F

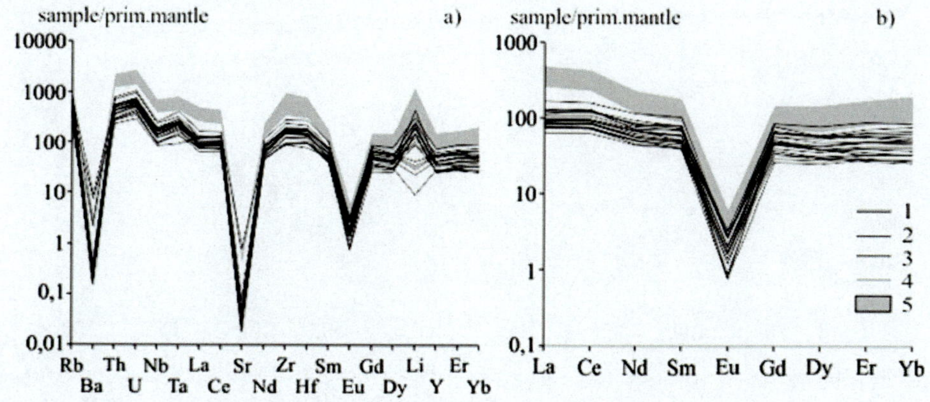

Fig. 3 Distribution patterns of trace and rare earth elements in comendite (1) and homogeneous glasses of melt inclusions in quartz of comendites of the Sant association (2–4) and pantellerites of the Dzarta-Khuduk association (5). Normalization of the relatively primitive mantle [15].

are 0.8–1.1 wt%, H_2O—0.8–2.2 wt%, while in the glasses of melt inclusions from sample SNT-2/8 the concentrations of no more than 0.13 wt% and 0.3 respectively.

5 Discussion

The results of the investigation of melt inclusions in quartz of the comendites of the Sant bimodal association allow us to consider the characteristics of the composition and evolution of these rocks and estimate the role of various processes in their formation and concentration of ore matter.

The tendencies of the behavior of rare and rare earth elements during the melt evolution are well illustrated on the diagrams of their mutual correlations. In Fig. 4 shows the dependencies in the distribution of various incompatible elements in the homogenized melt inclusions and rocks of Sant bimodal association, including their basic varieties. Nb was used as differentiation index, because the partition coefficient of Nb between melt and crystal is close to zero. The diagrams show a clear direct correlation between the concentrations of Nb and Zr, Nb and Hf in the whole range of melt and rock compositions. Similar dependences with Nb are preserved for other rare elements, in particular, Rb, Ta, Th, U, Hf, Y, Be and B. Concentrations of rare earth elements also correlate well (positive correlation) with Nb concentrations in both melts and rocks. At the same time, it should be noted that this primarily concerns medium and heavy rare-earth elements and, to a much lesser extent, light REE. This feature, apparently, is connected with crystallization of rare earth minerals in the process of evolution of comendite magma, which is recorded by detection of monazite and chevkinite among crystalline inclusions and accessory minerals.

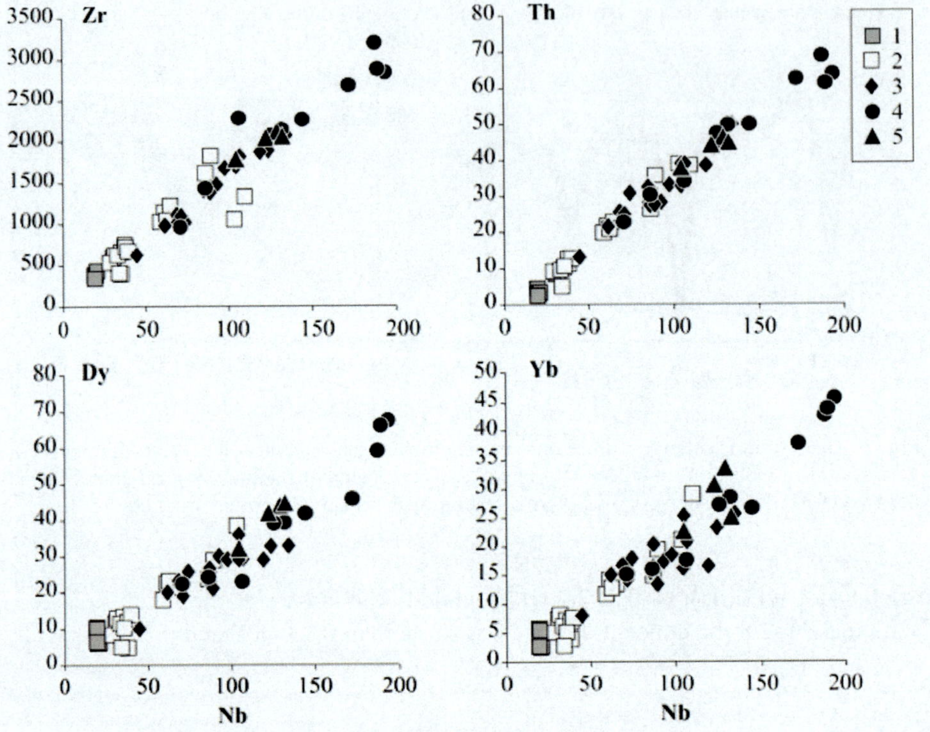

Fig. 4 Variations in trace element contents (ppm) versus Nb in rocks (1, 2) and glasses of melt inclusions in quartz of commendites (3–5) of the Sant association. 1—basalts, 2—rhyolites, 3–5—glasses of homogenized melt inclusions in quartz of comendite: 3—sample SNT-2/13, 4—sample SNT-2/9, 5—sample SNT-2/8

Thus, the presence of common trends between melts and rocks of different compositions gives us reason to believe that the basic and acid rocks of the Sant bimodal association are genetically related, and the process of crystallization differentiation of magma is responsible for the accumulation of many rare and rare earth elements in alkaline-salic melts.

As was noted above, the glasses of homogenized melt inclusions in the minerals of the two studied samples (SNT-13 and SNT-9) contain high concentrations of Li and F. The melts enriched with these elements were found earlier in our study of melt inclusions in minerals from trachydacite and pantellerite of the Dzarta-Khuduk bimodal association in Mongolia [8, 9]. It should be emphasized that both Dzarta-Khuduk and Sant magmatic associations are developed within the Mongolia-Transbaikalia Early Mesozoic zonal magmatic area, have a close age (210 and 208 million years, respectively) [4] and are characterized by a similar specifics of the rock composition. The spiderdiagrams (Fig. 3) show distinct similarity in the distribution of rare and rare earth elements in melt inclusions from phenocrysts of alkaline-salic rocks of both associations, one of the main features of which is the presence of Li maximum. An extremely important fact in the study of melt inclusions in phenocrysts of Dzarta-Khuduk pantellerites was the discovery of Li-rich fluoride melts along with silicate melts. This allowed to prove the principal possibility of generation of salt melts largely extracting Li at the final stages of alkaline magma differentiation. Such salt melts may later be separated from silicate magma and be a source of rare-metal mineralization. Evidence of the coexistence of a fluoride melts with silicate melts was obtained experimentally [10, 11] and by melt inclusions studies [12, 13].

As it follows from the Fig. 3, the compositions of the glasses of melt inclusions in quartz of comendites (sample SNT-8) does not obey the general trend with respect to Li. They, as well as the compositions of bulk rocks, are characterized by negative lithium anomaly. In addition, they contain the lowest concentrations of F (up to 0.2 wt%) and H_2O (up to 0.3 wt%). It is likely that these melts are residual magma after separation of the fluoride melt from them, and low concentrations of Li and volatile components (F and H_2O) are the result of their redistribution into the salt melt.

Thus, the revealed general regularities of melt evolution in the Sant and Dzarta-Khuduk massifs of the Mongolia-Transbaikalia zonal magmatic area, close age and composition of the rocks allow us to assume a similar formation mechanism associated with the dominant role of the process of crystallization differentiation of magma. High potential of alkalis as well as of volatile components (F and H_2O) in melts contributed to significant concentration of many rare elements in them. In the late stages of the evolution of comendite and pantellerite magmas, separation of the salt fluoride melt with the extraction of significant Li contents took place.

In general, the bimodal magmatic associations of the Mongolia-Transbaikalia zonal magmatic area can be considered as promising objects for rare metal mineralization.

Acknowledgements The work was carried out at the Laboratory of Rare Metal Magmatism of IGEM RAS with financial support from the Research Program and the Russian Foundation for Basic Research (Project #20-05-00306).

References

1. Kovalenko V.I.: Petrology and Geochemistry of Rare-Metal Granitoids, 250 p. Science, Novosibirsk (1977) (in Russian)
2. Schmitt, A.K., Trumbull, R.B., Dulski, P., et al.: Zr–Nb–REE mineralization in peralkaline granites from the AmisComplex, Brandberg (Namibia): evidence for magmatic pre-enrichment from melt inclusions. Econom. Geol. **97**, 399–413 (2002)
3. Trua, T., Daniel, C., Mazzuoli, R.: Crustal control in the genesis of Plio-Quaternary bimodal magmatism of the Main Ethiopian Rift (MER): geochemical and isotopic (Sr, Nd, Pb) evidence. Chem. Geol. **155**, 201–231 (1999)
4. Yarmolyuk, V.V., Kozlovsky, A.M., Kudryashova, E.A., et al.: Early Mesozoic rift magmatism of the Central Mongolia: geochronological evidence for evolution of

Mongolian-Transbaikalian zoned igneous province. DAN **475**(6), 669–675 (2017) (in Russian).

5. Sobolev, A.V.: Melt inclusions in minerals as a source of principle petrological information. Petrology **4**(3), 228–239 (1996)
6. Peretyazhko I.S., Savina, E.A., Karmanov, N.S.: Comendites and pantellerites of Nemrut volcano, Eastern Turkey: genesis and relations between the trachyte-comenditic, comenditic, and pantelleritic melts. Petrology **23**(6), 576–622 (2015)
7. Macdonald, R., Belkin, H.E., Wall, F., et al.: Compositional variation in the chevkinite group: new data from igneous and metamorphic rocks. Mineral. Mag. **73**(5), 777–7968 (2009)
8. Andreeva, I.A., Kovalenko, V.I., Yarmolyuk, V.V., et al.: Immiscibility of Silicate and Salt (Li, Na, F) Melts in Comendite at the Zaart Khudag Ore Occurrence, Central Mongolia: Evidence from Melt Inclusions. Dokl. Akad. Nauk **414**(4), 655–660 (2007)
9. Andreeva, I.A., Kovalenko, V.I.: Evolution of the trachydacite and pantellerite magmas of the bimodal volcanic association of Dzarta-Khuduk, Central Mongolia: investigation of inclusions in minerals. Petrology **19**(4), 363–385 (2011)
10. Gramenitsky, E.N., Shchekina, T.I., Devyatova, V.N.: Phase relations in fluorine-containing granite and nepheline syenite systems and distribution of elements between phases. M. GEOS, 188 p. (2005).
11. Veksler I.V., Dorfman A.M., Kamenetsky M., et. al.: Partitioning of lanthanides and Y between immiscible silicate and fluoride melts, fluorite and cryolite and the origin of the lanthanide tetrad effect in igneous rocks. Geochimica et Cosmochimica Acta **69**(11), 2847–2860 (2005)
12. Solovova, I.P., Girnis, A.V., Kovalenko, V.I.: Fluoride and chloride melts included in phenocrysts of agpaitic acid volcanic rocks from Pantelleria Island. DAN **433**(3), 390–393 (2010)
13. Vasyukova, O., Williams-Jones, A.E.: Fluoride-silicate melt immiscibility and its role in REE ore formation: evidence from the Strange Lake rare metal deposit, Quebec-Labrador, Canada. Geochimica et Cosmochimica Acta **139**, 110–130 (2014)
14. Yarmolyuk V.V., Kovalenko V.I.: Riftogenic magmatism of active continental margins and its ore content, 263 p. Nauka, Moscow (1991) (in Russian)
15. Sun, S.-S., McDonough, W.F.: Chemical and isotopic systematics of oceanic basalts: implications for mantle composition and processes. In: Saunders, A.S., Norry, M.J. (eds.) Magmatism in the Ocean Basins. Geological Society of London, Special Publication, vol. 42, pp. 313–345 (1989)

Composition and Parameters of Generation of Primary Magma of Ultramafic Massif Uitkomst: Thermobarogeochemical Study of Microinclusions

I. P. Solovova

Abstract

Primary melt inclusions in olivine from ultramafic rocks of the Uitkomst Massif, South Africa were studied. It was found that the parent magma has high levels of MgO (up to 22.5 wt%), FeO, SiO_2 and LREE, the content of CaO, alkalis, volatiles and fluid-mobile components. Melting of magma took place at a temperature of 1590 °C and a pressure of 2.5 GPa. The main process of magma evolution in the early stages was crystallization differentiation, which was complicated by contamination by crustal matter as the magma moved to the Earth's surface.

Keywords

Molten inclusions in Olivines · Ultramafic rocks · The Uitcomst Massif in South Africa · Evolution of Magma · Crystallization differentiation

1 Introduction

Petrological study of magmatic layered intrusions of the world has shown that their main volume was formed in precambrian period and they have a complex structure. The composition of the rocks composing them varies from ultrabasic rocks to granites. The largest deposits of platinum group elements, Cr, V and Cu-Ni sulfides are connected with the basic and ultrabasic layered complexes. Bushveld complex

I. P. Solovova (✉)
Institute of Geology of Ore Deposits, Petrography, Mineralogy and Geochemistry RAS, Staromonetnyy. 35, Moscow 119017, Russia
e-mail: solovova@igem.ru

N. Vladykin (ed.), *Alkaline Rocks, Kimberlites and Carbonatites: Geochemistry and Genesis*, Springer Proceedings in Earth and Environmental Sciences,
https://doi.org/10.1007/978-3-030-69670-2_11

is one of the oldest (2050 million years), with a maximum yield area. The question of the composition and genesis of Bushveld's parental magmas is still debatable [21, 22], because the compositions of fresh fine-grained boundary sills within the massif are extremely diverse and cannot be considered as analogues of the original magmas. Long-term study of the rocks of the complex allowed us to assume participation in the formation of the modern appearance of massif of various processes, such as metasomatic study of the mantle source, mixing of magmas of contrast composition, pulsing inflow of melts into the chamber and their contamination with crustal material, cumulative processes, late reactions and secondary changes in the rocks. In this connection, it is impossible to consider the bulk composition of rocks as analogues of original magmas. The first attempt to estimate potential Bushveld source magmas was made by G. Davis with co-authors [11]. Based on the compositions of fine-grained sills in the bottom of the complex, the authors assumed the existence of several source magmas, which, in their opinion, had an unusual composition and could be defined as boninite, magnesian andesite and basalt with a high content of SiO_2 and MgO. The researchers used computer modeling [1, 2, 13] to reconstruct the parent magmas of the Bushveld complex and its close to it composition of the Yoko-Dovyren stratified massif (Transbaikalia). At present, based on the compositions of the most primitive cumulates and quenched contacts, the possibility of contamination of crustal material with komatiitic magma was considered [18, 21, 22]. For Bushveld [21] believes that in this case the initial melt contained up to 19 wt% MgO. According to Balykin et al. [2], the primary melt also had high concentrations of MgO, but no more than 11% with a total alkali content of up to 1.5%. At the same time, the basalt magma of the Yoko-Dovyren massif, according to his calculations, should have contained up to 20% MgO.

Information on the results of the study of inclusions in minerals is of a single character. High volatiles in polyphase inclusions in the chromite of the Merensky reef of the Bushveld complex were interpreted as a result of metasomatic effects of fluid separated from the underlying rocks [17]. Two types of melt inclusions were studied in apatite from nelsonites and gabbro of the Bushveld Upper Zone with compositions of SiO_2—31 and 51, MgO—7.7 and 1.5, FeO 36 and 23 wt% [20]. According to Holwell et al. [15], some sulfide and sulfide-silicate inclusions in chromite of Bushveld may represent a trapped melt. Kottke-Levin and co-authors [16] studied inclusions in chromite crystals at 1300 °C and assumed that they contained two types of melt whose compositions they believed corresponded to B1 and B2 Bushveld magma. For the Yoko-Dovyren layered massif, data on the composition of two types of melt inclusions in the olivine from peridotite-dunite-troctolite-gabbro were also obtained: acid glass inclusions and crystallized basic inclusions [5]. Both types inclusions are enriched with volatile components (in acid melt—up to 7 wt% H_2O, 1 wt% F and 0.7 wt% Cl).

The Uitkomst Massif, due to its genetic, temporal and spatial relationship with the Bushveld Massif, is considered as part of the Bushveld Magmatic Province. The method of experimental study of microinclusions in minerals is a new approach to solving the problem and allows obtaining direct information on genesis, composition and physicochemical conditions of initial magma generation. It has been suggested that the satellite subvolcanic sill Uitcomst is the only medium where melt inclusions could be preserved in chemically primitive rocks such as dunites and peridotites.

2 Methods

Double-polished plates with thickness up to 0.3 mm were made for investigation of inclusions in minerals. Before conducting thermometric experiments, they were studied with the Olympus BX-51 optical microscope. The microinclusions detected were photographed, their species identity and tightness were determined.

A high-temperature micro furnace with a Pt—30% Rh heater was used in the work, which allows to heat samples up to 1500 °C. Thermocouple of the muffle was graduated by gold melting point (1063 °C) and NaCl (800 °C). For the melts of inclusions of the basic composition time of the exposure at the selected temperature did not exceed 15 min to avoid loss of volatile components. The use of high-temperature microcamera Linkam TS1500 allowed during heating to observe and record phase transitions under the microscope.

After the experiments the preparations with melt inclusions were pressed into polystyrene or epoxy resin. The analyses were performed on the electron-probe microanalyzer JXA-8100 of JEOL firm (IGEM RAS, analyst S. E. Borisovsky) and on the electron-probe microanalyzer Superprobe JEOL JXA-8230 at the Geological Department of MSU (Moscow State University of M. V. Lomonosov, analyst N. G. Zinovieva). The analyses were performed at accelerating voltage 15 and 20 kV, current 20 and 30 nA, beam diameter of electrons of 1, 2, 5 and 12 μ. Certified natural minerals and glasses were used as standards: Si, Al, Fe, Mg, Ca–USNM 111240/2 (basalt glass VG-2); Na–jadeite, K–orthoclase, Mn–rhodonite, Ti–TiO_2, Zr–ZrO_2, F–MgF_2, Cl–vanadinite; Cu, Fe, S–$CuFeS_2$, Ni–Ni.

The contents of trace elements, H_2O, Cl, F, S were determined by the method of secondary ion mass_spectrometry on the Cameca IMS_4f device (Yaroslavl Branch of the Institute of Physics and Technology of the Russian Academy of Sciences, Yaroslavl Branch of the Institute of Physics and Technology of the Russian Academy of Sciences, analyst S. G. Simakin). The primary O-beam with acceleration voltage of 10 kV and current of 3–6 nA was focused on the sample surface with the area of 15–20 μ. The content of the elements was calculated in relation to the measured isotopes of elements to 30Si, determined according to the certified standards. The accuracy was: <10 r.% for the contents >1 ppm, <20 r.% for the contents 0.1–1 ppm, <40 r.% for the contents <0.1 ppm.

Raman spectra of the phase composition of inclusions were obtained on the SENTERRA "Bruker" spectrometer (Institute of Physical Chemistry and Electrochemistry named after A. N. Frumkin, Institute of Physical Chemistry and Electrochemistry, Russian Academy of Sciences). Registration parameters: laser radiation wavelength (λ) 532 nm, power 2–10 mW depending on the object of investigation, resolution ~3 cm^{-1}, Olympus BX-51 microscope, ×20 and ×50 lenses. The time of accumulation of spectra was 500 s.

3 Petrography of Rocks

To carry out the work, we took samples of rock from the borehole of LM6 hole in the interval of depths from 628.6 to 665 m (Fig. 1), drilled to a depth of 1100 m in the zone of maximum thickness in the central part of the sill Uitcomst section (Main Harzburgite Layer). The studied ultramafic rocks consist of olivine without signs of serpentinization, orthopyroxene, chromite, rarer clinopyroxene and single grains of phlogopite. In the lower parts of the section there is a significant amount of intergranular material, the amount of which decreases to the upper parts of the studied hole interval.

Since olivine and orthopyroxene dominate the rock, their composition is examined in more detail. Two types of olivine are distinguished by morphological and genetic characteristics. The first type includes large grains (up to 1 cm) of cumulus olivine Ol I (Fig. 2a). The size of olivine grains of the second type (Ol II) does not exceed 500 μ, they are characterized by round or oval shape (Fig. 2b and c). The grains of olivine Ol II are cemented with ortho- and clinopyroxene, occasionally with phlogopite, which determines the poikilite structure of the rock.

Studies on electron and ion microprobes have shown that there is no zoning in both types of olivine. Thus, the content of a number of elements in the center and in the marginal zones of the crystal are: FeO 9.06 and 9.02 wt%, MgO 49.37 and 49.59 wt%, CaO 0.10 and 0.12 wt%, Ni—4140 and 4296 ppm, Sc—9 and 8 ppm, V—4 and 4 ppm. Cumulus olivine prevails in the lower parts of the studied section. The amount of olivine of the second type increases above the section.

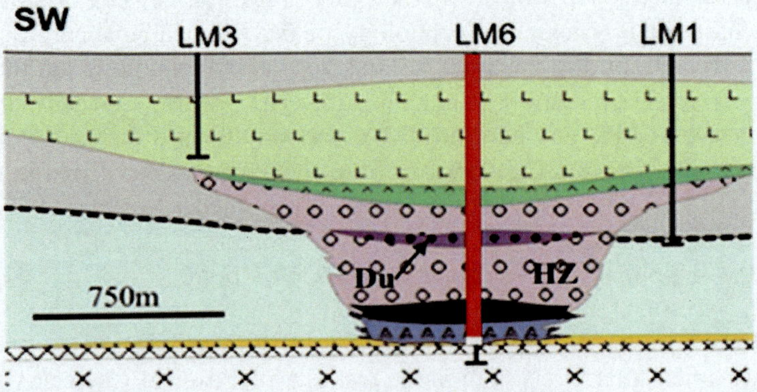

Fig. 1 LM6 borehole position in Uitkomst sill. HZ—Main Harzburgite Layer, Du—dunites

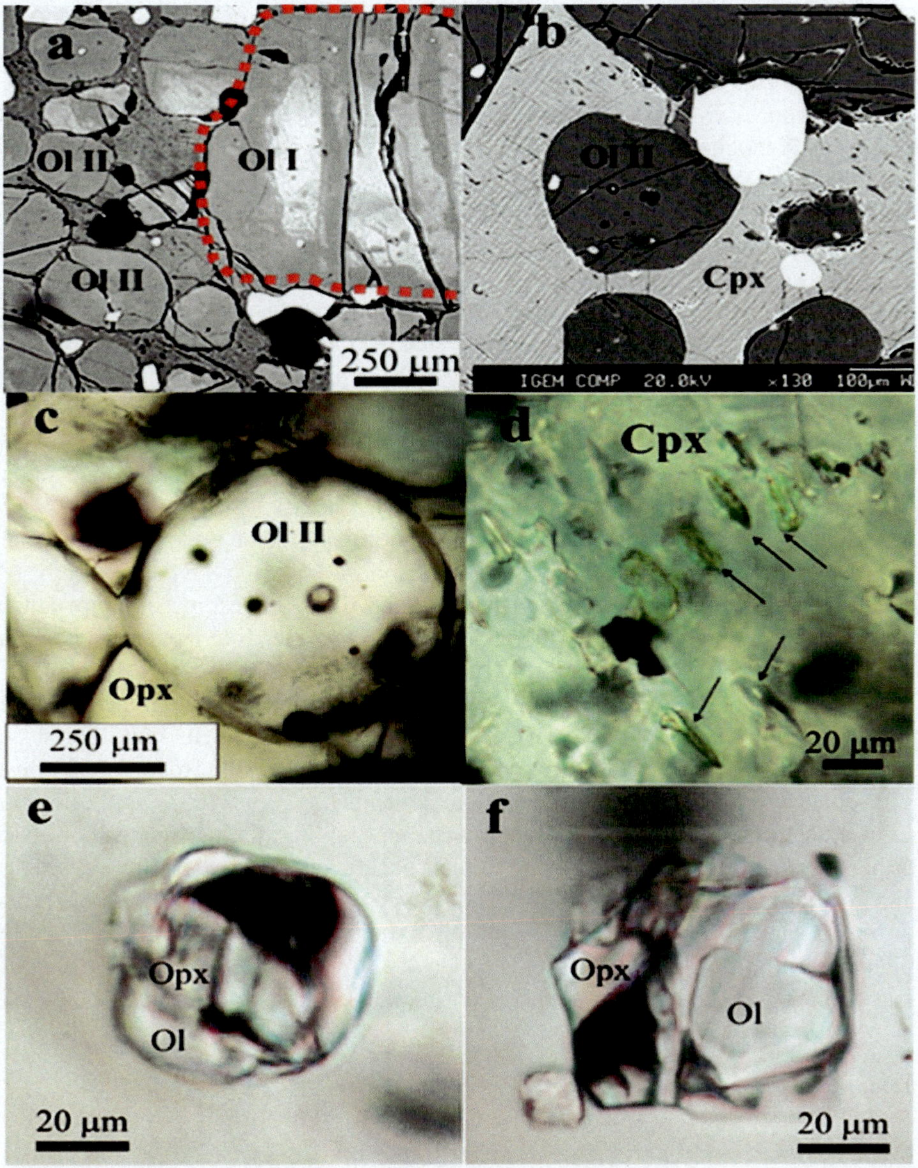

Fig. 2 Photomicrographs of minerals and inclusions in studied rocks. **a** Fragment of cumulus olivine (Ol I), contoured with red dotted line, and rounded olivine of later generation (Ol II). View in reflected light. **b** and **c** Phases of poikilitic structure: grains of olivine Ol II cemented by clinopyroxene and orthopyroxene. View in BSE (**b**) and in transparent light under microscope (**c**). **d** Wedge-shaped sub-parallel melt inclusions in resorbed clinopyroxene, view in transparent light. **e**, **f** Primary nature-crystallized melt inclusions in cumulus olivine (Ol I), containing daughter phases—olivine and orthopyroxene

Despite the morphological differences between olivines of the first and second types, their chemical composition is close and does not depend on the crystal separation form. Their magnesia varies in a narrow range (#Mg—0.90–0.91), CaO concentration does not exceed 0.15 wt%, amount of NiO—0.45–0.50 wt% (Table 1). Xenogenic polyphase inclusions are found in olivine of the first type.

Contents of rare and rare-earth elements in both types of olivines are also equal (Table 2, Fig. 3). The phases are equally enriched with HREE $(Gd/Yb)_{PM}$ = 0.085–0.110. The proximity of concentrations of basic and impurity components, the magnesia of type I and type II olivines, and the presence of crystalline spinel inclusions in them suggest that the poikilitic structure was formed during the late stage recrystallization of early cumulus olivine. At the same time, only single grains of first-generation large-crystalline olivine (Ol I) remain in the upper zones of the studied section.

At wide variations of magnesia of all studied types of orthopyroxene (#Mg =0.893–0.913), their earliest extractions and daughter orthopyroxene inclusions are characterized by lower concentrations of CaO (0.27–0.68 wt%) and Al_2O_3 (0.45–0.99 wt%) inclusions in comparison with orthopyroxene poikiloblast and their lamellas in clinopyroxene (CaO up to 1.97% and Al_2O_3 up to 1.81%, respectively) (Table 3, Fig. 4). For the last two types, the graph shows the estimated temperatures for Brey and Koehler [9]. In open conditions, its decrease leads to increase of alumina and decrease of CaO concentration in orthopyroxene composition.

Orthopyroxene type I and II is enriched with heavy rare earth elements and Y with respect to LREE. The total REE content of elements in poikiloblasts is noticeably higher than in the early orthopyroxene secretions (1.16 vs. 0.222 ppm). The same pattern is observed for rare microelements (395.7 vs. 105.7 ppm). For rare earth elements, there are certain differences and similarities between olivine and orthopyroxene of different types due to their genetic characteristics. Similar spectra and relations $(Gd/Yb)_{PM}$ are observed for type I and type II olivines and early orthopyroxene (Opx I) (Fig. 3). The orthopyroxene of poikiloblast characterized by high concentrations of rare earth elements and a lower HREE/LREE ratio (2) compared to the primary orthopyroxene of the rock (4) contrasts sharply with them.

Table 1 Compositions of olivine, wt%

	1	2	3	4	5	6	7	8
SiO_2	40.56	40.97	40.65	40.45	40.35	41.25	40.52	39.65
FeO	9.06	9.01	9.14	8.84	9.30	9.26	7.36	8.00
MnO	0.13	0.14	0.10	0.13	0.16	0.13	0.12	0.12
MgO	49.59	50.16	50.26	50.01	48.88	49.67	51.43	52.58
CaO	0.10	0.11	0.04	0.05	0.03	0.08	0.17	0.04
NiO	0.49	0.46	0.45	0.44	0.38	n.a.	n.a.	n.a.
Total	99.92	100.85	100.19	99.47	99.10	100.38	99.59	100.39
#Mg	0.907	0.908	0.907	0.910	0.904	0.905	0.926	0.921

Note 1, 2—cumulus olivine (Ol I); 3, 4—grains of poikilitic structure olivine (Ol II); 5–8—daughter phases in primary melt inclusions

Table 2 Concentrations of rare elements in rock minerals, ppm

	1	2	3	4	5	6	7	8
	Ol I	Ol I	Ol II	Ol II	Opx I	Opx II	Cpx-cl	Cpx-res
Sc	20.94	23.49	21.95	20.45	30.53	43.20	97.92	83.96
Nb	0.009	0.012	0.008	0.006	0.008	0.026	0.515	0.527
La	0.006	0.002	0.002	0.002	0.005	0.027	7.710	3.729
Ce	0.018	0.009	0.008	0.018	0.020	0.118	24.781	10.415
Nd	tr	tr	0.006	0.004	0.021	0.142	19.220	6.947
Sr	0.682	0.583	1.026	0.710	1.534	0.490	33.215	24.779
Sm	0.005	0.005	tr	tr	0.018	0.092	5.002	1.973
Hf	tr	0.024	tr	tr	0.006	0.094	1.401	0.600
Zr	0.306	0.445	0.298	0.548	0.349	1.211	23.178	8.697
Ti	78.69	110.96	51.83	114.02	72.87	347.35	1612	1171
Eu	0.000	0.003	0.003	0.002	tr	0.032	1.073	0.574
Gd	0.013	0.018	0.003	0.007	0.007	0.176	5.090	2.446
Dy	0.031	0.057	0.003	0.007	0.010	0.181	4.266	2.139
Y	0.393	0.643	0.099	0.148	0.197	1.155	22.374	11.318
Er	0.043	0.116	0.020	0.020	0.024	0.163	2.513	1.332
Yb	0.125	0.242	0.028	0.055	0.117	0.227	2.169	1.096
Lu	0.020	0.029	0.006	0.010	0.010	0.023	0.309	0.170

Note (1), (2)—cumulus olivine (Ol I), (3), (4)—olivine poikiloblasts (Ol II); (5)—early orthopyroxene (Orkh I), (6)—orthopyroxene poikiloblast (Opx II); (7), (8)—Cpx poikiloblasts, (7)—pure crystal (Cpx-cl), (8)—resorbed crystal with signs of melting (Cpx-res)

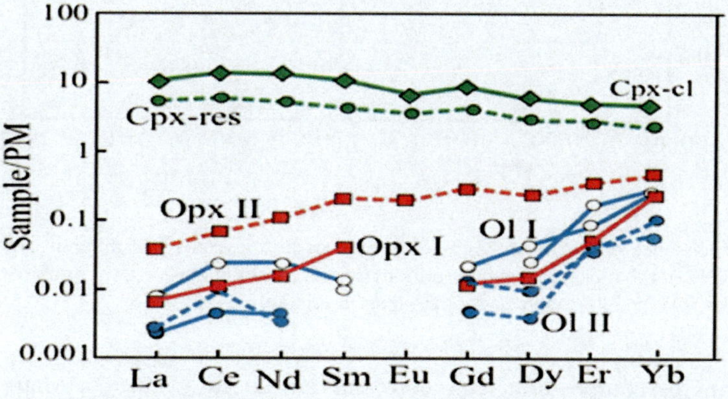

Fig. 3 Distribution of rare-earth elements in Ol I and Ol II, Opx I and Opx II, in pure (Cpx-cl) and resorbed (Cpx-res) clinopyroxene grains (the graph shows a dashed line). The concentrations are normalized to the primitive mantle

Table 3 Chemical compositions of pyroxenes and phlogopite, wt%

	1	2	3	4	5	6	7
	Opx I	Opx II	Opx III	Opx IV	Opx IV	Cpx	Flg
SiO_2	56.96	55.61	56.52	58.30	58.30	54.24	39.06
TiO_2	0.05	0.05	0.08	0.00	0.00	0.12	3.80
Al_2O_3	0.53	1.07	1.81	0.45	0.45	2.10	13.85
Cr_2O_3	0.06	0.58	0.00	0.10	0.10	1.31	1.19
FeO	6.37	5.94	7.15	6.26	6.26	3.01	3.46
MnO	0.14	0.12	0.17	0.18	0.18	0.06	0.00
MgO	34.57	33.74	33.74	35.10	35.10	18.69	22.62
CaO	0.68	1.63	0.97	0.27	0.27	20.48	0.00
Na_2O	0.00	0.05	0.04	0.03	0.03	0.49	0.73
K_2O	0.00	0.00	0.00	0.00	0.00	0.00	8.97
NiO	0.10	0.08	0.10	0.08	0.08	0.07	0.26
Total	99.45	98.87	100.59	100.70	100.70	100.49	93.68
#Mg	0.906	0.910	0.894	0.909	0.909	0.917	0.921

Note 1—early phenocrysts of rock; 2—Opx poikilocrystal; 3—Opx lamelles in poikilocrysts of Cpx; 4, 5—daughter phases in primary melt inclusions; 6—clinopyroxene poikilocryst; 7—phlogopite poikilocryst, additionally containing Cl—0.18, F—0.31 wt%

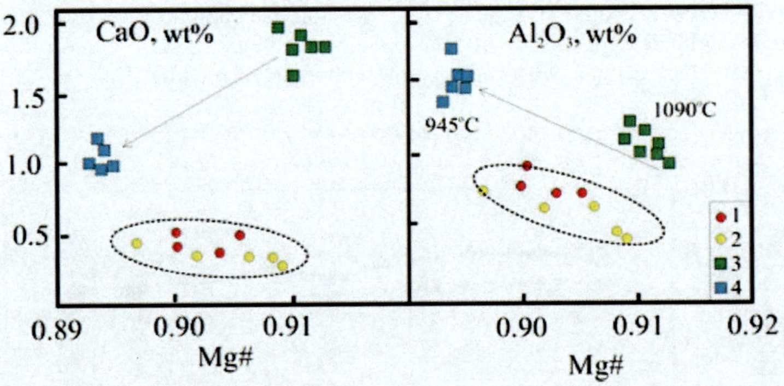

Fig. 4 Dependence of CaO and Al_2O_3 from Mg# of orthopyroxene of different types. (1) early orthopyroxene of the rock, (2) daughter orthopyroxene in melt inclusions, (3) orthopyroxene of the poikiloblasts, (4) lamelles of the orthopyroxene in clinopyroxene

In terms of composition, clinopyroxene are augite (Table 3). Many grains of clinopyroxene are characterized by the resorption of marginal zones and the presence of numerous sub-parallel fine melt inclusions of wedge-shaped shape (Fig. 2d). The distribution spectra of rare-earth elements in fresh and resorbed grains are shown in Fig. 3. Clinopyroxene is enriched with all rare elements, the value of $(Gd/Yb)_{PM}$ is considerably higher, and the HREE/LREE ratio is lower than 1. The element contents in clinopyroxene with inclusions are somewhat lower in comparison with pure grains, but the shape of the spectra is the same. The location and appearance of melt inclusions, the identity of the REE spectra shapes of clinopyroxenes with and without inclusions, but with different levels of accumulation, can be related to pyroxene melting in situ, possibly as a result of local temperature increases leading to a redistribution of impurities in the melt.

Phlogopite is rare phase and contain up to 1 wt% Na_2O and Cl, F in equal quantities. #Mg of phlogopite reaches 0.93, the sum does not exceed 93.7 wt%.

4 Results of Studying Inclusions

Single primary fully crystallized inclusions 30–50 μ in size were found in cumulus olivine. They are located azonally in the volume of olivine and are characterized by oval or partially cut form (Fig. 2e, f). Since the olivine phenocrysts contain numerous crystalline spinel inclusions, they often form combined melt inclusions together with the melt. The silicate component of such inclusions on the border with the olivine host has a meniscus-shaped border, which indicates its originally liquid state. The low-density deformed gas bubble is located between the daughter phases.

Based on the experiments, it was found that olivine is the first crystallizing daughter phase from the melt. Its magnesia in some inclusions reaches #Mg = 0.927. The concentration of CaO varies from 0.04 to 0.25 wt%. The second crystallizing daughter phase is represented by orthopyroxene. Its #Mg varies from 0.896 to 0.909. In terms of chemical composition, the orthopyroxene daughter phase is close to the orthopyroxene of the rock (Fig. 4). When heated, the first signs of melting of the daughter phases were observed at 1050 °C, which led to the appearance of melt. The low-density gas phase acquired rounded forms. The presence of the melt was confirmed by the movement of the gas bubble during the temperature rise of the experiments. At the temperature of 1290 °C, a mixture of molten crystals of olivine and orthopyroxene, as well as remnants of the ore phase, was observed in the inclusions. The increase in temperature of the experiments led to further melting of crystalline phases. Homogenization of inclusions with dissolution of all phases, including fluid phase, was not achieved at heating up to 1450 ° C (Fig. 5c). In some cases, decrepition of inclusions was observed, which resulted in partial loss of substance and heating stopped.

Most inclusions contained glass (melt at test temperature) and a gas bubble. However, the possibility of the presence of rims of a daughter olivine deposited on the walls of the vacuoles during crystallization of the melt in natural conditions and optically indistinguishable from the host mineral is not excluded. To obtain the bulk composition of the melt and to reduce the error, the analyses were performed from the scanning area of 5 × 5, 10 × 10 and 15 × 15 μ.

To reconstruct the composition of the trapped melt the inverse fractionation method was used, the essence of which is to recalculate the obtained analyses into equilibrium with the host olivine, whose magnesia varies in the range from 0.907 to 0.911 with an average value of 0.910. An important role in such calculations is played by the Mg–Fe distribution coefficient between the olivine and the melt

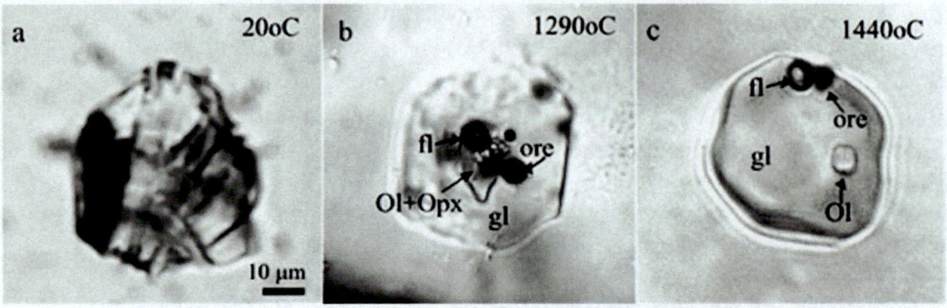

Fig. 5 Photomicrographs of phase changes in primary melt inclusion at temperatures indicated in the photographs. Transparent light. gl—glass (at experience temperature), fl—fluid phase, ore—ore phase, Ol + Opx—melting of daughter phases (olivine and orthopyroxene)

(Kd = $(Fe/Mg)^{Ol}/(Fe^{2+}/Mg)^{Gl}$). For basic and ultrabasic rocks of normal alkalinity Kd is taken equal to 0.3–0.33. In our calculations we used the distribution coefficient Mg–Fe, equal to 0.315, previously adopted in the works devoted to the study of Bushveld rocks [21, 22]. The most primitive melt in equilibrium with Fo91 contains 22.5% MgO at 12.6 wt% FeO. It is characterized by a SiO_2 concentration of 51.9 wt%, Al_2O_3 concentration of 8.3 wt%, low CaO content of 2.5 wt% and alkali amount of 2.1 wt% at Na_2O/K_2O—3.8. The melts are low titanium (TiO_2 is no more than 0.08–0.3 wt%). The number of volatile components in the melt, determined by the method of secondary ion mass spectrometry (SIMS), corresponds: 168 ppm H_2O, 264 ppm Cl, 258 ppm F, 114 ppm S. Fractionated melts with MgO concentrations up to 15.5 wt% contain up to 56.5% SiO_2, 10 wt% Al_2O_3, 4% CaO, and 4% Na_2O, up to 549 ppm F and 2288 ppm S. They are visibly enriched with H_2O– to 7665 ppm and Cl– to 1728 ppm.

Rare metal composition of melts is characterized by close concentrations of the most incompatible elements. Their concentrations are 10–350 times higher than in the primitive mantle. The spectra have clearly expressed minimums of Nb and Ti.

The total content of rare earth elements varies from 47 to 144 ppm. Melts of inclusions are characterized by the predominance of LREE over HREE (Fig. 6). The LREE content varies in a narrow range (La_{PM}—from 15 to 35). The greatest variation of values is observed for HREE, for example, the value of Yb_{PM} changes almost 6 times (from 0.5 to 2.9). The melt spectra with maximum MgO and FeO content have the steepest negative slope ($(La/Yb)_{PM}$ = 29 vs. 5.9) for a low-magnesia melt. The spectra of the fractionated melts with reduced MgO and enriched H_2O and Cl content have a clearly marked negative anomaly of Eu. A review of data on REE spectra of mafic and ultramafic volcanogenic rocks, hypabbysal and plutonic rocks [6] shows that the negative anomaly of Eu is not typical for them.

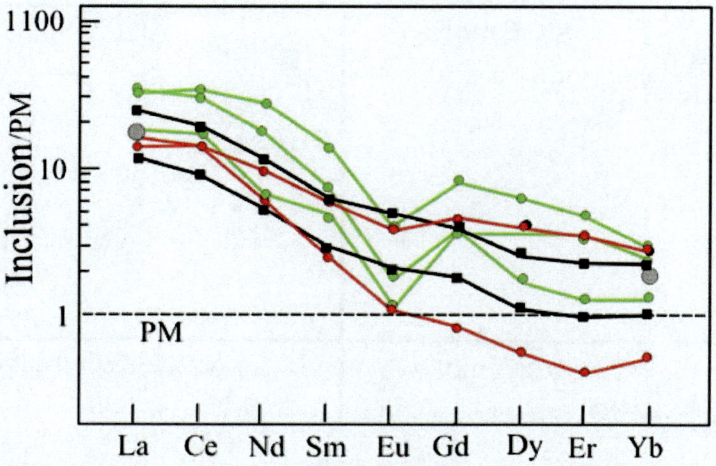

Fig. 6 Concentrations of rare earth elements in the melts of inclusions normalized to the primitive mantle and recalculated to equilibrium with olivine Fo91. The red spectra correspond to the earliest melts; the green spectra with the negative anomaly Eu correspond to the later magmatic differentiates. For comparison, the black solid line diagram shows REE concentrations in the model melts B1 and B1-UM [8] and gray circles for komatiitic-like magma [21]

The spectra of the model boninite melts on the spidergram are in an intermediate position, but differ from the melts we obtained for the initial melts with higher concentrations of heavy rare-earth elements. The same is true for komatiitic-like magma.

5 Discussion

5.1 Chemical Composition of the Parental Magma and Its Evolution

Despite the limited number of primary inclusions found in the olivine, their study allowed us to obtain information on the composition of the original melt independent of secondary changes in the rocks and to identify the main regularities of mafic magma evolution. The experiments showed that olivine is a liquidus phase. Removal of its main components (FeO, MgO and MnO) from magma should systematically increase the share of other components. On the correlation diagrams of FeO + MgO + MnO—oxide (in mol%) the melts form negative trends confirming the dominating role of olivine and prevalence of crystallization differentiation process (Fig. 7). The exception is alkali metals, for which the trend direction changes are observed.

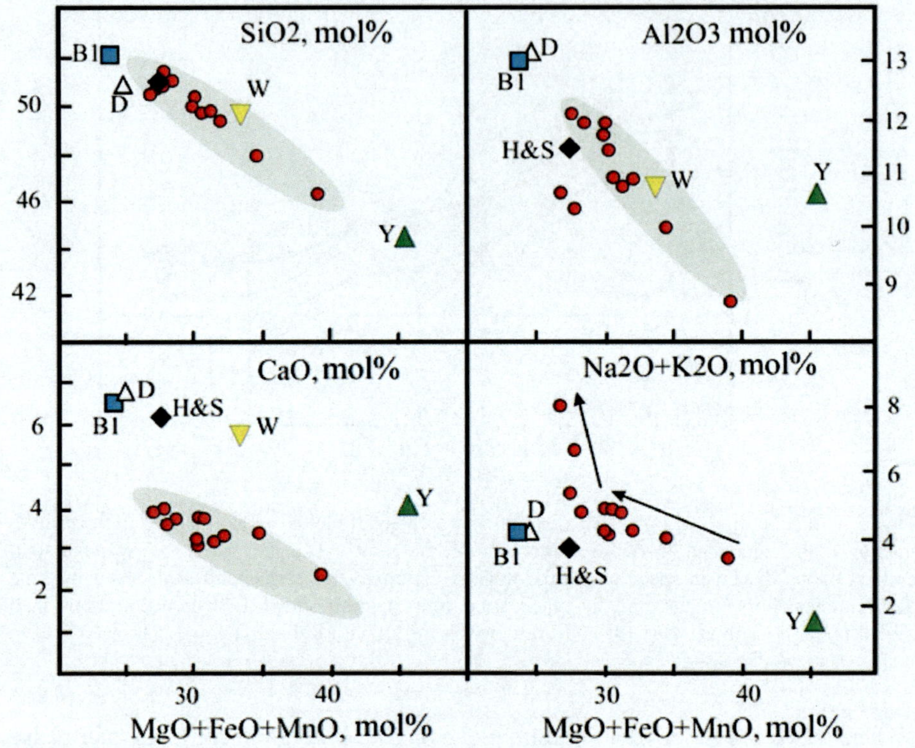

Fig. 7 Dependence of concentrations of components on MgO + FeO + MnO in melts (in mol%). Red circles—our data; Model melt: Boninite magma B1—Barnes et al. [8] and D—Davis [12]; H&S—Harmer and Sharpe [14]; komatiite-like magma: W—Wilson [21] and Y—Yudovskaya et al. [22]

As mentioned above, the composition of the inclusion melts was recalculated to be in equilibrium with the host olivine Fo91. However, the absence of zoning in their grains suggests that the magnesia of olivines changed during the diffusion of Fe–Mg. Under subliquidus conditions, phases with a higher concentration of fayalite component and lower value of #Mg should have crystallized. On the other hand, it was previously shown [21] that early olivine in Bushveld ultramafites could correspond to Fo 92.2. A higher value of #Mg olivine up to 0.93, was theoretically determined when evaluating the composition of primitive magma using one of the samples of harzburgite as its chemical analogue [22]. In equilibrium with olivine Fo 92.2, the melt will contain up to 26 wt% MgO.

Surprisingly the constancy of the content of the most incoherent elements with wide variations of the coherent. This contradicts the role of crustal contamination, which should cause significant variations in the left side of the diagram with insignificant variations of HREE. We can assume that this is due to the dominant crystallization of olivine and orthopyroxene, phases characterized by high HREE content (Fig. 3). As a result, the melt depleted with heavy HREE while maintaining

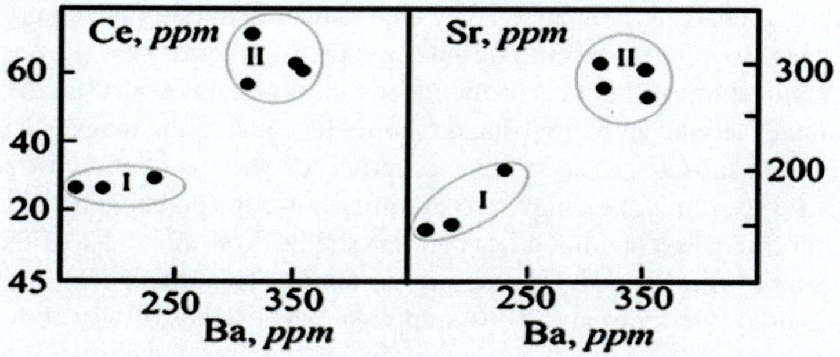

Fig. 8 Dependence of the contents of fluid-mobile elements Ce and Sr from the concentration of Ba in the melts of inclusions

the light element concentration at the same level. On the other hand, this phenomenon can be connected with the shift of mantle melting area from garnet-peridotite facies to spinel or even plagioclase peridotite facies.

Analyses have shown that melts with MgO up to 22.5 wt% contain low concentrations of fluid-mobile elements (Ba, Sr, Th, Ce, U) (Fig. 8, field I), which is consistent with the low role of volatile components in the magma generation zone and the absence of primary fluid inclusions in the olivine.

Such melts are characterized by the maximum concentration of ore-forming elements (3078 ppm Ni), which becomes especially important when solving the question of the source of ore components. As MgO decreases in the melt, the concentrations of fluid-mobile components are increasing (Fig. 8, Field II), including alkalis. In melts with 17.0 wt% MgO and below their concentration sharply increases from 2.9 to 5.1 wt% and the trend slope changes. At this stage of magma evolution, such a trend bend (Fig. 7) may indicate that the fluid of the melt enriched with water and chlorine enters the system. The appearance of such a fluid is associated with the interaction of hot magmatic differentials with cow rocks. Evidence of this is microxenolites in the studied rocks and in late olivines.

Melt of inclusions with maximum MgO is characterized not only by low content of heavy rare-earth elements, but also by steep slope of the spectrum, which indicates the residual garnet in the source. Enrichment of LREE magma indicates its melting from the metasomatically enriched mantle substrate.

5.2 Water in the Melt

In considering the sources of ultramafic magma and their evolution, researchers (e.g., Sobolev et al. [7]) pay great attention to the behavior of volatiles, including water, during the melt crystallization. As mentioned above, the H_2O concentration in inclusions from early olivine crystals is low and corresponds to 168–283 ppm. However, in the later inclusions, its concentration reaches 7665 ppm. This prompted us to analyze more attentively the results of studying the compositions of both the daughter phases and the olivines containing them.

At crystallization of the melt of inclusions in olivine phenocrysts, the magnesia of the daughter olivine at the first stages should be equal to the magnesia of the host olivine. But the subsequent crystallization proceeds with its continuous enrichment with iron. Subsequently, at a high rate of diffuse Fe–Mg exchange between the host mineral and the daughter mineral, which crystallized on the walls of the vacuole, the magnesia of the boundary phases is equalized. This effect was considered in detail by Danyushevsky et al. [10]. As applied to volcanites, it was shown that the redistribution of these components between an olivine phenocryst and daughter crystals enriched with the latter Mg can lead not only to an increase in magnesium oxide in the homogenized melts of inclusions, but also will affect the concentrations of other elements. In our case, at the initial completely crystallized of the melt in the inclusions, it was found that the magnesia of the daughter olivine in inclusions not only was not equalized, but was higher than that of the host olivine (Table 1). This phenomenon is very rare and is related to the loss of hydrogen from hermetic inclusions, especially at high iron content in melts. It was previously studied in melt inclusions in the olivine boninites of New Caledonia [19].

The interaction of water with FeO in the inclusion melt is described by the reaction $2FeO + H_2O = Fe_2O_3 + H_2$. The resulting hydrogen diffuses beyond the inclusions, and Fe_2O_3 is preserved in the melt and does not take part in the formation of daughter olivine at lower temperatures. Thus, MgO concentration in relation to FeO artificially increases at constancy of total Fe, which leads to crystallization in inclusions of daughter olivine with increased magnesia. Calculations show that removal of 0.1 wt% H_2O should lead to increase of #Mg olivine by 0.007–0.01 units. High temperatures and high iron content in magma are favorable for this phenomenon. Calculations show that inclusions with MgO content up to 15.6 wt% could lose up to 0.22 wt% H_2O. Considering that such inclusions have a maximum H2O concentration of 7665 ppm, they could contain at least 1 wt% water.

5.3 PT-Parameters of Initial Magma

Our high-temperature experiments at 1 atm. showed that the homogenization temperature of inclusions exceeds 1450 °C. The question of the influence of external and fluid pressure, as well as physical and chemical properties of the mineral and the melt on the behavior of inclusions during heating was considered by us earlier [4]. The work showed that the homogenization temperature of inclusions may differ from the crystallization temperature of a mineral depending on the presence or absence of volatile components in the melts. To clarify the PT-parameters of magma formation and melt evolution, we used the method of mineral thermometry. The olivine-orthopyroxene-melt is an association suitable for solving the problem of conditions for the generation of *parental* magma in the mantle. Using the program developed based on the results of more than 3600

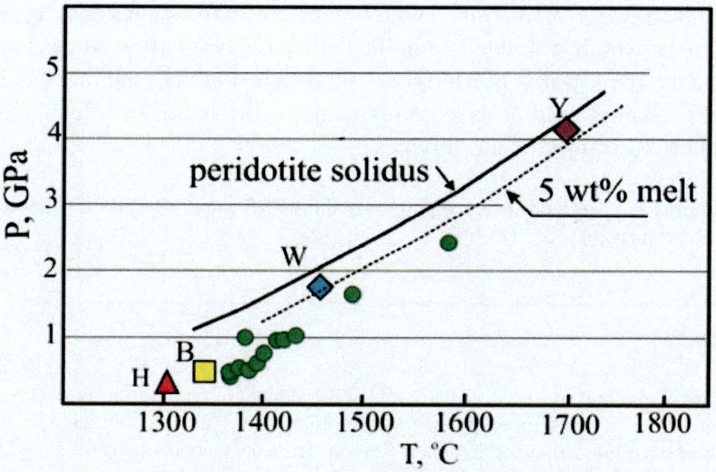

Fig. 9 Temperature and pressure primary and intermediate stages of melt evolution. Round symbols are our data. For comparison, the following model compositions of Bushveld magma are placed on the diagram: B-1—composition of boninite-like magma [8], komatiite-like magma [21, 22]

experiments with high-magnesia compositions [3], we evaluated the parameters for the most primitive melt with 22.5 wt% MgO as 1590 °C and 2.5 GPa (Fig. 9).

The primitive komatiite-like magma, according to Wilson [21] and Maier et al. [18], is located near a weakly differentiated parent magma that we have identified. Its PT-parameters are slightly lower than the values we obtained. Another group of inclusions contains melts with MgO concentrations of 15.4–17.4 wt%. They are enriched with alkalis, H_2O, Cl and have a distinct negative Eu anomaly. Since these inclusions also contain the mineral association Ol-Opx-melt, the melt crystallization parameters estimated for them are 1380–1400 °C and ≤ 1 GPa. The model magma B-1 of the boninite type falls into the region of their compositions [8].

6 Conclusion

The melts studied in hermetically sealed melt inclusions, isolated from secondary processes, belong to the ultramafic type. Primary magma is characterized by high concentrations of SiO_2 and LREE, low CaO, alkalis, volatile and fluid-mobile components. The source of such melt could be mantle garnet-containing harzburgite, metasomatically enriched just before magma generation. The generation of magma took place at extremely high temperature and pressure—1590 °C and 2.5 GPa.

The main process of magma evolution in the early stages was crystallization differentiation, which was later complicated by contamination of crustal matter as the magma moved towards the surface. The conclusion is based on a sharp increase of H_2O, Cl, alkalis, fluid-mobile components in the inclusion melts and the presence of microxenolithes in the olivines.

Acknowledgments The work was supported by the RFBR grant 18-05-00291a and the Russian Academy of Sciences (0136-2019 0014).

References

1. Ariskin, A., Konnikov, E.G., Kislov, E.V.: Modeling of the equilibrium crystallization of ultramaficites in the appendix to the problems of phase stratification formation of Dovernensky Pluton (Northern Baikal Region, Russia). Geochemistry 2, 113–155 (2003)
2. Balykin, P.A., Lavrenchuk, A.V., Petrova, T.E.: Formation types of the peridotite-gabber arrays of the intracontinental mobile belts, compositions of the initial melts and their evolutionary trend. Lithosphere **4**, 3–20 (2004)
3. Girnis, A.V.: Equilibrium olivine-orthopyroxene-melt as a thermobarometer for mantle magmas. Petrology **11**(2), 115–127 (2003)
4. Girnis, A.V., Solovova, I.P.: Petrogenetic informativeness of the melt microinclusions in the deep rocks minerals. Geochemistry **1**, 20–28 (1989)
5. Konnikov, E.G., Kovyazin, S.V., Nekrasov, A.N., Simakin, S.G.: Fluid-magmatic interaction of mantle magmas with rocks of the lower crust: data on study of inclusions in minerals of the intrusion. Geochemistry 10, 1027–1046 (2005)
6. Lesnov, F.P.: Rare earth elements in ultramafic and mafite rocks and their minerals (in Russian). In Book: Novosibirsk. Acad, 401 pp. "Geo" Publishing House (2007)
7. Sobolev, A.V., Sobolev, S.V., Kuzmin, D.V., Malich, K.N., Petrunin, A.G.: The mechanism of Siberian meimechites formation and nature of their connection with traps and kimberlites. Geol. Geophys. **5**(12), 1293–1334 (2009)
8. Barnes, S.-J., Maier, W.D., Curl, E.A.: Composition of the marginal rocks and sills of the rustenburg layered suite, bushveld complex, South Africa: implications for the formation of the platinum-group element deposits. Econ. Geol. **105**, 1491–1511 (2010)
9. Brey, G., Koehler, T.: Geothermobarometry in four-phase lherzolites II. New thermobarometers, and practical assessment of existing thermobarometers. J. Petrol. **31**, 1353–1378 (1990)
10. Danyushevsky, L.V., Della-Pasqua, F.N., Sokolov, S.: Re-equilibration of melt inclusions trapped by magnesian olivine phenocrysts from subduction-related magmas: petrological implications. Contrib. Mineral. Petrol. **138**, 68–83 (2000)
11. Davies, G., Cawthorn, R.G., Barton, J.M., Morton, M.: Parental magma to the Bushveld complex. Nature **287**, 33–35 (1980)
12. Davies, G.: The petrogenesis of the peripheral zone of the Rustenburg Layered Suite and associated sills between Harteebestpoort and Buffelspoort Dams Johannesburg: Western Bushveld complex. Ph.D. thesis. 444 p. University of the Witwatersrand (1982)
13. Eales, H.V., Costin, G.: Crustally contaminated komatiite: primary source of the chromitites and marginal, lower, and critical zone magmas in a staging chamber beneath the Bushveld complex. Econ. Geol. **107**, 645–665 (2012)
14. Harmer, R.E., Sharpe, M.R.: Field relations and strontium isotope systematics of the Eastern Bushveld complex. Econ. Geol. **80**, 813–837 (1985)
15. Holwell, D.A., McDonald, I., Butler, I.B.: Precious metal enrichment in the Platreef, Bushveld complex, South Africa: evidence from homogenized magmatic sulfide melt inclusions. Contrib. Mineral. Petrol. **161**, 1011–1026 (2011)

16. Kottke-Levin, J., Gauert, C., Merkle, R., Sommer, H.: Pilot study of melt inclusions in chromites of the Merensky Reef reflecting melt compositions at Karee Mine, western Bushveld Complex, South Africa. SAO/NASA Astrophysics Data System (ADS), Abstract #V21A, 2085 p. (2008)
17. Li, C., Ripley, E.M., Sarkar, A., Shin, D., Maier, W.D.: Origin of phlogopite-orthopyroxene inclusions in chromites from the Merensky Reef of the Bushveld omplex, South Africa. Contrib. Mineral. Petrol. **150**, 119–130 (2005)
18. Maier, W.D., Barnes, S.J., Karykowski, B.T.: A chilled margin of komatiite and Mg-rich basaltic andesite in the western Bushveld omplex, South Africa. Contrib. Mineral. Petrol. **171**, 57–79 (2016)
19. Solovova, I.P., Ohnenstetter, D., Girnis, A.V.: Melt inclusions in Olivine from the Boninites of New Caledonia: postentrapment melt modification and estimation of primary magma compositions. Petrology **20**(6), 529–544 (2012)
20. Wang, M., Veksler, I., Cawthorn, G.: Apatite-hosted melt inclusions in the upper zone of the Bushveld complex. Abstact IMA, p. 236 (2014)
21. Wilson, A.H.: A chill sequence to the Bushveld complex: insight into the first stage of emplacement and implications for the parental magmas. J. Petrol. **53**(6), 1123–1168 (2012)
22. Yudovskaya, M.A., Naldrett, A.J., Woolfe, J.A.S., Costin, G., Kinnaird, J.A.: Reverse compositional zoning in the uitkomst chromitites as an indication of crystallization in a magmatic conduit. J. Petrol. **10**, 1–21 (2015)

New Mineralogical Type of Mantle Substrate (Nakyn Kimberlite Field, Yakutia)

S. M. Sablukov, L. I. Sablukova, and Yu B. Stegnitskiy

Abstract

Detailed study of kimberlites and mantle xenoliths from pipes of Nyurbinskaya, Botuobinskaya and Mayskaya Nakyn field revealed their unusual, interesting and important mineralogical features, which indicate the reduced nature of asthenospheric impact on the mantle substrate of the area (absence of megacrystic picroilmenites of is compensated by presence of large orange-red titanous pyropes of "megacryst" type), development of subduction processes (picroilmenite, chrome-omfacite) and crystallization of mantle melts (chrome-diopside), as well as on the impact of the processes of "calcium" (*chromium-calcium* and *titanium-chromium-calcium)* mantle metasomatosis (green garnet). The exotic nature of the mantle substrate emphasizes the presence of MARID and alcremite xenoliths. Pipes of Nakyn field are unique in terms of mineralogical features of kimberlites and mantle xenoliths, both among other pipes of Yakutia Diamond-bearing Province, and among kimberlite tubes of the World.

Keywords

Kimberlite · Nakyn field · Mantle xenoliths · Omphacite · Green garnets

S. M. Sablukov (✉) · L. I. Sablukova
RUSGEO Limited, Moscow, Russia
e-mail: sablukoff@rambler.ru

Y. B. Stegnitskiy
NIGP AK ALROSA (PAO), Mirny, Russia

N. Vladykin (ed.), *Alkaline Rocks, Kimberlites and Carbonatites: Geochemistry and Genesis*, Springer Proceedings in Earth and Environmental Sciences,
https://doi.org/10.1007/978-3-030-69670-2_12

1 Introduction

Diamond-bearing kimberlites of the ***Nakyn field***, opened in 1994 [5], are located separately, at a significant distance (~300 km) from other fields of the Yakutia diamondiferous province (Fig. 1), and in terms of physical composition so far have no analogues among the kimberlites of Yakutia (and not only Yakutia) [1, 2]. The later discovered diamondiferous kimberlites of the ***Menda-Barylaiskaya area*** (2007), Central Yakutia, the right bank of the Lena River, 100 km to the south of Yakutsk [40]) and the ***Seldyukar Field*** (2015, Western Yakutia, 60 km to the north of the city Mirny [17]), as a whole (at this, the initial stage of study) do not differ much in composition from the kimberlites of other, long known fields of Western Yakutia (in any case, belong to the most common type of ilmenite-bearing kimberlites—Group-1 kimberlites (by Smith et al. [41]).

The Nakyn field kimberlites stand out sharply among other diamondiferous kimberlites of Yakutia with a very low content of all incoherent elements (except potassium, rubidium, and phosphorus), older age: 399 million years, early Devon [32] rather than 320–370 million years, late Devon—early Carbon [3], and unusual isotope characteristics.

In terms of sharp predominance among the deep minerals of pyrope and chrome spinel in the complete absence of picroilmenite "megacrystic" association [15, 48] kimberlites of the Nakyn field stand out sharply among other diamondiferous kimberlites of Yakutia and similar to the kimberlites of Group 2 of South Africa and ilmenite-free kimberlites of the alumina series (Al-series) of the M.V. Lomonosov deposit in the Arkhangelsk diamond province [9, 19, 27]. However, by Nd-Sr

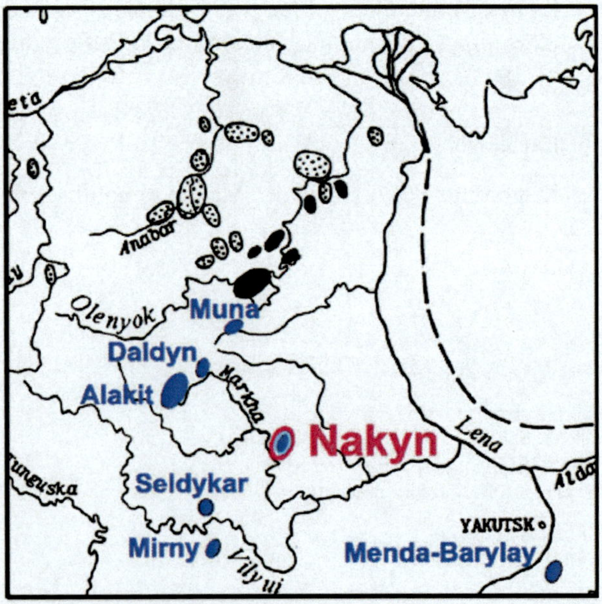

Fig. 1 Layout of kimberlite fields on the Siberian platform

isotopic characteristics approaching the BSE (eNd = +1; eSr = +25 [33]) kimberlites of Nakyn in general close to other kimberlites of Yakutia and kimberlites of Group 1 of South Africa. This uniqueness of the kimberlites of the Nakyn field is determined by an unusual set and composition of some mantle minerals, as well as an unusual structure of the mantle substrate of the Nakyn field as a whole.

2 Geological Objects, Samples and Methods of Research

In 2006–2016, we studied in detail and in a comprehensive manner in the Nakyn field the Nyurbinskaya kimberlite pipe (together with the pipes dyke), the Botuobinskaya pipe and the Mayskaya pipe (85 boreholes, more than 10,000 linear meters of core, the walls of the quarries, the material of ore yards), as well as the Nyurbinskaya and Botuobinskaya placers (more than 120 boreholes).

Totally more than 1100 mantle xenoliths (about 1000 pcs. in Nyurbinskaya pipe, about 50 pcs. in Botuobinskaya pipe and about 50 pcs. in Mayskaya pipe) were studied in details. Mineralogical analysis of about 100 kimberlites and about 400 core samples of placers was carried out. X-ray spectral microanalysis of minerals (about 4000 analyses) was carried out in FSUE Gintsvetmet laboratory, on Camebax-microbeam device at accelerating voltage of 20 kV, current 15–20 nA, linear probe size of about 3–4 μm. Study of microelement composition of mantle xenolite garnets was carried out in Isotope Center of FSUE VSEGEI. Method—LA-ICPMS, laser ablation system (laser sampling) DUV-193, with excimer laser COMPex-102, mass spectrometer Element-II. Beam diameter $\sim$200 microns, laser pulse repetition rate 10 Hertz. Totally 30 garnet grains from different types of mantle xenoliths of Nyurbinskaya tube were investigated. The determination of TR-parameters of mineral paragenesis formation was carried out with the help of "PT-Quick" program.

3 The Features of the Set and Composition of Deep (Mantle) Minerals

The set of basic, predominant deep minerals in kimberlites (and mantle xenolites) of Nakyn field is typical for kimberlites of group-2 of South Africa [41]—it is olivine, red-violet pyrope and chrome spinel.

3.1 Olivine

Macrocrystals and megacrystals of olivine in size 0.5–2 cm (up to 6 cm) constitute up to 40% of the volume of kimberlites of all three studied pipes. However, all olivine is completely replaced by a secondary minerals unit (serpentine, carbonate).

Fresh olivine is found only in the one sample of kimberlites Nyurbinskaya pipe at a depth of 896 m. The olivine grains are colorless, with very weak greenish or yellowish tint, sometimes with inclusions of purple pyrope. Olivine has a composition common to the chrome mineral association of kimberlites: 6.8–9.4% FeO, 0.30–0.50% NiO, 0.01–0.07% Cr_2O_3, <0.05% CaO.

3.2 Pyrope of the Red-Purple Colours

The most common "fresh" deep mineral of Nakyn kimberlites is pyrope. Both in kimberlites and in mantle xenoliths pyrope of usual red-violet color of different shades and tones sharply prevails. The composition of the majority of pyrope is also quite usual for chrome mineral association of kimberlites of different types: 0.5–14% Cr_2O_3, 7.5–9.0% FeO, 1–9% CaO, 0.05–0.6% TiO_2.

3.3 Chrome Spinel

Chrome spinel are much less common in mantle xenoliths of Nyurbinskaya pipe than garnet. In terms of composition peculiarities, they are quite diverse; there are chromites of different facies of depth [45] from spinel-pyroxene facies (B) to diamond-pyrope facies (D), with a noticeable prevalence of chromites of high-baric facies. In kimberlites, the share of chrome spinels of diamond paragenesis has been increased [43], sometimes up to 20–25% of their total amount. The composition of most chrome spinels is quite common for the chrome mineral association of diamondiferous kimberlites of different types.

The associations of other deep minerals of Nakyn kimberlites are quite unusual.

3.4 Picroilmenite

In Nakyn kimberlites there is no pikroilmenit "megacryst" association at all. This is in contradiction with their Nd-Sr isotopic values, as close to BSE, as in other kimberlites of Yakutia and kimberlites of Group-1 of South Africa, saturated with picroilmenite, as with the isotopic and mineralogical side is determined by the impact on their mantle substrate asthenospheric source.

This lack of picroilmenite of the "megacryst" association in the Nakyn kimberlites is all the more surprising because they contain a significant number of orange-red titanium pyrope in appearance and composition corresponding to the pyrope of the "megacryst" association of paragenesis with picroilmenite, typical for Group-1 kimberlites. These orange-red titanium pyropes are found in Nakyn kimberlites in the form of typical "megacrystals"—large nodules and garnetites (size from 1 to 5.5 cm!) (Fig. 2). There are also serpentinized olivine megacrysts (up to 6 cm in size) with inclusions of titanium orange pyrope of "megacryst" association and xenoliths of garnet peridotite and garnet clinopyroxenites (up to 10 cm in size) with orange pyrope of "megacryst" association [18]. However, there is no picroilmenite at all in all these mantle rock samples.

But there is a picroilmenite in kimberlites of all pipes of Nakyn field. It is rare, its maximum number is noted in the Nyurbinskaya pipe.

Picroilmenite is present only in garnet-containing mantle xenoliths (peridotites, pyroxenites, eclogites) or in metasomatic rocks (marids, glimmerites, protokimberlites of lamprophyre type).

In all mantle xenoliths (peridotite, pyroxenite, eclogite) picroilmenite is an accessory mineral present as rare grains up to 1 mm in size. The content of pycroilmenite sometimes reaches 0.1–0.5% of volume, and then elements of sideronite microstructure of the rock appear. Picroilmenite is found only in garnet peridotites and garnet clinopyroxenites with oriented, directive textures: chain, linear, plane-parallel, including in distinctly parallel-zonal (stratified, layered) rocks (Fig. 3A–C). In peridotites and pyroxenites of massive texture (and allotriomorphic, panidiomorphic, and protogranular structure (Fig. 3D–F)) picroilmenite was not found.

Inclusions of metasomatic mantle rocks (marids, glimmerites) picroilmenite is also found in the form of small grains and in accensual amounts. The maximum amount of picroilmenite (up to 5%) is found in inclusions of protokimberlites of lamprophyre type, sometimes with elements of sideronite structure. Rocks of this type both externally and by their structure and composition are similar to

Fig. 2 External appearance of orange-red titanium pyrope megacrysts from kimberlites of the Nyurbinskaya pipe

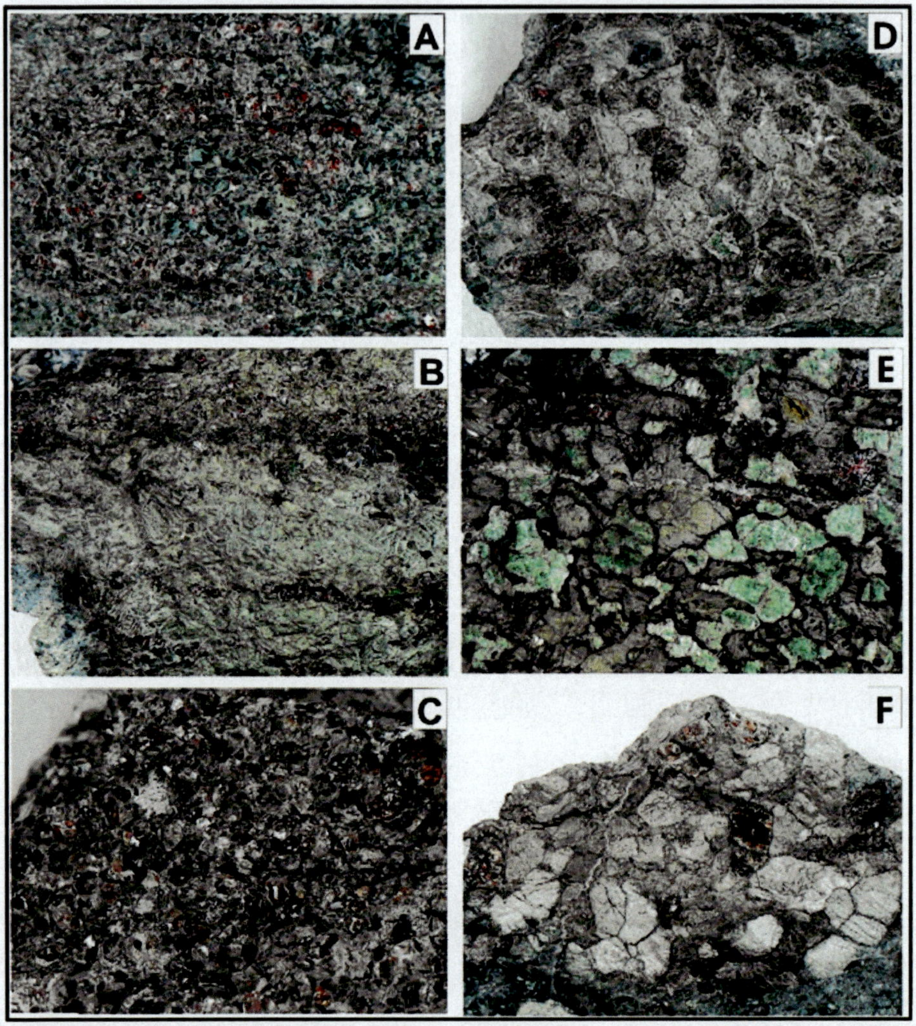

Fig. 3 Structural types of mantle xenoliths containing clinopyroxene in the form of omphacite and chromium-omphacite (A, B, C) and in the form of chrome-diopside (D, E, F). **A**, **B**—garnet peridotites with oriented directive structure; **C**—eclogites. **D**, **E**—garnet peridotites: **D**—protogranular (panidiomorphic) texture, **E**—allotriomorphic texture; **F**—titanium-garnet peridotites of the "megacryst" association of the panidiomorphic texture. Photo field 3 * 4 cm

MARID-type rocks and glymmerites with which they may have a genetic connection. Then, by analogy with the nature of the term "MARID" formation (Mica-Amphibole-Rutile-Ilmenite-Diopside), rocks of this type can be called MAOIte (Mica-Apatite-Olivine-Ilmenite).

The chemical composition of picroilmenite from the most different mantle rocks of the Nyurbinskaya pipe (garnet peridotites, garnet clinopyroxenites and eclogites)

is of the same type and equally unusual, that underlines the supposed genetic unity of these types of mantle rocks (Table 1). It is characterized by an increased content of TiO_2, MgO and Cr_2O_3, a very low content of hematite component, and in the classification diagram [50] (Fig. 4) its composition goes beyond the range of compositions of typical kimberlite pycroilmenites of megacrystic association. Small differences between ilmenites of garnet peridotites on the one hand, and garnet clinopyroxenites and eclogites on the other hand, are observed mainly in the increased content of chromium and magnesium in peridotites ilmenites (0.5–2.2% vs. 0.25–0.89% Cr_2O_3 and 11.9–15.3% vs. 9.98–13.75% MgO).

Picroilmenite from inclusions of protokimberlites of lamprophyre type (MAOItes) and MARIDS differs appreciably from picroilmenite from xenolites of garnet peridotite-pyroxenites by lower magnesium content (7.0–7.7% MgO), The extremely low content of chromium and aluminum (0–0.08% Cr_2O_3 and 0–0.10% Al_2O_3) and the increased content of manganese (0.85–0.95% MnO), but at the same time shows similarity to it in terms of high titanium content and low hematite component. Its composition also goes beyond the typical ilmenite from kimberlites and practically forms a single trend of changes in compositions with ilmenite from garnet peridotite and pyroxenites of the Nyurbinskaya pipe. This underscores to some extent the genetic unity of these very different types of rock inclusions.

Table 1 Chemical composition of picroilmenites

No*	TiO_2	Al_2O_3	Cr_2O_3	FeO	MnO	MgO	Total
1	58.52	0.49	1.38	24.60	0.40	14.55	99.94
2	57.82	0.65	1.32	25.78	0.28	13.85	99.70
3	57.92	0.21	1.38	25.54	0.31	14.24	99.60
4	59.24	0.08	1.30	23.39	0.42	15.30	99.73
5	56.90	0.51	1.46	27.29	0.64	12.66	99.46
6	57.43	0.20	0.65	29.74	0.34	11.58	99.94
7	58.71	0.51	0.42	25.81	0.59	13.80	99.84
8	57.16	0.42	1.38	28.13	0.52	11.92	99.53
9	57.77	0.35	1.78	26.37	0.55	13.21	100.03
10	58.80	0.23	0.69	27.27	0.29	12.88	100.16
11	56.54	0.49	0.72	29.04	0.33	12.31	99.43
12	57.73	0.29	0.63	28.13	0.28	12.55	99.61
13	58.00	0.34	0.61	28.30	0.32	12.71	100.28
14	56.78	0.50	0.50	31.15	0.16	10.83	99.92
15	52.13	0.08	0.03	38.93	0.87	7.20	99.24
16	53.26	0.02	0.06	37.95	0.95	7.23	99.47
17	51.91	0.05	0.00	39.34	0.90	7.05	99.25

Note No*: 1–5—garnet peridotite; 6–14—garnet clinopyroxenite; 15–17—lamprophyre type protokimberlites

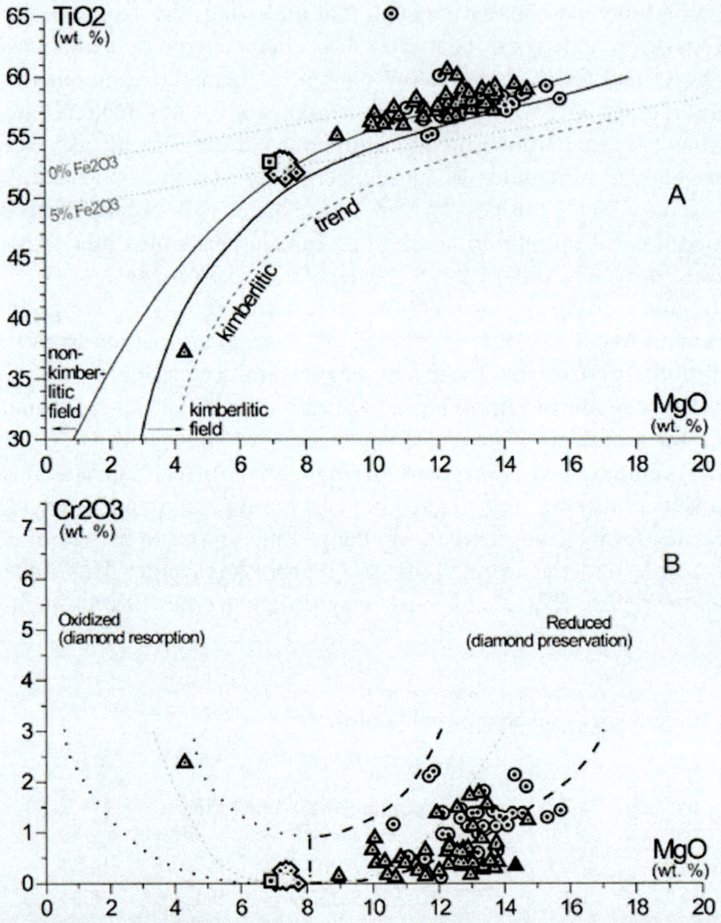

Fig. 4 Composition of picroilmenite from mantle xenoliths of the Nyurbinskaya pipe: diagram [50] (top); diagram [11, 12] (below). Legend: points of the compositions: circles—garnet peridotites; triangles—garnet clinopyroxenites and eclogites; rhombuses and squares are marids and protokimberlite of the lamprophyric-types

In contrast to the kimberlites of the pipe, in the placer pipe **Nyurbinskaya** found typical in composition picroilmenite megacrystic association. The primary root source of these picroilmenites is still unknown.

Single picroilmenites from inclusions of garnet peridotites of the Botuobinskaya pipe and the Mayskaya pipe fully correspond both in structural position and in chemical composition to picroilmenites from inclusions of garnet peridotites of the Nyurbinskaya pipe.

Wide spread of orange-red titanium pyrope of "megacrystic" association with complete absence of picroilmenite of "megacrystic" association, and also **originality of present picroilmenite composition** is one of the most interesting mineralogical features of kimberlites of Nakyn field pipes.

3.5 Orthopyroxene

Another bright mineralogical feature of kimberlites of Nakyn field is almost complete absence of **orthopyroxene** in kimberlites and mantle xenoliths, even in the form of pseudomorphoses. Usually less distinct pseudomorphoses on clinopyroxene are easily diagnosed in Nakyn mantle xenoliths, thus much more expressive and easily defined pseudomorphoses on orthopyroxene in mantle xenoliths of Nakyn are practically absent. From approximately 1000 mantle xenoliths studied only 4 xenoliths with orthopyroxene were found: 3 xenoliths of orthopyroxenite and 1 xenolith of websterite. Therefore, overwhelming part of peridotite of mantle inclusions from Nakyn kimberlites is "lerzolites" by peculiarities of garnet composition (by Sobolev [45]), and by modal mineral composition it is "verlites". Probably, the unusual increased content and variety of garnet composition of the verlite paragenesis in some pipes of the Nakyn field (see below) corresponds to it.

This unusual mineralogical feature of Nakyn kimberlites is **almost complete absence of orthopyroxene**, until it finds an explanation.

3.6 Clinopyroxene

Clinopyroxene secretions are widespread in kimberlites of the Nakyn field. Due to the very high degree of kimberlite change, fresh clinopyroxene in kimberlites is quite rare, but with depth, the proportion of its fresh secretions increases.

The overwhelming part of clinopyroxene grains from kimberlites and mantle xenoliths of Nakyn has typical bright emerald-green color (Fig. 5).

For clinopyroxene from kimberlites and different types of deep inclusions Nyurbinskaya pipe (from the overwhelming number of inclusions) are characterized by unexpectedly similar features of the composition (Table 2 and Figs. 6,7,8): high calcium content (Ca/(Ca + Mg) 47.6–50.6%), increased sodium content (2–5% Na_2O) and aluminum content (3.5–5.5% Al_2O_3), reduced iron content (1.3–2.4% FeO) and reduced magnesium content (12.6–15.0% MgO) with variable chromium content (0.07–4.5% Cr_2O_3). The composition of clinopyroxenes of only a few samples comes out of these narrow limits. "Tail" in high-sodium and high-alumina region, where compositions of large number of grains (Fig. 6) are located, is not characteristic for chrome-diopsides from kimberlites and mantle xenoliths of other regions.

Component analysis of minerals shows that the vast majority of chromium-containing clinopyroxenes in the Nyurbinskaya pipe are chrome-omfacites rather than chrome-diopsides as in other kimberlite regions. For

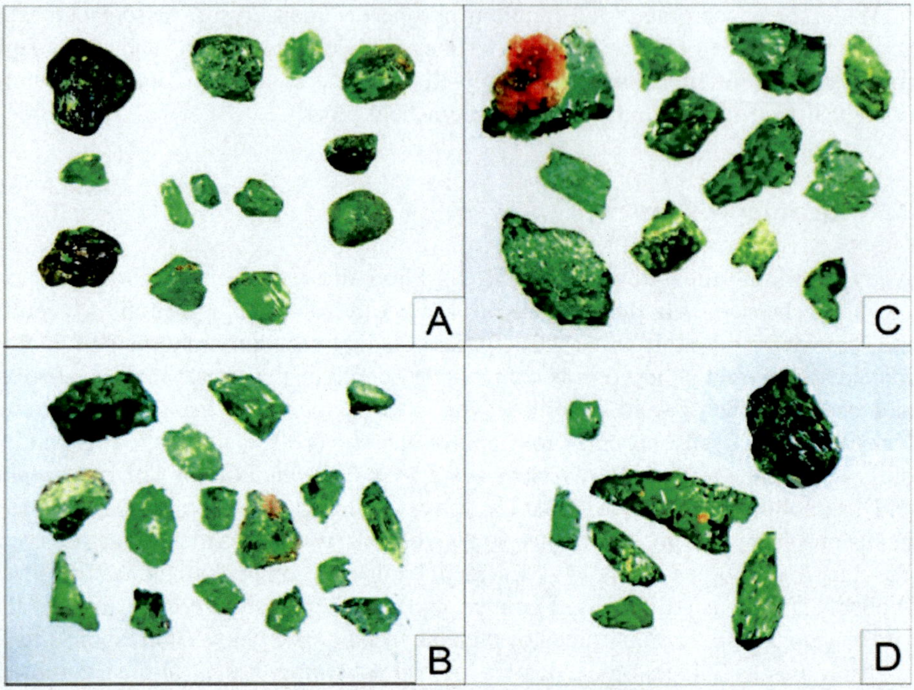

Fig. 5 External appearance of chromium omphacites: **A**—Nyurbinskaya pipe, **B**—Botuobinskaya pipe, **C** and **D**—Mayskaya pipe

identification of clinopyroxene species as omphacite, the following conditions must be met (all values—in the formulas) (according to [49]):

$0.15 < Na <= 0.75$; Al^{IV} 0.0–0.15; Fe^{3+} 0.0–0.12; $(Na\text{-}Fe^{3+})/Fe^{3+} > 1$; $Al^{IV}/(Al^{IV} + F^{e3+)} > 0.15$.

Based on these conditions, the proportion of omphacites among those selected to study chromium-containing clinopyroxenes in the Nyurbinskaya pipe and rocks of some other regions was estimated (Table 3). Thus, in the Zimny Bereg kimberlite region (Arkhangelsk diamondiferous province), the proportion of omphacites among clinopyroxenes is 3–4 times less than in the Nyurbinskaya pipe [29, 30]. Such a high share of omphacites in the Nyurbinskaya pipe is due to a large number of garnet peridotites with oriented, directive textures. It is very well seen at comparison of figures of clinopyroxenes composition for peridotites and eclogites-clinopyroxenites (Figs. 6 and 7).

Clinopyroxenes of garnet peridotite differ by unusual composition in the pipe. The discriminative diagram [47] (Fig. 7) shows that clinopyroxenes from garnet peridotites as a whole are less magnesial than clinopyroxenes from eclogites and garnet clinopyroxenites. This is a very unusual fact.

Table 2 Chemical composition of clinopyroxenes

No*	SiO_2	TiO_2	Al_2O_3	Cr_2O_3	FeO	MnO	MgO	CaO	Na_2O	K_2O	Total
1	54.71	0.37	5.02	2.21	1.55	0.06	13.69	18.35	3.31	0.00	99.27
2	54.00	0.34	4.48	1.66	1.67	0.05	14.09	19.23	3.19	0.00	98.71
3	53.81	0.21	4.02	2.93	1.42	0.04	13.67	19.25	3.15	0.00	98.50
4	54.01	0.42	4.66	1.61	2.08	0.08	13.56	18.91	2.84	0.01	98.18
5	54.18	0.36	4.32	1.71	1.63	0.05	14.50	19.72	3.12	0.00	99.59
6	54.27	0.27	4.22	1.86	2.50	0.03	13.32	19.29	3.25	0.00	99.01
7	54.84	0.40	4.87	1.92	1.66	0.03	13.77	18.98	3.55	0.00	100.02
8	54.68	0.30	3.61	1.48	2.10	0.08	14.47	20.39	2.49	0.00	99.60
9	54.75	0.36	4.73	0.91	1.75	0.06	13.91	19.56	2.77	0.00	98.80
11	54.49	0.39	4.81	0.94	1.67	0.07	14.17	19.78	2.88	0.01	99.21
13	54.42	0.36	4.49	0.55	1.67	0.04	14.50	20.78	2.49	0.00	99.30
12	54.87	0.34	3.52	0.22	1.69	0.11	16.00	20.24	2.02	0.02	99.03
10	55.01	0.39	4.29	1.13	2.14	0.03	13.95	20.04	3.27	0.00	100.25
14	54.17	0.33	4.51	0.09	2.75	0.09	13.83	20.71	2.30	0.02	98.80
15	54.59	0.30	3.97	0.18	1.95	0.02	15.51	20.39	2.18	0.00	99.09
16	53.40	0.40	4.72	0.17	1.96	0.02	15.01	20.51	2.61	0.00	98.80
17	54.98	0.19	3.94	0.39	5.06	0.06	14.47	17.75	2.15	0.34	99.33
18	55.32	0.05	0.86	1.35	2.06	0.05	17.81	21.44	1.05	0.12	100.11
19	54.71	0.04	1.28	2.63	1.79	0.10	16.09	21.09	1.74	0.03	99.50
20	54.72	0.18	1.82	1.79	2.27	0.05	16.30	20.09	1.77	0.06	99.05
21	54.24	0.10	1.30	1.69	2.52	0.11	16.41	20.90	1.56	0.07	98.90

Note No*: 1–8—garnet peridotite; 9–14—garnet clinopyroxenite; 15–16—eclogite; 18–21—garnet peridotite of protho-granular texture, massive structure

So, clinopyroxenes from sharply prevailing types of mantle xenolites of Nyurbinskaya pipe are omphacites: either simply omphacites (eclogites), or chromium-containing omphacites (garnet clinopyroxenites), or chrome-omphacites (according to [49]) (almost all garnet peridotite and chromium spinel peridotite). The difference between clinopyroxenes from peridotites and clinopyroxenites (including eclogites) is clearly shown only by chromium content (more than 1% and less than 1% Cr_2O_3, respectively).

On this general high sodium background are sharply distinguished and contain not omphacite but, as it should be, kimberlites—chrome-diopside, single mantle xenoliths of 4 types: (1) primary-magmatic garnet peridotite of protogranular, panidiomorphic and (partially) allotriomorphic structure; (2) garnetites chromic association; (3) garnet clinopyroxenites with red-orange garnets of titanium, "megacrystic" association; (4) metasomatic rocks of MARID series.

Unusuality of Nyurbinskaya tube mantle xenoliths clinopyroxenes composition is also clearly seen in diagram [25] (Fig. 8). The considerable part of clinopyroxenes from garnet peridotites and garnet clinopyroxenites with directive textures falls into the field "Off-craton garnet peridotites and spinel peridotites". In the field

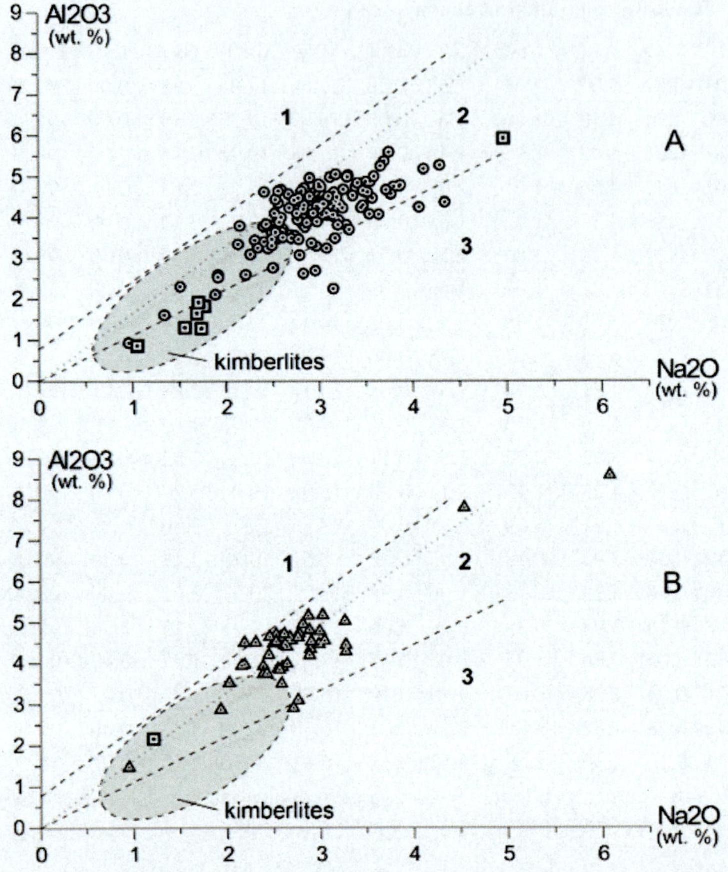

Fig. 6 Composition of clinopyroxene from mantle xenoliths of the Nyurbinskaya pipe (diagram [13, 26, 28]): **A**—From garnet peridotites with oriented structure (circles) and garnet peridotites of protogranular textures (squares); **B**—From garnet clinopyroxenites, eclogites (triangles), and marids (square)

"On-craton garnet peridotite" all chrome-diopsides from above mentioned 4 types of mantle xenoliths, and also chrome-omphacites from a part of garnet peridotite get.

In the **Botuobinskaya pipe**, the studied clinopyroxenes were divided into chrome-omfacites and chrome-diopsides in about half. The clinopyroxenes from the **Mayskaya pipe** are characterized by even higher levels of Al_2O_3 (3.1–6.2%), Na_2O (4–6.8%) and Cr_2O_3 (4–7.2%), lower FeO (1.7–2.3%) and MgO (less than 12%). They correspond to chromium-omfacites of the high-baric depth facies with an unusually high fraction of the ureyite (cosmochlore) component, according to which some points in their compositions go far beyond the usual discriminatory diagram [25].

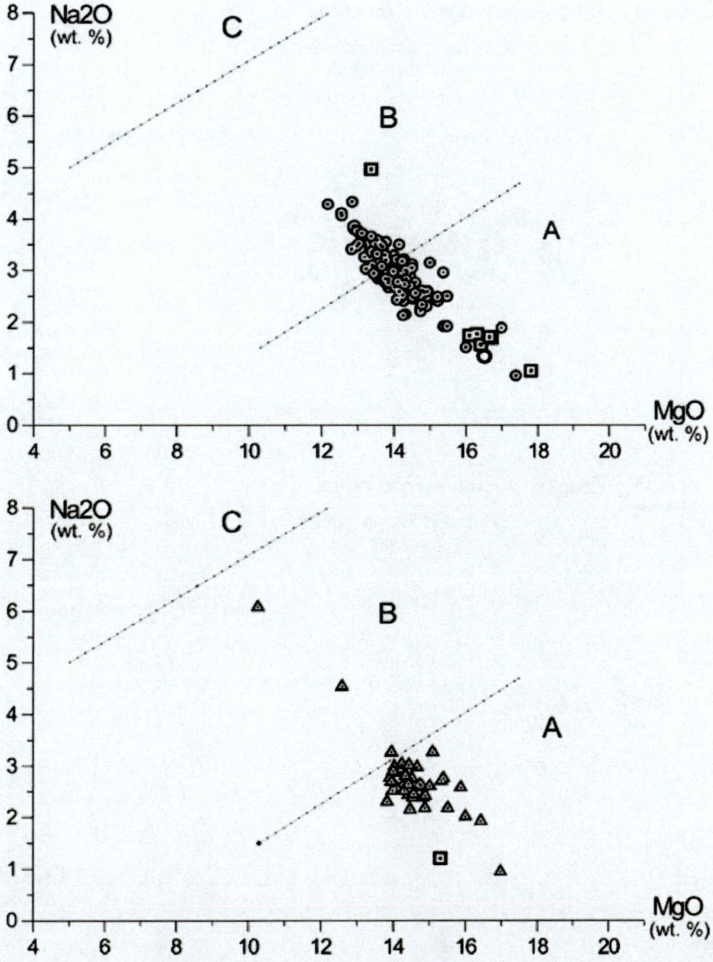

Fig. 7 Composition of clinopyroxene from mantle xenoliths of the Nyurbinskaya pipe (diagram [47]: top—from garnet peridotites with oriented structure (circles) and garnet peridotites of protogranular textures (squares); bottom—from garnet clinopyroxenites, eclogites (triangles) and marids (square). **A**, **B**, **C**—types of eclogites according to [7]

In general, the kimberlites of the Nyurbinskaya pipe (as well as those of Botuobinskaya and Mayskaya pipe) differ significantly from the kimberlites of other fields of the Yakutia province, kimberlites of the Arkhangelsk diamondiferous province, and other regions of the world in their unusual composition. Unusual composition of clinopyroxenes from kimberlites and mantle xenoliths, probably, is connected with unusual conditions of formation and structure of mantle substrate of Nakyn field.

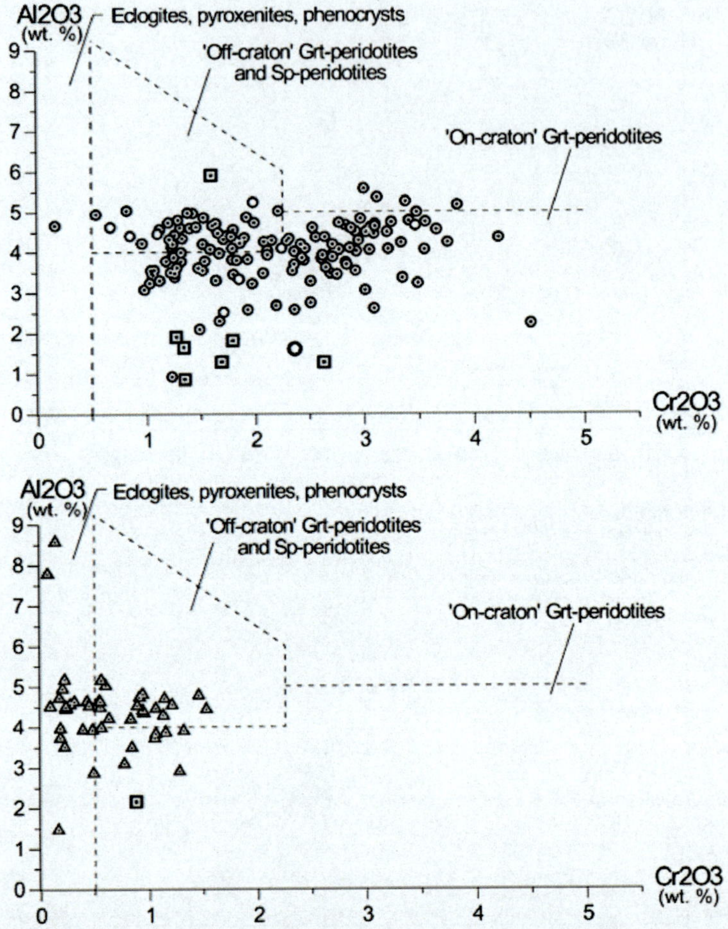

Fig. 8 Composition of clinopyroxene from mantle xenoliths of the Nyurbinskaya pipe (diagram [25] above—from garnet peridotites with oriented structure (circles) and garnet peridotites of protogranular textures (squares); below—from garnet clinopyroxenites, eclogites (triangles) and marids (squares)

3.7 Pink Garnets

Among the overwhelming mass of red-violet pyrope in kimberlites of the Nakyn field there are pale pink low chromium, low titanium pyrope, which correspond in composition to garnets from a rare type of mantle inclusions—alkremites [24]. From studied ~1000 mantle xenoliths only one alkremite xenolith was found in Nyurbinskaya pipe (Fig. 9) [36]. However, in kimberlites of the Nyurbinskaya pipe dyke a high content of garnets of this very type and composition is noted (Table 4), which may indicate a wider development of these rare rocks—alcremites in the mantle of the region.

Table 3 The ratio of omphacite and chromdiopside in different types of rocks

Objects	Quantity of grains			Share omphacite, %
	Total	Omphacite	Diopside	
Nyrbinskaya pipe kimberlites	58	54	4	**93**
Nyrbinskaya pipe mantle xenoliths	195	171	24	**88**
Nyrbinskaya pipe total	253	225	28	**89**
Arkhangelsk district, Al-series (deposit name Lomonosov)*	245	60	185	24
Arkhangelsk district, Fe–Ti-серия (deposit name Grib et al.)*	208	67	141	32
Arkhangelsk district, panned samples from alluvium	376	319	57	**85**

Note *Al-series, Fe–Ti-series of kimberlites rocks after [27].

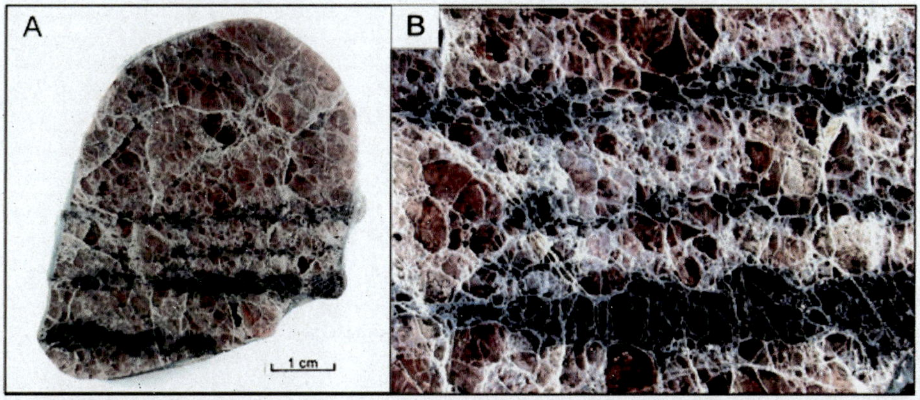

Fig. 9 External appearance of xenlite alkremit from the Nyurbinskaya pipe, sample no. H-678: **A**—General view; **B**—Layered structure of alcremite, polished surface, field 2.5 cm

3.8 Green Garnets

One of the most interesting mineralogical features of kimberlites in the Nakyn field is the presence of green garnets.

Green garnets (grossular-uvarovite-pyrope) are very rare in concentrates of diamondiferous kimberlites of Yakutia, South Africa, Venezuela and Canada [6, 14, 39]. For the first time green garnets in concentrates of kimberlites of Yakutia, xenoliths of grospydites and inclusions in diamonds are described by Sobolev [42–44]. Detailed study of green garnet ***xenoliths*** from kimberlites of dyke Newlands (South Africa) has shown that they are metasomatic formations of olivine-uvarovite-chromite veins type at moderate depths of upper mantle

Table 4 Chemical composition of pale pink and green color garnets

No*	SiO_2	TiO_2	Al_2O_3	Cr_2O_3	FeO	MnO	MgO	CaO	Na_2O	Total
1	41.63	0.07	23.72	0.16	7.83	0.22	18.77	7.51	0.00	99.91
2	41.99	0.07	23.78	0.11	7.48	0.29	20.76	4.8	0.02	99.30
3	42.01	0.12	23.82	0.12	7.38	0.35	20.79	4.92	0.03	99.54
4	41.82	0.31	22.75	0.61	7.15	0.41	21.92	3.83	0.08	98.88
5	38.25	0.34	11.79	13.85	7.72	0.45	10.61	15.80	0.03	98.84
6	38.22	0.30	13.96	10.56	7.43	0.42	11.15	16.14	0.04	98.22
7	37.55	1.33	10.24	14.50	5.94	0.45	10.74	17.30	0.04	98.09
8	37.08	1.77	9.41	14.86	5.47	0.38	7.96	22.26	0.30	99.49
9	37.99	1.57	10.42	13.74	5.68	0.48	8.13	21.49	0.03	99.53
10	37.75	1.50	10.05	14.12	5.80	0.39	7.92	21.61	0.06	99.20
11	37.25	1.64	9.24	14.93	5.68	0.52	7.89	22.26	0.24	99.65
12	37.52	1.59	9.35	14.88	5.50	0.34	8.05	22.06	0.00	99.29
13	37.71	1.66	10.82	12.30	6.92	0.43	7.37	22.07	0.00	99.28
14	38.46	1.52	10.84	12.85	5.77	0.44	7.99	21.27	0.03	99.17
15	38.96	0.44	13.91	9.46	5.94	0.36	5.32	25.34	0.03	99.75
16	38.82	0.58	15.87	7.56	6.13	0.38	7.14	22.18	0.06	98.72
17	39.15	0.63	15.98	7.64	6.08	0.44	7.26	22.37	0.05	99.59
18	40.06	0.59	17.67	6.78	7.38	0.51	11.44	15.43	0.10	99.95
19	40.28	0.51	17.02	7.61	6.87	0.45	11.75	16.06	0.04	100.59
20	40.13	0.38	17.96	6.67	7.33	0.53	12.16	14.88	0.07	100.09
21	39.38	0.10	17.59	6.66	6.23	0.48	11.33	17.01	0.04	98.80
22	39.30	0.11	17.73	6.35	6.26	0.45	11.60	16.89	0.02	98.70
23	40.15	0.07	17.97	6.86	6.33	0.40	10.98	17.02	0.03	99.79
24	40.28	0.14	17.92	6.84	6.23	0.45	11.05	17.03	0.04	100.00

Note No*—kimberlites pipes: 1—pipe Nyrbinskaya; 2–4—dyke Nyrbinskaya; 5–13—pipe Botuobinskaya; 14–24—pipe Mayskaya
No*—color: 1–4 pale pink; 5–24 green

(80–90 km) [35, 37]. A new type of "modal" mantle metasomatosis was identified —"calcium" metasomatosis (with varieties: *chromium-calcium* and *titanium chromium-calcium mantle metasomatosis*).

In a detailed study of kimberlites and placers of the Nakyn field, we first found green garnets in such a significant amount, which had previously been noted only in the kimberlites of dykes Newlands (South Africa) [6].

Green garnets were found by us in kimberlites of Botuobinskaya and Mayskaya pipes, as well as in the Nyurbinskaya placer, in the Nyurbinskaya pipe green garnets were not found [16].

Macrocrystals of green garnets up to 4–8 mm in size are found in the samples of kimberlites of Botuobinskaya and Mayskaya pipes ("in situ"), the grains of these garnets are strongly cracked and changed by the cracks, so in the ceiling samples they are usually presented as tens and hundreds of fragments of different sizes (from class −4 + 2 mm to class −0.25 mm). Rare garnets with chrome spinel are noted. In some samples from kimberlites of the Botuobinskaya pipe content of green garnets is from 0.2–0.4 to 14% of the heavy fraction.

Green garnets of the Nakyn field are diverse in color shades (Figs. 10 and 11): there are grains of green, emerald-green, gray-green, swampy and bluish-green to blue color, the most diverse in color is green garnets from the Mayskaya pipe (Fig. 11). They are characterized by a rather wide range of composition (Table 4 and Fig. 12) in terms of CaO (10–26%), Cr_2O_3 (4–15%), TiO_2 (0.1–1.86%), FeO (5.5–9%) content and belong to uvarovit-grossulars and uvarovit-pyrops [8]. Of microelements, they are characterized by an increased content of LREE, etc., but especially Zr—up to 652–1042 ppm [37].

Fig. 10 External appearance of green garnet monofractions from kimberlites of the Botuobinskaya pipe: **A**—sample with one-color garnets; **B**—sample with multi-colored garnets. Size class −1 + 0.25 mm

Fig. 11 External appearance of green garnet monofractions from kimberlites of the Maiskaya pipe: **A**—size class −4 + 1 mm; **B**–**E**—monofractions of color varieties of garnet, size class −1 + 0.25 mm

For the grains of green garnets of various shades, some features of the chemical composition are outlined. For example, in the tube May, garnets of *emerald-green* color are characterized by high content of CaO (23–25.5%), Cr_2O_3 (9–11.0%) and reduced content of TiO_2 (0.3–0.6%) and FeO (5.5–6.5%), while *bottle-green* garnets are characterized by lower CaO (15–16%), Cr_2O_3 (6–8%) and TiO_2 (0.3–0.6%) content and higher FeO (6.0–8.0%) content. Differences in characteristics of green garnets are also observed on the level of separate tubes of Nakyn field. As for

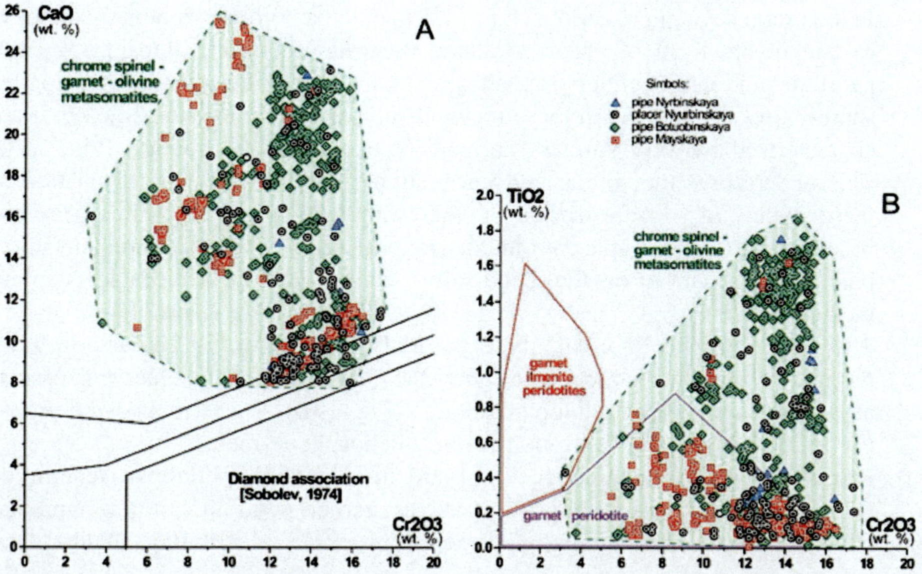

Fig. 12 Composition of garnet of blue-green color scheme from the kimberlites of the Nakyn field. **A**—diagram [45]; **B**—diagram [38]

composition, green garnets of the Mayskay pipe as a whole noticeably differ from green garnets of the Botuobinskaya pipe in reduced content of chromium and titanium. And if to take into account that in a pipe Nyurbinskaya green garnets are not found at all, it is possible to consider green garnets as one of the basic typomorphic minerals-indicator for kimberlites of Nakyn field [38], on which features separate pipes differ both on quantitative level (on a principle "more—less"), and on qualitative level (on a principle "is—no").

Such a high content of green garnets, as in the pipes Mayskaya and Botuobiskaya, is typical only for kimberlites dyke Newlands (South Africa). But such quantity and ***diversity*** in color and chemical composition of green garnets (and garnets of verlite paragenesis in general) as in the Botuobinskaya and Mayskaya pipes is not in any other pipes of the world.

Green garnets are the brightest, typomorphic mineralogical markers of kimberlites in the Nakyn field.

4 Distribution of Trace Elements in Garnets

For garnets from main types of mantle xenoliths the content of microelements (micro impurities) was determined by LAICP-MS method. It allowed specifying some peculiarities of their origin.

1. In Nyurbinskaya pipe various types of mantle xenoliths containing clinopyroxene in the form of chrom-omfacite (instead of chrome-diopside)—garnet peridotite and garnet clinopyroxenites with directive textures, and also eclogites sharply prevail. Garnets in these rocks also have different composition, but are characterized by very similar "normal" type of REE distribution (Fig. 13A). This underscores the similar and fairly simple conditions for the formation of these types of mantle rocks in the composition of the initial layered basite-hyperbasite complexes (and/or oceanic crust), and only their metamorphic transformation in the subduction zone (without noticeable, explicit metasomatosis).
2. The same simple, "normal" type of REE distribution is characterized by red-orange titanium garnets of megacrystals and rocks of "megacrystic association". (titanium-garnet clinopyroxenites) (Fig. 13B). These rocks were probably formed during crystallization from asthenospheric melts.
3. Rocks with primary characteristics of structure containing clinopyroxene in the form of "usual" chrom-diopside, such as garnet peridotite of protogranular structure have "protruding" character of REE distribution with clear

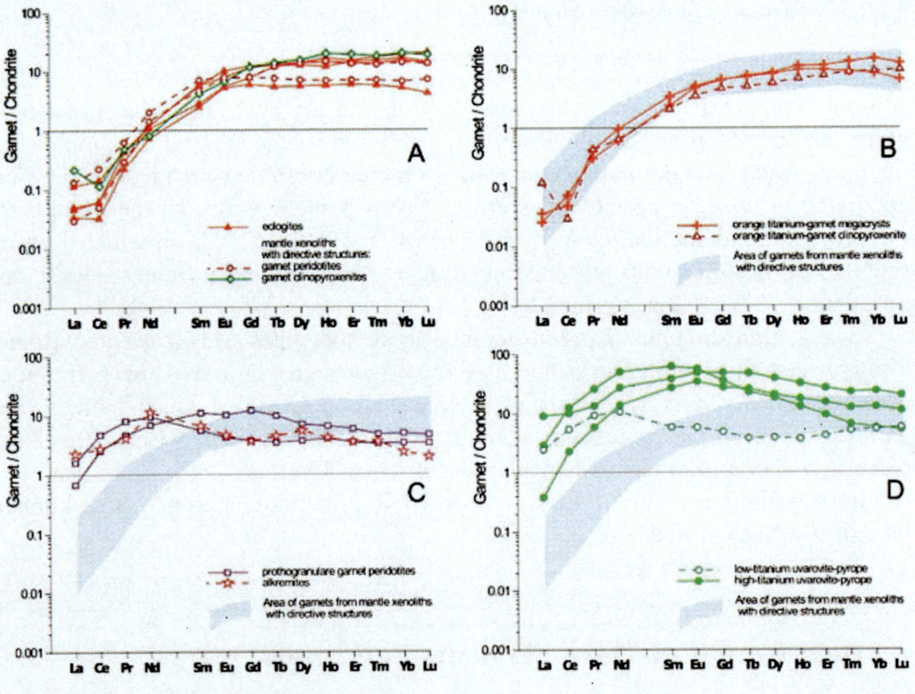

Fig. 13 Chondrite-normalized REE distribution in garnets from mantle xenoliths of the Nyurbinskaya pipe. **A**—garnet peridotites and garnet clinopyroxenites with directive structure, eclogites; **B**—orange titanium-garnet megacrysts and titanium-garnet clinopyroxenites; **C**—garnet peridotites of protogranular textures and alcremites; **D**—uvarovite-pyropes [35, 37]

metasomatic enrichment in the part of light REE (Fig. 13C). Alcremite garnets also have approximately the same "settled" nature of REE distribution.

4. The most abnormal character of REE distribution is fixed in green garnets of uvarovite metasomatites (Fig. 13D). And if low titanium uvarovite by the nature of REE distribution ("protruding", with clear enrichment by light REE) is similar to protogranular garnet peridotites, then high titanium uvarovites sharply differ from all other studied garnets by "raised" type of REE in the region of medium and light REE. Of other microelements, they are characterized by an increased content of Y, etc., but especially Zr—up to 652–1042 ppm [37]. These features fix intensive influence of metasomatic processes on rocks containing these garnets.

5 TP-Parameters of Nodule Formation

The available set of deep minerals and mineral paragenesis is extremely limited by the high degree of change in Nakyn kimberlites. Only garnet and clinopyroxene are preserved from substitution. These are far from the best minerals for reliable determination of TP-parameters of mantle rock formation.

In total, more than 20 geothermometers and geobarometers were used and the number of their combinations could not be taken into account. As a result, any intelligible values of TP-parameters were obtained in the calculation of 6 monomeral thermometers and barometers (gar, cpx) and a single bimineral geothermobarometer (gar + cpx).

The use of the monomeral clinopyroxene thermobarometer NT00T—NT00P [21] has shown that the points of the calculated parameters, though forming a rather compact field, are characterized by clearly underestimated pressure values (and to a large extent unrealistic values of temperature), which can hardly be considered close to reality. Perhaps, this is due to the unusual composition of both garnet and clinopyroxene, as mentioned above.

For some exotic mantle rock samples (alcremitic and uvarovite metasomatites) the TP-parameters of formation were calculated using a thermometer C99T [4] and a geo barometer GLM06P [10].

The most comprehensible and, probably, the closest to the real picture of the location of points of TR-parameters of mantle xenoliths formation is given by the mineral paragenesis most widespread in different types of nodules: clinopyroxene + garnet (Cpx + Ga). The temperature was determined using NS94T geothermometer, pressure—using NS94P barometer [20]. The small dispersion of the obtained results and their consistency with petrographic features of xenoliths confirm the maximum possible correctness of application of this method in such difficult conditions (Fig. 14).

The overwhelming majority of various garnet peridotites and garnet clinopyroxenites with derective textures, and also eclogites on sizes of TP-parameters form the uniform, very compact field with parameters 800–950 °C and 25–37 kbar,

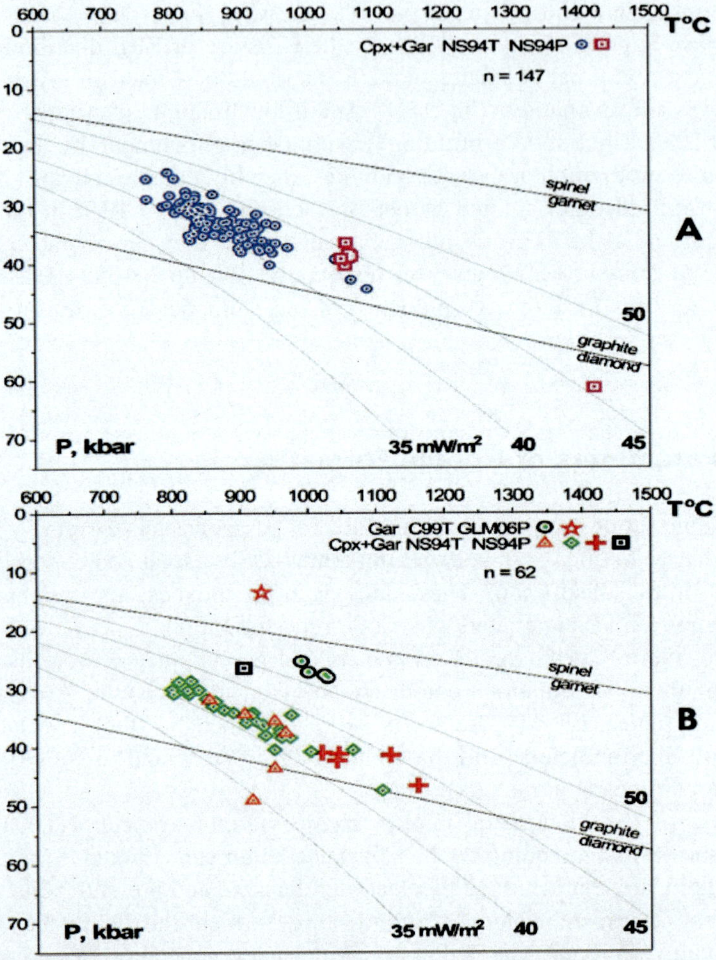

Fig. 14 TP-formation conditions of mantle xenoliths of the Nyurbinskaya pipe. **A**—garnet peridotites of protogranular (panidiomorphic) and allotriomorphic texture of massive structure (squares), garnet peridotites of oriented directive structure (circles); **B**—eclogites (triangle), garnet clinopyroxenites of oriented, directive structure (rhombuses), titanium-garnet clinopyroxenites of the "megacryst" association (crosses), alcremites (star), marids (square), chromite-uvarovite metasomatites (circle)

obviously oriented along conductive geotherm approximately 41 mW/m^2 [23]. Only the so-called *primary magmatic* garnet peridotites of protogranular, panidiomorphic and allotriomorphic structure (1000–1070 °C and 35–40 kbar) are sharply distinguished by higher temperature and high barometric conditions of formation (Fig. 14A) and garnet clinopyroxenites of iron-titanium series (900–1150 °C and 33–47 kbar) (Fig. 14B) with clinopyroxene-chromdiopside, but these rocks do not enter the zone of diamond paragenesis. Only 2 samples of very unusual, exotic rocks correspond to the zone of diamond paragenesis: primary-magmatic garnet peridotite of allotriomorphic structure (with garzburgite pyrope) and eclogite of "I B" group, which contain the most high sodium and high alumina omphacites. Two more samples of eclogite and clinopyroxenite also correspond to the region of diamond stability, but they are located practically at the boundary of graphite-diamond transition. The other nodules correspond to conditions of grospidite (C_2) and coenzytic (C_3) subfacion of graphite-pyrope facies of C depth [45]. Such exotic types of inclusions as marids and chromite-uvarovite metasomatites are characterized by lower formation pressures (24–27 kbar and 900–1030 °C), and alkremites proved to be the lowest-baric of all investigated samples (13.6 kbar and 930 °C).

As a whole, character of change of parameters of formation of studied nodules and peridotites and eclogites of different types well corresponds to conductive geotherm 41 mW/m^2 [23].

6 Discussion

Present in kimberlites of Nakyn field deep mineralogical material (mantle xenoliths, megacrysts and xenocrysts), its large number, variety and originality of composition emphasize the complex structure and character of diverse processes of formation of mantle substrate of Nakyn field.

The apparent contradiction between the signs of the impact of asthenospheric diapir (plume) on the mantle substrate of the Nakyn field (Nd-Sr isotopic characteristics of kimberlites, approaching BSE) and the absence in kimberlites of the main typical mineralogical feature of such impact (megacrystals of picroilmenite), is explained by the mineralogical features of kimberlites of the Nakyn field. The presence in kimberlites of typical megacrystals of orange-red titanium pyrope and inclusions of garnet peridotite-clinopyroxenites with the same pyrope of megacrystic association is in this case a mineralogical sign of the process of asthenospheric impact on the mantle substrate of the Nakyn field. This process was clearly reduced - either because of the unusual material composition of the asthenospheric diapyre in this area, or because of the location of the Nakyn field in the "marginal" zone of exposure to such a diapyre—similar to the same "marginal" location of kimberlites of the Lomonosov deposit, Zimny Bereg, Arkhangelsk province [31].

Unusual composition of picroilmenite, which is present in xenoliths of mantle rocks of Nakyn field kimberlites, is caused by development of deep processes, not related to the influence of asthenospheric diapir. Picroilmenite in kimberlites of Nakyn is present in inclusions of mantle rocks of definite type, namely in eclogites, and also in garnet peridotites and clinopyroxenites (often zonal) with medium-fine grain structure and oriented, directive, layered textures. In the same types of mantle rocks, the composition of clinopyroxene—omphacite, chrome-omphacite (but not chrom-diopside)—is unusual. For these same rocks is not uncommon the presence of a number of grains of pale purple and orange-pink pyrope with a difference in the content of chromium up to 2 times, which emphasizes the nonequilibrium nature of the mineral paragenesis of rocks. And, at last, the character of REE distribution in garnets from this type of mantle rocks is “normal”, which does not allow speaking about any serious influence of mantle metasomatosis on them. Peculiarities of structure, mineral composition and mineral composition of inclusions of eclogites, garnet peridotite and clinopyroxenites with oriented textures allow to assume an important participation in their formation by subduction processes [34]. The important role of subduction processes in the formation of mantle substrate in the Nakyn field is also noted for diamondiferous eclogites and pyroxenites [46]. As for similar composition of picroilmenite from metasomatic mantle rocks inclusions, these rocks were formed due to melting of mantle substrate of eclogite-peridotite-pyroxenite (mentioned above), so their minerals inherit some characteristic features of mineral composition of initial rocks (in particular—picroilmenite).

The development of omphacite in eclogites, garnet peridotite and clinopyroxenite with oriented textures is probably due to the fact that these rocks were formed not as a result of crystallization of mantle smelting, but have, in fact, metamorphic origin (in the subduction zone) and their mineral composition and composition reflect the material composition of the original rocks of layered basite-hyrbasite complexes and/or oceanic crust.

In contrast, the rocks formed as a result of crystallization of mantle melts—garnet peridotites of protogranular and allotriomorphic structure, garnet chromium association and garnet clinopyroxenites of iron-titanium (“megacrystic”) series contain “usual” for kimberlites clinopyroxene–chrom-diopside.

As a result of crystallization of mantle smelting, rocks of alcremitic type—parallel-banded coarse-crystalline plutonic rocks with elements of magmatic separation (liquation?) into garnet and spinel layers—were formed. The pink alcremite pyrope, widely distributed in the kimberlites of the Nyurbinskaya pipe dyke, indicates a possible broader development of these rare rocks in the Nakyn field mantle.

One of the most interesting features of Nakyn kimberlites is the presence of green grenades in them. Such quantity and variety in color and chemical composition of green garnets (and garnets of werlite paragenesis as a whole) as in pipes Botuobinskaya and Mayskaya is not in any other tube of the world, The unusually wide development of green garnets in kimberlites of the Nakyn field indicates an intensive impact of the processes of “*calcium*” (*chromium-calcium* and *titanium-chromium-calcium*) mantle metasomatosis on rocks of mantle substrate [35, 37]. Green garnets with such a high content of titanium (up to 1.86% TiO_2) were found in kimberlites for the first time. It is possible that the increase in the content of incompatible elements (LREE, Ti, Y, Zr, etc.) in high chromium garnets and in high calcium high chromium garnets (uvarovite-pyrophies) is also due to the influence of titanium-chromium-calcium metasomatic processes of “diffusion” type on “primary” mantle rocks of protogranular structure.

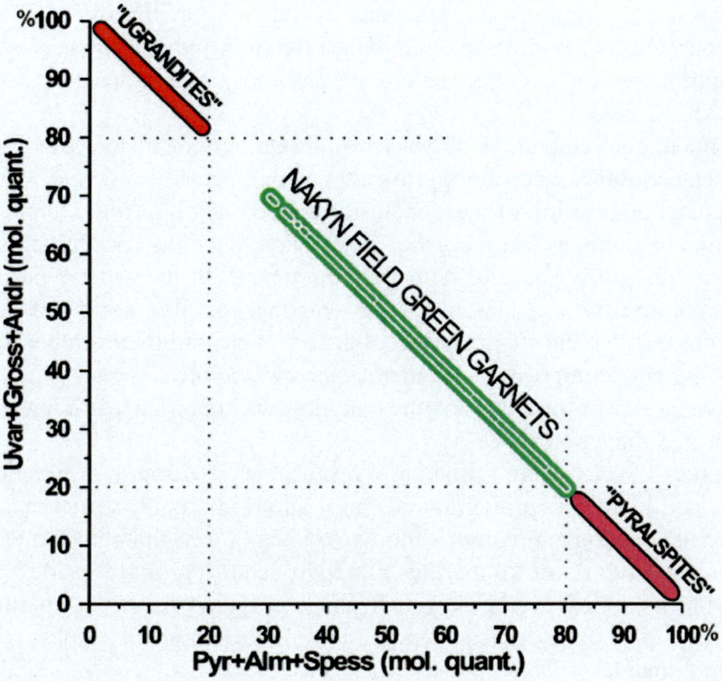

Fig. 15 Position of the compositions of green garnets from the kimberlites of the Nakyn field in relation to two main groups of garnets: pyralspites and ugrandites

The compositions of various green (blue-green) garnets and other garnets of werlite paragenesis from Nakyn kimberlites form a kind of continuous "bridge" between two rows of the most widespread, contrasting and usually not mixed types of garnets—"pyralspites" and "ugrandites" (Fig. 15), which underlines the unusual conditions of green garnet formation in Nakyn field. (Previously it was noted only for single green garnets from different regions of the world [22]).

7 Conclusion

Detailed study of kimberlites and mantle xenoliths of Nakyn field pipes (Nyurbinskaya, Botuobinskaya and Mayskaya pipes) revealed their unusual, interesting and important mineralogical features.

Set of basic, predominant deep minerals in kimberlites and mantle xenolites of Nakyn field is typical for kimberlites of group-2 of South Africa—olivine, red-violet pyrope and chrome spinel. Orthopyroxene is practically absent, which does not find explanation yet.

The absence of megacrist picroilmenite in Nakyn kimberlites is compensated by the presence of large nodules of orange-red titanium pyrope "megacrist" association, emphasizing the reduced nature of asthenospheric impact on the mantle substrate of the area.

Picroilmenite in kimberlites of Nakyn is present only inclusions of eclogites, and also garnet peridotites and clinopyroxenites with oriented, directive textures. The picroilmenite composition of these inclusions differs sharply from the picroilmenite composition of a typical "megacrystic" association by an increased titanium content and an extremely low share of hematite component. In the same types of mantle rocks, the composition of clinopyroxene—omphacite, chrome-omphacite (but not chrom-diopside)—is unusual. The peculiarities of the structure, mineral composition and mineral composition of inclusions of eclogites, garnet peridotites and clinopyroxenites with oriented textures suggest an important participation in their formation of subduction processes.

Depth rocks, which were formed as a result of crystallization of mantles melting —garnet peridotite of protogranular and allotriomorphic structure, garnetites chromium association and garnet clinopyroxenites of iron-titanium ("megacrystic") series contain "usual" for kimberlites clinopyroxene—chrom-diopside.

The pink alcremitic pyropes, which are widespread in the kimberlites of the Nyurbinskaya pipe dyke, indicate a possible wide development of these rocks in the Nakyn field mantle.

Unusually wide development in kimberlites of Nakyn field of various color and composition of green garnets indicates an intensive impact of the processes of "calcium" (*chromium-calcium* and *titanium-chromium-calcium*) mantle metasomatosis on rocks of mantle substrate. Green garnet is an ideal, easily diagnosable search mineralogical marker.

As a whole, the conducted works allowed to reveal uniqueness of pipes of Nakyn field on mineralogical features of kimberlites and mantle xenoliths, both among other pipes of Yakutia Diamond-bearing Province, and among kimberlite pipes of the World.

References

1. Bobrievich, A.P., Ilupin, I.P., Kozlov, I.T., and others: Petrography and Mineralogy of Kimberlite Rocks of Yakutia, M., Nedra (1964) (in Russian)
2. Bogatikov, O.A., Kononova, V.A., Golubeva, Yu., Zinchuk, N.N., Ilupin, I.P., Rotman, A. Ya., Levsky, L.K., Ovchinnikova G.V., Kondrashov I.A.: Petrogeochemical and isotopic variations of composition of kimberlites of Yakutia and their causes. Geochemistry (9), 915–939 (2004) (in Russian)
3. Brachfogel, F.F.: Geological aspects of kimberlite magmatism of the North-East of the Siberian platform. Yakutsk, 128 p. (1984) (in Russian)
4. Canil, D.: The Ni-in-garnet geothermometer: calibration at natural abundances. Contrib. Miner. Petrol. **136**, 240–246 (1999)
5. Cherny, S.D., Fomin, A.S., Yanygin, Yu.T., Banzeruk, V.I, Kornilova, V.P.: Geology and composition of the Nakyn field kimberlite pipes and diamond properties (Yakutia)/7-th IKC. Cape Town. Extended Abstracts, pp. 9–10 (1998)

6. Clarke, D.B., Carswell, D.A.: Green garnets from the Newlands kimberlite, Cape Province, South Africa. Earth Planet. Sci. Lett. **34**, 30–38 (1977)
7. Coleman, R.G., Lee, D.E., Beatty, L.B., Brannock, W.W.: Eclogites and eclogites: their differences and similarities. Bull. Geol. Soc. Amer. **76**(3), 483–508 (1965)
8. Dawson, J.B., Stephens, W.E.: Statistical analysis of garnets from kimberlites and associated xenoliths. J. Geol. **83**, 589–607 (1975)
9. Golubeva, Yu., Pervov, V.A., Kononova, V.A.: Low titanium diamondiferous kimberlites—new petrogeochemical type. Comparison of Nakyn (Yakutia) and Zolotitsky (Arkhangelsk Region) varieties. Region Geol Metal (27), 15–25 (2006) (in Russian)
10. Grutter, H., Latti, D., Menzies, A.: Cr-saturation arrays in concentrate garnet compositions from kimberlite and their use in Mantle Barometry. J. Petrol. 1–20 (2006)
11. Gurney, J.J., Moore, R.O.: Kimberlite garnet, chromite and ilmenite compositions, applications to exploration. In: Sheahan, P., Chater, A. (eds.) Diamonds: Exploration, Sampling and Evaluation, pp. 109–146. Prospectors and Developers Association of Canada, Toronto (1991)
12. Haggerty, S.E.: The chemistry and genesis of opaque minerals in kimberlites, pp. 295–307. In: Ahrens, L.H., Dawson, J.B., Dunkan, J.B., Erlank, A.J. (eds.). Physics and Chemistry of the Earth, vol. 9. Pergamon Press, Oxford, 940 pp. (1975)
13. Ilupin, I.P.: Findings of "kimberlite" minerals in nonkimberlite erupted rocks. In: Proceedings of TsNIGRI, vol. 188, pp. 46–51 (1984) (in Russian)
14. Kharkiv, A.D.: Finding of ore wherlite xenolithe in kimberlite pipe "Zimnyaya" (Verkhne-Munskiy diamond-bearing region, Yakutia). Reports of USSR Academy of Sciences. Volume 239, 1, pp. 181–184 (1978) (in Russian)
15. Kornilova, V.P., Fomin, A.S., Zaitsev, A.I.: A new type of diamondiferous kimberlite rocks on the Siberian platform. Region Geol Metal (13–14), 105–117 (2001) (in Russian)
16. Kornilova, V.P., Rogova, V.V., Ivanov A.S.: Paragenetic associations of garnets from indigenous bodies and panned halos of the Nakyn kimberlite field. In: Problems of Forecasting and Prospecting of Diamond Deposits in Closed Territories: Materials of the Conference devoted to the 40th Anniversary of YANIGP TSNIGRI JSC "ALROSA".—Yakutsk: YANTS SO RAS Publishing House, pp. 183–190 (2008) (in Russian)
17. Maltsev, M.V., Tolstov, A.V., Fomin, V.M., Starkova, T.S.: New kimberlite field in Yakutia and typomorphic features of its minerals-indicators. Vestnik VSU. Series: Geol. (3), 86–94 (2016) (in Russian)
18. Marakushev, A.A.: Peridotite nodules in kimberlites and basalts as indicators of the deep structure of the lithosphere In: 27 IYC. Petrology. Section P.09.Reports. T.9. Moscow: Science, pp. 153–161 (1984) (in Russian)
19. Mahotkin, I.L., Zhuravlev, D.Z.: Isotopic composition of Sr- and Nd diamondiferous kimberlites and melilitites of the Arkhangelsk region. DAN SSSR, vol. 332, no. 4, pp. 491–495 (1993) (in Russian)
20. Nikitina, L.P., Simakov, S.K.: TERRA nova. In: Fifth International EMPG Symposium, Vol. 6, p. 4 (1994)
21. Nimis, P., Taylor, W.R.: Single clinopyroxene thermobarometry for garnet peridotites. Part 1. Calibration and testing of a Cr-in-Cpx barometer and an enstatite-in-Cpx thermometer. Contrib. Mineral. Petrol. **139**, 541–554 (2000)
22. Platonov, A.N., Matsyuk, S.S., Simonenko, K.V.: Crystallochemistry and spectroscopy of Cr3 + ions in green garnets from mantle wherlites. Mineralogichniy zhurnal **T. 29**(3), 25–38 (2007) (in Russian)
23. Pollack, H.N., Chapman, D.S.: On the regional variation of heat flow, geotherms, and lithospheric thickness. Tectonophysics **38**, 279–296 (1977)
24. Ponomarenko, A.I.: Alkremites—a new type of alumina hyperbasites (xenolites from kimberlite pipe "Udachnaya"). Reports of AS USSR. T. 225(4), 928–931 (1975) (in Russian)
25. Ramsay, R.R.: Geochemistry of diamond indicator minerals. Ph.D. Thesis, University of Western Australia, Perth, Australia (1992)

26. Sablukov, S.M., Budkina, L.I.: Deep inclusions and peculiarities of the chemical composition of the coexisting minerals. Proc. of Intern. TSNIGRI. Episode. **229**, 16–23 (1988) (in Russian)
27. Sablukov, S.M.: About petrochemical series of kimberlite rocks. Reports of the USSR Academy of Sciences, T.313(4), 935–939 (1990) (in Russian)
28. Sablukov, S.M., Sablukova, L.I., Shavirina, M.V.: Mantle xenoliths from kimberlite deposits of round diamonds of Zimneberezhny district (Arkhangelsk diamondiferous province). Petrology, T.2(5), 525–554 (2000) (in Russian)
29. Sablukov, S.M., Sablukova, L.I., Verichev, E.M.: Types of mantle substrate of Zimneberezhny district in connection with the formation of kimberlites with rounded and flat-sided diamonds (Arkhangelsk diamondiferous province). In: Proceedings of the II International Workshop "Deep Magmatism, Magmatic Sources and the Plume Problem" Irkutsk-Vladivostok, IrSTU Publishing House, pp. 215–234 (2002)
30. Sablukova, L.I., Sablukov, S.M., Verichev, E.M., Golovin, N.N.: Petrography and chemistry of minerals of mantle xenoliths and xenocrists from kimberlites of a tube named after V. Grib, Zimny Bereg, Russia. In: Proceedings of III International Seminar "Plumes and the problem of deep sources of alkaline magmatism" (Khabarovsk). Irkutsk, IrSTU Publishing House, pp. 65–95 (2003)
31. Sablukov, S.M., Sablukova, L.I.: Degree of asthenospheric influence as one of the determining factors of kimberlite rock diversity. In: Proceedings of the VI International Workshop "Deep Magmatism, its Sources and Plumes" Irkutsk-Mirny, Institute of Geography SB RAS, pp. 145–158 (2006)
32. Sablukov, S.M., Banzeruk, V.I., Sablukova, L.I., Stegnitsky, Yu.B., Bogomolov, E.S., Lebedev, V.A.: Ancient age of kimberlites of Nakyn field (Yakutia) is one of the reasons of their material composition originality. In: VIII International Conference "New ideas in the Earth sciences". Moscow, Russia. Reports, T. 5. pp. 209–212 (2007) (in Russian)
33. Sablukov, S.M., Sablukova, L.I., Stegnitsky, Yu.B., Banzeruk, V.I.: Litospheric mantle of Nakyn kimberlite field (Yakutia) by mantle studies of mantle xenoliths and basalts of Nyurbinskaya tube. In: Proceedings of VII International Seminar "Alkaline magmatism, its sources and plumes" Irkutsk-Naples, Institute of Geography of SB RAS, pp. 140–156 (2007)
34. Sablukov, S.M., Sablukova, L.I., Stegnitsky, Yu.B., Karpenko, M.A., Spivakov S.V.: Volcanic rocks of the Nyurbinskaya tube: reflection of the upper mantle of the region from reef to carbon, geodynamic effects. In: Proceedings of VIII International Seminar "Deep Magmatism, its Sources and Plumes" Vladivostok, Institute of Geography of SB RAS, pp. 71–103 (2008)
35. Sablukova, L.I., Sablukov, S.M., Stegnitsky, Yu.B., Karpenko, M.A.: Green garnets of the "wehrlite" mantle xenoliths from kimberlites: their composition and origin (Newlands dike, South Africa; Nyurbinskaya pipe, Nakyn field, Yakutia) In: 10th IKC, India, Bangalore. Extended Abstracts Reg. No. 10IKC-142. Pp. 1–5 (2012)
36. Sablukov, S.M., Sablukova, L.I., Stegnitsky, Yu.B., Karpenko, M.A.: Banded alkremite xenoliths from Nyurbinskaya kimberlite pipes (Nakyn field, Yakutia) In: 10th IKC. India, Bangalore, Extended Abstracts Reg. No. 10IKC-140, pp. 1–5 (2012)
37. Sablukov, S.M., Sablukova, L.I., Stegnitsky, Yu.B., Karpenko, M.A.: Origin of mantle xenoliths with green garnets from kimberlites (Newlands dyke, South Africa and Nyurbinskaya tube, Yakutia) In: Proceedings of XIII International Seminar "Deep magmatism, its sources and plumes". Irkutsk, pp. 178–202 (2013)
38. Sablukova, L.I., Sablukov, S.M., Stegnitsky, Yu.B., Karpenko, M.A.: Use of unusual mineralogical features of the kimberlites of the Nakyn field for identification of the indigenous sources of placid halos In: Proceedings of the IV Regional Scientific and Practical Conference of ALROSA. Peaceful, pp. 187–190 (2014) (in Russian)
39. Schulze, D.J.: Green garnets from South Afrikan kimberlites and their relationship to wehrlites and crustal uwarowites. In: Proceeding 4th IKC, Pert, Vol. 2 pp. 820–826 (1986)

40. Smelov, A.P., Andreev, A.P., Altukhova, Z.A., Babushkina, S.A., Bekrenev, K.A., Zaitsev, A.I., Izbekov, E.D., Queen, O.V., Mishnin, V.M., Okrugin, A.V., Oleynikov, O.B., Surnin, A. A.: Kimberlites of Manchara pipe: new Kimberlite field of Central Yakutia. Geol. Geophys. **51**(1), 153–159 (2010) (in Russian)
41. Smith, C.B., Gurney, J.J., Skinner, E.M.W., Clement, C.R., Ebrahim, N.: Geochemical character of Southern African kimberlites. A new approach based on isotopic—contents. Trans. Geol. Soc. S. Afr. **88**, pp. 267–280 (1985)
42. Sobolev, N.V., Jr., Kuznetsova, I.K., Zyuzin, N.I.: The petrology of grospydite xenoliths from the Zagadochnaya kimberlite pipe in Yakutia. J. Petrol. **9**(2), pp. 253–280 (1968)
43. Sobolev, N.V.: On mineralogical criteria of diamond-bearing kimberlites. Geol. Geophysics. **3**, 70–80 (1971). (in Russian)
44. Sobolev, N.V., Lavrent'ev, YuG., Pokhilenko, N.P., Usova, L.V.: Chrome-rich garnets from the kimberlites of Yakutia and their paragenesis. Contrib. Mineral. Petrol. **40**, 39–50 (1973)
45. Sobolev, N.V.: Depth inclusions in kimberlites and the problem of the upper mantle composition. Novosibirsk, Nauka, pp. 264 (1974) (in Russian)
46. Specius, Z.V., Taylor, L., Valey, D.: Chemistry and $d^{18}O$ garnets in diamondiferous xenoliths from Nyurbinskaya pipe. In: Proceedings of VI International Seminar "Deep magmatism, its sources and plumes". Irkutsk-Mirny, pp. 59–78 (2006)
47. Taylor, L.A., Neal, C.R.: Eclogites with oceanic crustal and mantle signatures from the Bellsbank kimberlite, South Africa, Part I: mineralogy, petrography and whole rock chemistry. J. Geol. **95**, 551–567 (1989)
48. Tomshin, M.D., Fomin, A.S., Kornilova, V.P., Cherny, S.D., Yanigin, Yu.T.: Especially the magmatic formations of the Nakyn kimberlite field in Yakutsk province. Geology and Geophysics, pp. 1693–1703 (1989) (in Russian)
49. Udovkina, N.G.: Eclogists of the USSR. Moscow. Science, p. 286 (1985) (in Russian)
50. Wyatt, B.A, Baumgartner, M., Ancar, E., Grutter, H.: Compositional classification of "kimberlitic" and "non-kiberlitic" ilmenite. In: 8th International Kimberlite Conference Selected Papers, Volume 2. Lithos 77, pp. 819–840 (2004)

Petrology of the Yubileynaya Kimberlite Pipe: Application to the Variation of Kimberlites Composition with the Depth

Z. V. Spetsius, A. Ivanov, and A. V. Aminov

Abstract

The article attempts to generalize the available data on the chemical composition, petrographic and mineralogical features of kimberlite rocks of the upper and deep levels of the Yubileynaya pipe. The results contain on original data of petrography, content of mineral phases in the groundmass and distribution of indicator minerals in kimberlites drill cores of upper and deep levels of the pipe. Obtained data indicate that Yubileynaya kimberlites are characterized by a predominant garnet association of indicator minerals, high content of pyropes of lherzolite paragenesis, low prevalence of titanic garnets, two types of ilmenites and spinelides compositions. We can assume that these parameters are typical for the medium-grade kimberlite bodies. The study of garnet chemistry and data on the ratio of eclogitic and ultrabasic paragenesis garnets in the concentrate of this pipe indicate the prevalence of eclogitic garnets in the composition of indicator minerals. This suggests a significant contribution of eclogitic diamond paragenesis to the formation of diamonds population in this pipe that. In our opinion, that also caused an increased content of large diamond crystals in kimberlites of the Yubileynaya pipe.

Keywords

Kimberlites · Petrology · Yubileynaya pipe

Z. V. Spetsius (✉)
Diamond and Precious Metal Geology Institute Siberian Branch Russian Academy of Sciences, Yakutsk, Russia
e-mail: spetsiuszv@gmail.com

A. Ivanov · A. V. Aminov
Geo-Scientific Geology Enterprise (NIGP), ALROSA OJSC, Mirny, Yakutia, Russia

N. Vladykin (ed.), *Alkaline Rocks, Kimberlites and Carbonatites: Geochemistry and Genesis*, Springer Proceedings in Earth and Environmental Sciences,
https://doi.org/10.1007/978-3-030-69670-2_13

1 Introduction

The Yubileynaya kimberlite pipe differs from many other industrial pipes of Yakutian kimberlite province their big dimension and because it produces many of the largest and rather valuable diamonds. Deposit of the Yubileynaya was discovered in 1975 by geologists of the Amakinsky expedition: I. J. Bogatyx, V. P. Roschin, Y. M. Erinchek and V. I. Lugowski. Quarrying of the deposit was started in 1989 and continues to the present day. At these days, the depth of opening mining is more than 450 m from the surface. The pipe is located in the upper part of the Markha River, 15 km to the NW from the town of Aikhal [3]. Mineralogy and petrography of kimberlites as well as a geology of the Yubileynaya pipe were discussed in some books and papers from the discovery until present days [3, 4, 7–9]. But in connection with high variation of peculiarities of kimberlites with the depth some questions of geology and properties of different kimberlite types still are debatable.

The results given in the paper contain on original data of petrography, content of mineral phases in the groundmass and distribution of indicator minerals in kimberlites drill cores of prospecting levels: upper—(+) 440—(+) 530 m and deep (−) 280—(−) 680 m of absolute mark of the ocean level. The drilling work was particularly done to define the different kimberlite types and to delineate the pipe morphology and variation of diamond grade.

The results are based on original data of petrography, content of mineral phases in the ground matrix and distribution of indicator minerals in kimberlites drill cores of prospecting levels: from (+) 440 to (−) 680 m of absolute mark of the ocean level that means from near to the surface of the pipe to the depths approximately 1500 m.

The variability of the petrography-mineralogical composition of the Yubileynaya pipe kimberlites is discussed according to the comparative analysis of kimberlites from core drills of the upper exploration levels (+) 440—(+) 530 m (2001 year), and according to the results of the study of kimberlites from the core drills of the deep levels (−) 280—(−) 660 m, obtained in recent years (2018–2019). The results based on original data of petrography, content of mineral phases in the ground matrix and distribution of indicator minerals in kimberlites of drill cores of prospecting levels.

2 Geology of the Pipe

Yubileynaya is the largest pipe in the Yakutia diamond province. In plan, it has an elongated, close to pear-shaped shape with an azimuth extension of 72° (Fig. 1). Adjacent to the pipe is a system of dikes oriented in the same direction. In the immediate vicinity of Yubileynaya pipe is located the Ozernaya satellite pipe. On the NW flank of the deposit lays a kimberlite body of Ottorzhenets, which, according to a number of facts, is a fragment of the Yubileynaya pipe [3]. The size of the ore body under the overlying rocks was 1293 × 741 m, with an area of

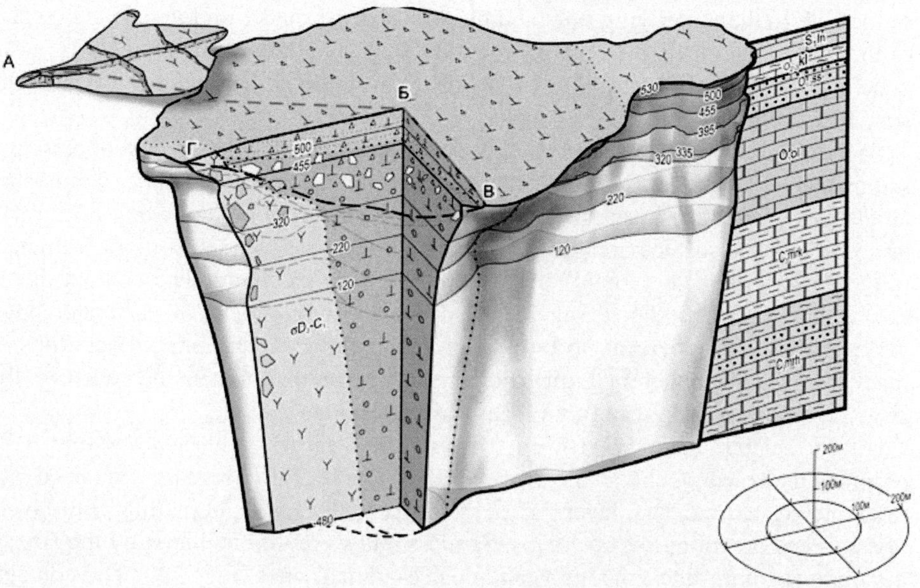

Fig. 1 Volumetric model of the Yubileynaya pipe (view from the south), according to [4]

560.000 m^2. With the depth of the pipe, size is substantially reduced. At the same time, its narrowing occurs more moderately in the Eastern part than in the Central and Western parts, which are most extended in the upper levels.

In the Eastern part of the pipe, the bell is observed only in the near-surface part of the ore body up to a depth of 130 m (abs. mark + 420 m) with the drop of contacts to the center of the body at angles of 50–70°, deeper angles of incidence increase to 75–85°. At a depth of 430 m from the surface of the ore body, the pipe turns into a dike-like body with a sub-vertical deposition and steeply falling (85–90°) contacts.

In the structure of the pipe, rocks of two main phases of intrusion are distinguished: the flanks are composed of massive porphyry kimberlites (coherent) of the first phase, and the Central channel, starting from the base of the bowl-shaped extension of the crater part of the pipe (from a depth of 300 m and below), is made of autolith breccias (pyroclastic kimberlites) of the second phase. Kimberlite rocks of the first phase of intruding, which compose the Western and Eastern flanks of the deposit, are similar in texture and structural features, petrography and mineralogy of the rocks and diamond content. These are dense, gray-green colored rocks with a distinct porphyry structure of the ground mass. Transitions between kimberlite rocks of the first and second intruding phases are gradual, through a mixing zone with a capacity of 5–30 m.

The most complex pipe structure and heterogeneous rock composition are located in the central part of the crater extension. It is made by various volcanogenic rocks—from volcanogenic-sedimentary formations composing the upper crater part of the bell, to the underlying tuffs and tufobreccias injected with kimberlite breccias of the final stage of the diatrema formation [3]. At the base of the crater part was a zone saturated with large blocks of the surrounding sedimentary rocks—the so-called "xenolith belt".

Within the upper levels of the crater extension of the pipe, subhorizontal sedimentary-volcanogenic rocks had a predominant development composing almost completely the near-surface part of the central "ore column" and occupying up to 85% of the surface area of the pipe. Their thickness varies from a few meters in the edge parts of the "bowl" up to 100–150 m in its center. The content of the debris material of the enclosing locally-derived country rocks in the crater bowl varies from the first percent up to 60% of their volume. Occasionally, xenoliths of mantle rocks and coherent kimberlite fragments were found in the rocks of the bowl, composing the Western and Eastern "ore pillars".

At a depth of 70–150 m from the earth's surface within the sedimentary-volcanogenic thickness, zones were discovered composed of rhythmically-sedimentary layers of pelitomorphic deposits, alternating with layers of sand- and gravelite-like rocks [3]. Sandy strata were differentiated by the size of the fragments into fine-, medium- and coarse-grained ones (Fig. 2A). The contacts between the layers are usually sharp. The layers are collected in zones with a thickness from the first tens of centimeters to 1 m. In the Eastern part of the crater, these zones were inclined to the west at an angle of 20°, in the Western part—to the South-East at an angle of 20–25°, i.e. their fall at the opposite ends of the crater was directed to the center of the pipe [3].

Fig. 2 Examples of different kimberlite types of the Yubileynaya pipe. **A** Volcano-clastic layered kimberlites (depth—75 m); **B** Coherent kimberlites (depth—405 m); **C** Sharp contacts of CK and KPK kimberlites (depth—1387 m); **D** Nuclear and porphyry magmoclasts in the kimberlites of the deep levels (depth—1420 m). Photo in transmitted light at 25x magnification

3 Methods of Research and Samples

Major element compositions of indicator minerals were determined with a Superprobe JXA-8800R electron microprobe at the ALROSA *PJSC* (Mirny). Natural minerals and synthetic were used as standards. Analytical conditions included an accelerating voltage of 15 keV, a beam current of 20 nA, beam size of 5 µm, and 20 s counting time for all elements. All analyses underwent a full ZAF correction.

The mineral composition of the kimberlite groundmass was established by the X-ray and thermography analysis. Shooting conditions were as follows: ARLX'-TRA diffractometer, CuKα radiation, V = 40 kV, J = 40 mA and thermal analyzer DTG60AH (Shimadzu), heating speed 20°/min, T = 1000 °C. In all cases for analyses of groundmass were used a special preparation. Kimberlites were crushed and from shaving class −1.0 +0.5 mm from groundmass were taken all pieces of sedimentary rocks and any micro-xenoliths as well as a big grains (>1.0 mm) xenocrysts. Kimberlites from core drills of the upper exploration levels (+) 440—(+) 530 m and kimberlites from the core drills of the deep levels (−) 280—(−) 660 m have been used. About 150 samples of kimberlites have been investigated.

4 Results

4.1 Petrographic Characteristics and Variability of the Kimberlites Composition

The variability of the petrography-mineralogical composition of the Yubileynaya pipe kimberlites is discussed according to the comparative analysis of kimberlites from core drills of the upper exploration levels (+) 440—(+) 530 m (2001 year), and according to the results of the study of kimberlites from the core drills of the deep levels (−) 280—(−) 660 m, obtained in recent years (2018–2019). The results are based on original data of petrography, content of mineral phases in the ground matrix and distribution of indicator minerals in kimberlites drill cores of prospecting levels: (−) 280—(−) 680 m of absolute mark of the ocean level.

Among the kimberlite rocks of the upper levels: (+) 440, 470, 500, 530 m, porphyric kimberlites or coherent kimberlites (CK) according to [5] and autolith kimberlite breccias that following the classification of Scott Smith et al. [5] classified as Kimberley-type pyroclastic kimberlites (KPK) are distinguished.

Porphyric coherent kimberlites (CK) of the upper levels had a wide distribution within the Eastern ore column, and sometimes found as xenoliths in autolith kimberlite breccia (KPK). Porphyry kimberlite is characterized by gray colors sometimes with a brownish and greenish tint, small-medium-porphyry structures, sometimes taxite and microtaxite textures of the groundmass. The content of olivine macrocrysts pseudomorphoses is 12–16%, rarely up to 20% and olivine phenocrysts 8–12%, sometimes up to 25–30% (see Fig. 2B).

Taxitic textures are caused by a sharply uneven distribution of pseudomorphoses over olivine macrocrysts and phenocrysts and appear as linear or irregular zones that differ in the number and dimension of pseudomorphoses. Micro-taxite textures occur due to the uneven distribution of the ground mass, ore minerals, and phlogopite, localized either near pnenocrysts, or in clusters, often together with carbonate and serpentine.

Porphyric macrocrysts are mainly represented by pseudomorphoses of olivine macrocrysts (>1 mm), which are partially or completely replaced by serpentine, to a lesser extent by calcite. Pseudomorphoses of olivine macrocrysts have an angular, rounded-angular shape, with a size >1 mm, usually no more than 5–6 mm, in rare cases up to 8–10 mm. They are composed with serpentine and carbonate. The ratio of these minerals depends entirely on the degree of superimposed secondary conversion processes. A distinctive feature of pseudomorphoses for olivine phenocrysts is the idiomorphic form and maximum dimensions (up to 0.8–1.0 mm).

Microscopically, porphyric (coherent) kimberlites have a clastic medium-fine-porphyric structure and consist of pseudomorphoses of olivine macrocrysts (usually rounded-angular shape) of 1–5 mm in size, composed of fibrous and crypto grainy serpentine, carbonate, and a scattered in it ore mineral (magnetite). The rock contains pseudomorphoses of serpentine and carbonate based on idiomorphic grains of olivine phenocrysts (15–20%), as well as rare plates of chloritized phlogopite (4–5%) with a size of 0.05–0.2 mm, idiomorphic crystals of opaque ore mineral and perovskite (total content of 4–5%) with a size less than 0.05 mm.

Autolith kimberlite breccias or pyroclastic kimberlites (KPK) form mainly the Central ore column. They are characterized by gray, bluish-gray and dark gray colors with all sorts of green shades, medium-small-fragmental, medium-fragmental and rarely large-fragmental structures. A distinctive feature of the pyroclastic kimberlites is the high content of pyrope grains (0.6–0.9%), their larger sizes, a slightly increased content of picroilmenite (up to 0.1–0.4%), as well as rare relicts of olivine macrocrysts. This variety also characterized by an increased content of deep rock fragments (up to 1–1. 5%).

The content of pseudomorphoses after olivine macrocrysts is in the range of 10–30%, fragments of sidimentary carbonate rocks 2–12%, sometimes reaching up 15–18 consisting %, the content of pseudomorphoses after olivine phenocrysts varies within 8–15%. The autolith (magmoclast) texture of kimberlite groundmass is clearly displayed. Magmoclasts, as a rule, have a nuclear structure, and consist of a core—(minerals, xenoliths) and a concentric zonal shell of a main fine-grained mass with an orbital arrangement of kimberlite components in it. In addition to autoliths, rounded fragments of porphyry kimberlites were observed in the autolith kimberlite breccia. Near the border with the porphyry kimberlites of the Eastern ore column, their content increases sharply. Pseudomorphoses after olivine macrocrysts have an angular-rounded, rounded and sometimes irregular shape. Their sizes vary between 1–8 mm, rarely more than 10 mm. They are usually composed of serpentine and carbonate with fine ore inclusions. Pseudomorphoses on olivine phenocrysts have subidiomorphic and idiomorphic forms with a size of 0.1–0.8 mm and are composed of a carbonate-serpentine or serpentine-carbonate substance.

Microscopic examination of the pyroclastic kimberlite revealed the following composition: pseudomorphoses of serpentine and carbonate on olivine macrocrysts were (about 15%), 1–8 mm in size; pseudomorphoses for olivine phenocrysts were (8–14%), subidiomorphic and idiomorphic forms. Autoliths (magmoclasts) of rounded shape made up (5–7%), usually of nuclear structure (shell thickness up to 0.7 mm). Kernel were present by olivine pseudomorphoses, sometimes by large leists of phlogopite. The composition of the shells was carbonate-serpentine with inclusions of ore mineral, phlogopite and rare pseudomorphoses of olivine.

The content of xenoliths of deep rocks in kimberlites of the upper levels of the pipe is uneven—from single finds in kimberlite breccias of the crater facies to the maximum content in autolith kimberlite breccias (0.5 vol.%). Along with the xenogenic material of mantle rocks, fragments of crystalline basement rocks were observed in the kimberlites of the pipe. Their content in varieties of kimberlite rocks was very uneven from single fragments in kimberlite breccias of crater facies up to 6 vol. % in the central ore column.

According to the results of X-ray and thermographic analysis of the thin fraction of the kimberlites groundmass of the upper levels, the following content of mineral phases in the matrix of kimberlites was established (Fig. 3). Dominated minerals were serpentine (35%) and calcite (33.4%), while dolomite and phlogopite were common, with a significant number of other components (21.3%). Among the latter were present, unevenly distributed: barite, montmorillonite, gypsum, anhydrite, pyrite and magnetite.

As noted above, we attempted to identify the vertical zoning of the Yubileynaya pipe kimberlites by comparison the petrography and mineralogy of the upper and deep levels of the pipe. In the latter case, the data obtained during the mineral and petrographic study of kimberlites from the drill core of deep levels from −280 to −660 m (or 600 to 1200 m from the surface) have been used.

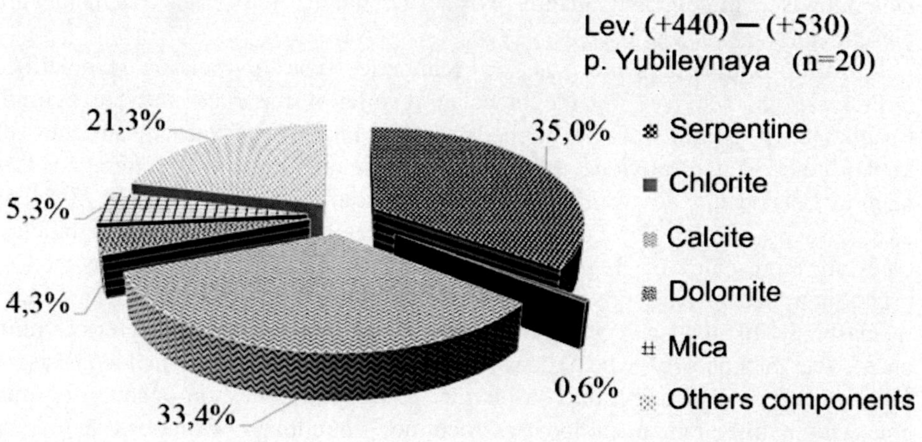

Fig. 3 Mineral content of the groundmass of kimberlites of the upper levels: (+)440—(+)530 m, Yubileynaya pipe

Coherent kimberlites (CK) of the deep levels of the Yubileynaya pipe are rocks of dark gray color with a dirty green tint, sometimes colored in brown tones and are characterized by a massive texture of the rock. Microscopically, coherent kimberlites have a medium-fine porphyry structure. In the rock, the grains of porphyry phenocrysts and of the groundmass are contrastingly separated. Porphyry phenocrysts occupy the main part of the rock and are presented by pseudomorphed olivine, less often there are relics of fresh olivine. Olivine pseudomorphoses are composed of fibrous or cryptogranular serpentine, fine-grained carbonate, sometimes with a fine rash of an ore mineral (goethite) or magnetite and have a different shape: from rounded to clastic-angular. The content of I-st generation (>1 mm) pseudomorphoses is 22–24%, the size of which is in average 5 mm, there are present sometimes olivine macrocrysts up to 14 mm. As porphyry phenocrysts, there are also present sometimes plates of phlogopite of rounded and irregular shape, up to 4 mm in size and content (2–3%). In the micrograin groundmass, there are distributed grains of garnet (1–2%), ore minerals (2%) and single grains of apatite. The groundmass of coherent kimberlite is composed of finely dispersed serpentine, pseudomorphoses of serpentine after olivine II generation, fine-grained calcite, dolomite, phlogopite, and ore minerals that vary in quantity and ratios.

The contents of various components in ore samples of kimberlites according to the results of macropetrographic calculations are given in Table 1. Note that the main component is pseudomorphed olivine; the number of mantle rock xenoliths in porphyry kimberlite is less than 4% on an average. Fragments of sedimentary rocks with a low number of xenoliths of metamorphic and ultrabasic rocks are common in pyroclastic KPK kimberlite (Table 1).

The composition of the minerals of the groundmass of coherent porphyry kimberlites is established by the X-ray and thermographic analysis (Fig. 4). As follows from the obtained data, serpentine (52.6%) dominates in the ground mass of deep levels kimberlites, with a subordinate amount of calcite, mica and chlorite. Dolomite is common, and apatite, pyroaurite, halite, pyrite, and magnetite are present.

Autolith kimberlite breccias or Kimberley-type pyroclastic kimberlites (KPK) are characterized by the breccian texture of the rock and the autolith (magmoclast) structure of the kimberlite groundmass. The autolith structure of kimberlite-cement is obvious. Magmoclasts have a porphyry structure. Nuclear magmoclasts occupy an average of 7% of the area of the sample (see Fig. 2C). The size of such secretions is about 5 mm, sometimes up to 9 mm. The magmoclast cores are represented by fragments of sedimentary rocks (most often), as well as pseudomorphed olivine, grains of ore minerals, and rare grains of garnet. Magmoclasts are distinguished by the dark hue of the shells (saturation with ore minerals), the thickness of which does not exceed one mm. The shell consists of pseudomorphoses of serpentine on olivine, phlogopite plates, an opaque ore mineral, and a fine-grained carbonate-serpentine groundmass. Composition of the groundmass of the shell in some cases, calcite (55%), dolomite (15%), opaque ore mineral (15%), serpentine (5%), and phlogopite (5%) were found in the sifts.

Table 1 The average component composition of different petrographic types of kimberlites of the pipe (results of counting components in ores, lin.%)

Drill #	Rock type	# of samples	Xenoliths			Magmo-clasts	Gt	Ilm	Ol pseudo-morph.
			Sediment rock	Umaf. xen.	Metam xen.				
438.1.1	KPK	3	9.69	0.45	1.03	4.58	0.12	0.07	16.77
438.1.1	CK	2	5.82	0.00	0.93	0.00	0.13	0.06	19.76
438.1.2	KPK	8	7.96	0.60	0.95	3.85	0.03	0.13	16.02
438.1.2	CK	1	4.78	0.00	0.00	0.00	0.06	0.12	27.85
438.2.1	KPK	4	8.51	0.41	0.62	3.04	0.15	0.18	19.14
438.2.1	PK	4	3.83	0.00	0.00	0.22	0.02	0.12	25.21
438.2.2	KPK	3	8.21	5.43	4.39	2.97	0.02	0.06	19.93
438.2.2	CK	6	5.73	0.15	0.00	0.00	0.10	0.03	23.77
438.3.2	KPK	3	16.70	0.59	0.54	15.26	0.09	0.02	5.92
438.3.2	CK	7	4.03	0.00	0.32	0.00	0.08	0.01	21.67
438.3.2	KPK	3	16.70	0.59	0.54	15.26	0.09	0.02	5.92
438.4.2	KPK	5	17.32	0.00	0.00	7.92	0.05	0.00	13.37
438.4.2	CK	1	1.19	0.00	0.00	0.00	0.00	0.00	23.81
438.5.2	KPK	1	5.89	0.00	0.00	7.06	0.07	0.00	13.96
438.5.2	CK	4	3.44	0.00	0.47	0.03	0.05	0.01	21.24
439.2.1	KPK	1	28.96	0.00	0.00	0.00	0.00	0.00	4.71
439.2.1	CK	5	4.81	0.00	0.00	0.00	0.02	0.00	20.59
Average on types	KPK	28	11.53	0.92	1.00	5.64	0.06	0.08	14.93
	CK	30	4.44	0.03	0.20	0.03	0.06	0.04	22.48

Notes **a** Accuracy of counting 0.01; **b** Results of components counting in linear %; **c** contents of Gt and Ilm, in % from the area of sample

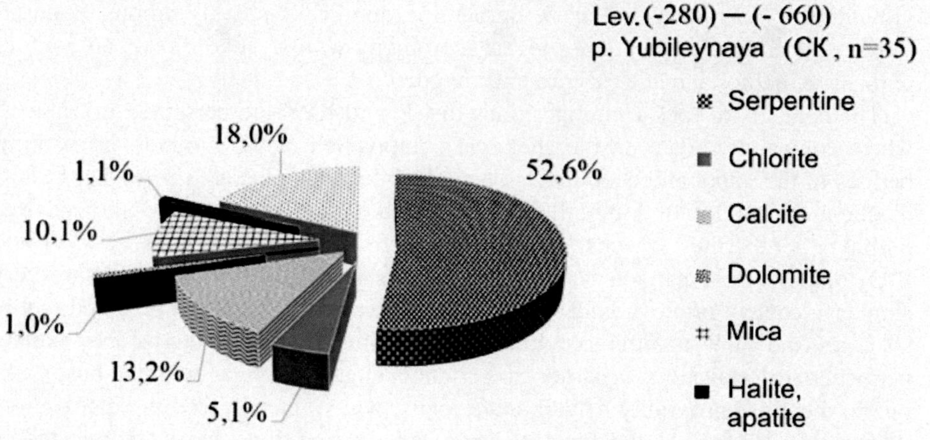

Fig. 4 Mineral content of the Yubileynaya porphyry kimberlites groundmass at deep levels (−) 280—(−)660 m

Magmoclasts with porphyry structure are also present in the KPK kimberlites of the Yubileynaya pipe (see Fig. 2D). Porphyry magmoclasts occupy an average of 4% of the area of the thin section and have a rounded-oblong shape. The size of porphyry secretions on average is less than 3 mm, maximum 6 mm. In the inner part of such formations, there is a section of rock similar in appearance to the CK. Porphyry macrocrysts are represented by serpentine pseudomorphoses on olivine, sometimes partially replaced by fine-grained carbonate. Their sizes are up to 2 mm. There are plates of phlogopite up to 1.2 mm, and rare garnet grains with a kelyphite rims. There is no opaque ore mineral in porphyry macrocrysts. A fine-grained groundmass of the inner part of porphyry magmoclasts is composed of calcite microlites, fine serpentine, phlogopite, and an opaque ore mineral.

The porphyry macrocrysts of the KPK kimberlites have a similar structure and composition to the porphyry macrocrysts of the CK. Pseudomorphosed olivines are replaced by serpentine, partially by carbonate (relics of fresh olivine are rare), and micro-inclusions of ore minerals are present. The average content of olivine pseudomorphoses in the KPK kimberlites is less than in the CK and is 15–18% of the area of the thin section. Various forms of ore mineral grains have sizes up to 3.8 mm (2%). Rounded grains and phlogopite plates with a maximum length of 2.9 mm occupy 1–2% of the area of the slot. Garnet grains in the kelyphite rims make up less than 1.0% of the total volume of macrocrysts. The groundmass of the macrocrysts is composed of fine-grained calcite, serpentine, phlogopite, and an opaque ore mineral. Rarely there are present grains of apatite.

The mineral composition of the groundmass of the KPK is close to those of the CK. In contrast to CK, the KPK kimberlites have a deep substitution (up to sulfates) of large olivine macrocrysts. KPK has a similar low content of deep xenoliths and xenoliths of metamorphic rocks, but the content of sedimentary rock fragments is significantly higher than in CK and reaches about 8%.

The data of X-ray and thermographic analysis of the composition of the KPK groundmass are shown in Fig. 5. Serpentine dominates sharply among minerals (60%), mica, calcite and chlorite are common, with a subordinate amount of pyroaurite, halite, hematite, pyrite and magnetite.

The contents of rock-forming oxides in CK and KPK are presented in Table 2, where comparative data on the chemical composition of the groundmass of kimberlites of the upper and deep levels of the Yubileynaya pipe are given. The results of chemical studies of kimberlite-cement of the Yubileynaya pipe showed that porphyry kimberlites are characterized by increased contents of TiO_2, K_2O and P_2O_5 oxides in comparison with kimberlite breccias (Table 2), however, are very similar in content of most oxides, which was noted in [3]. It should be noted that in all cases, the same textural-structural type of kimberlite, characterized by certain petrochemical indicators, is subject to certain changes in contact with the host rocks due to the close proximity of carbonate rocks, which is expressed in calcitization, dolomitization, etc. At the level of chemical composition, this is manifested in significant fluctuations in the main rock-forming oxides: CaO, MgO, SiO_2, TiO_2, etc.

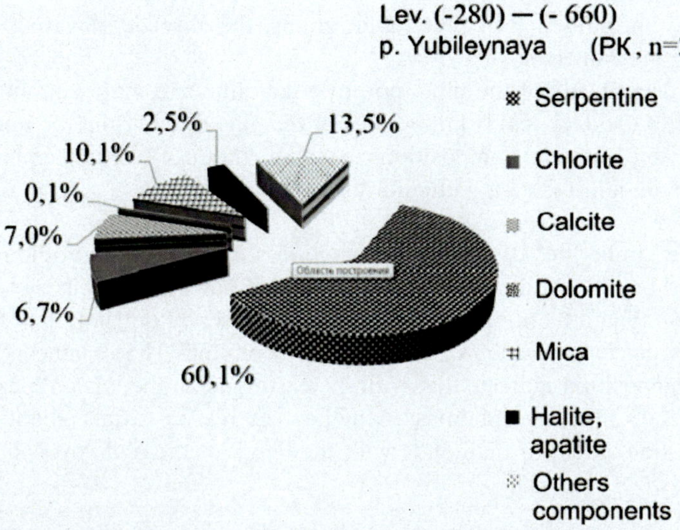

Fig. 5 The content of mineral phases in KPK groundmass of the Yubileynaya pipe (deep levels—280–660 m)

Table 2 Bulk chemical composition of the Yubileynaya pipe kimberlites, in wt%

	SiO_2	TiO_2	Al_2O_3	Fe_2O_3	CaO	MgO	MnO	K_2O	Na_2O	P_2O_5	S	p.p.p.
1	28.95	1.02	1.94	7.59	9.08	29.18	0.11	0.61	1.11	0.38	0.14	20.06
2	27.72	1.23	2.2	6.94	10.66	28.64	0.10	0.85	0.37	0.47	0.11	20.16
3	29.48	1.29	1.72	6.30	10.95	29.47	–	0.27	0.01	0.49	–	20.11
4	28.55	1.01	1.88	7.06	11.17	28.81	–	0.32	0.11	0.39	–	–

**Notes* 1—KPK (deep levels, average on 27 samples); 2—CK (deep levels, average on 35 samples); 3—average composition of kimberlites after [3]; 4—average composition of kimberlites on 27 samples (upper levels, 2001)

We emphasize that the kimberlites of deep levels in contrast to the previously studied rocks of the upper levels are characterized by high contents of K_2O and Na_2O, with approximately a constant number of TiO_2. This pattern is according to the authors of work [9] indicates an increase of the depths of formation of parental melts and, possibly, produces increasing their diamond.

As a result of the studying kimberlites of the upper and deep levels of the Yubileynaya pipe, their characteristic patterns of composition variability with the depth were established:

(a) With increasing depth, relics of fresh olivine are more common;
(b) In the deeper levels showed a decrease of garnet grains in the kelyphitic rims;

(c) Autoliths (magmoclasts) kernel in the upper levels differ from those in the deep ones by the presence of phlogopite grains, the absence of garnet grains and opaque ore minerals;
(d) In the deep levels of the pipe, porphyry autoliths are more common;
(e) Matrix of the deep levels kimberlites of the pipe differs from the upper ones in a more serpentinized composition, a smaller amount of carbonate minerals, and a large amount of mica and chlorite.

Thus, the kimberlites of deep levels were less exposed to hydrothermal effects, as evidenced by a lower degree of carbonation and other facts. This is confirmed by data from X-ray and thermographic analysis and observations in thin sections. This is evidenced by the presence of relics of fresh olivine. The obtained data on the chemical composition indicate that with increasing depth there is an increase in the content of K_2O in the groundmass, which gives reason to talk about a possible increase in the content of diamonds with the depth of the Yubileynaya pipe.

4.2 Mineralogical Features of the Yubileynaya Pipe Kimberlites

Indicator kimberlite minerals (IKM) in Yubileynaya pipe kimberlites are represented by garnets (average content ~0.02%), picroilmenite (~0.001%), and chromspinelides (~0.01%). Olivine and clinopyroxenes are rare. In this respect, Yubileynaya kimberlites differ significantly from those of the nearby Zarya pipe, where ilmenite predominates (Spetsius et al. 2014).

Garnets significantly predominate over other indicator minerals and are represented by several color varieties. Grains size usually do not exceed 5 mm in diameter. The difference between grain color ratios between porphyry CK and KPK is insignificant (Figs. 6 and 7). Crimson and orange garnets are dominate on both levels. Interestingly, if the upper levels of pink garnets are marked in the CK, but at deep levels they are recorded only in KPK. The obtained data show that porphyry

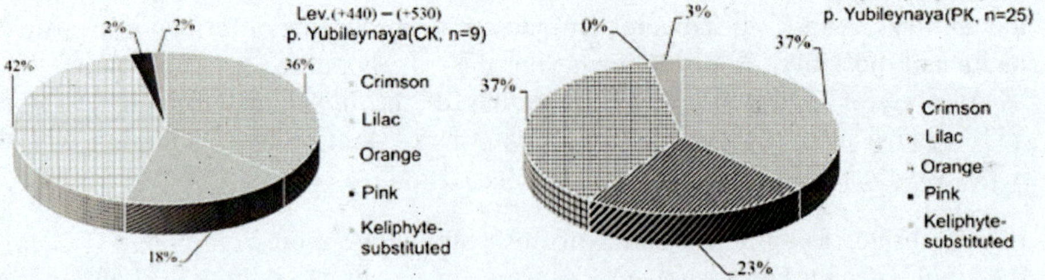

Fig. 6 The ratio of garnets color in kimberlites of the upper levels of the Yubileynaya pipe

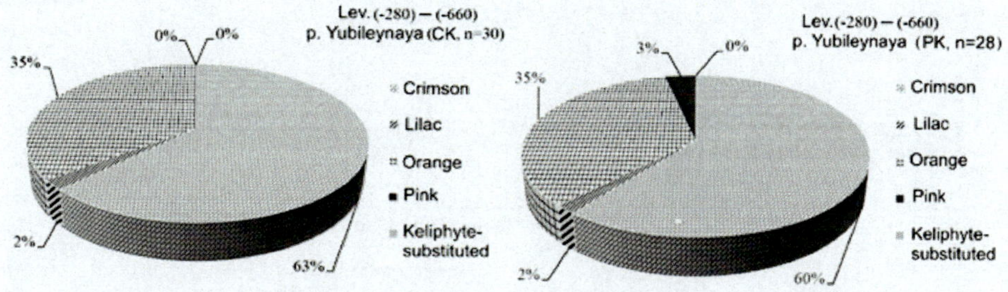

Fig. 7 The ratio of garnets color in kimberlites of deep levels of the Yubileynaya pipe

coherent kimberlites are characterized by slightly better indicator minerals preservation than KPK and an increase in the number of garnets.

The results of the study of mineralogical features of indicator minerals from the Yubileynaya pipe kimberlites are considered on the set of data from the microprobe analysis of garnets, ilmenites and chromespinelides. Comparison of the obtained results of garnets compositions from the deep levels of the Yubileynaya pipe with similar data on the upper levels of the pipe is shown in Figs. 8 and 9.

The garnets composition from kimberlites of the Yubileynaya pipe is diverse and varies widely on the upper and lower levels. In both cases, the garnets is dominated by lherzolite paragenesis of 50%, and most of them are grains with Cr_2O_3 content from 2.0 to 8.0 wt%. According to the classification of J. Dawson and V. Stephens [1] among garnets, chrome pyropes (group G9) and low-calcium chrome pyropes (group G10) predominate, their number is more than 20% on both upper and deep levels. About 15% grains are titanium and high titanium garnets with a TiO_2

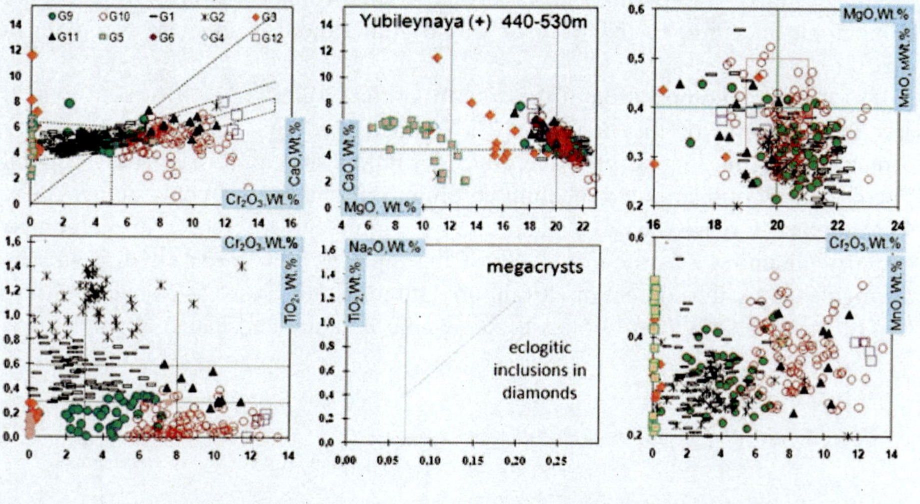

Fig. 8 Complex diagrams of garnets composition from upper levels of the Yubileynaya pipe

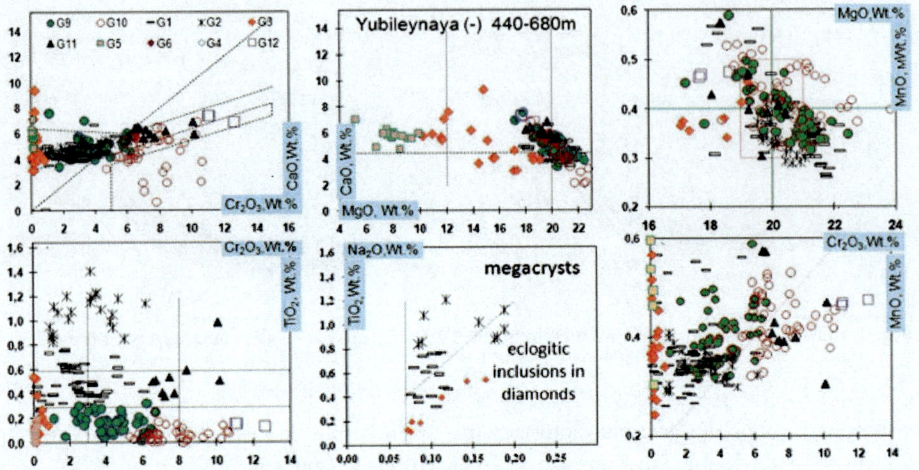

Fig. 9 Complex diagrams of garnets composition from deep levels of the Yubileynaya pipe

content of more than 0.6 wt%. Garnets of dunite-harzburgite paragenesis account for about 15% and eclogite paragenesis garnets less than 15%. The quantity of garnets of the diamond dunite-harzburgite association is about 10%, which is reflected in the well-known diagram of Sobolev [6]. It should be noted the increase with the depth of group G10 garnets according to the classification [1, 2]. It was also established that in the deep levels of the pipe, there are on 6% more pyropes that belong to a group of eclogitic paragenesis and the enhance presence of eclogitic paragenesis garnets of the diamond association is noted.

Ilmenite in kimberlites of the Yubileynaya pipe is relatively rare. The size of the ilmenite grains varies from 0.2 mm to 5.0 mm, the class −1 + 0.5 mm prevails. The predominant morphological variety is angular-rounded grains, without crystallographic faceting, in the form of single grains there is ilmenite of aggregate structure.

The chemical composition of ilmenite from kimberlites is very diverse (Figs. 10 and 11). Ilmenite of the first group is characterized by high MgO and Cr_2O_3 content, while the second group has low chromium and high manganese content. The most widely distributed medium-chromium picroilmenites containing 9–13 wt % MgO and 1–5 wt% Cr_2O_3. There are no significant differences in the composition of ilmenites of different levels of the pipe: at both levels, two groups are distinguished by the content of chromium with maximums of 1.5–2.0 and ~4 wt% Cr_2O_3, that probably corresponds to xenogenic ilmenite and matrix ilmenites.

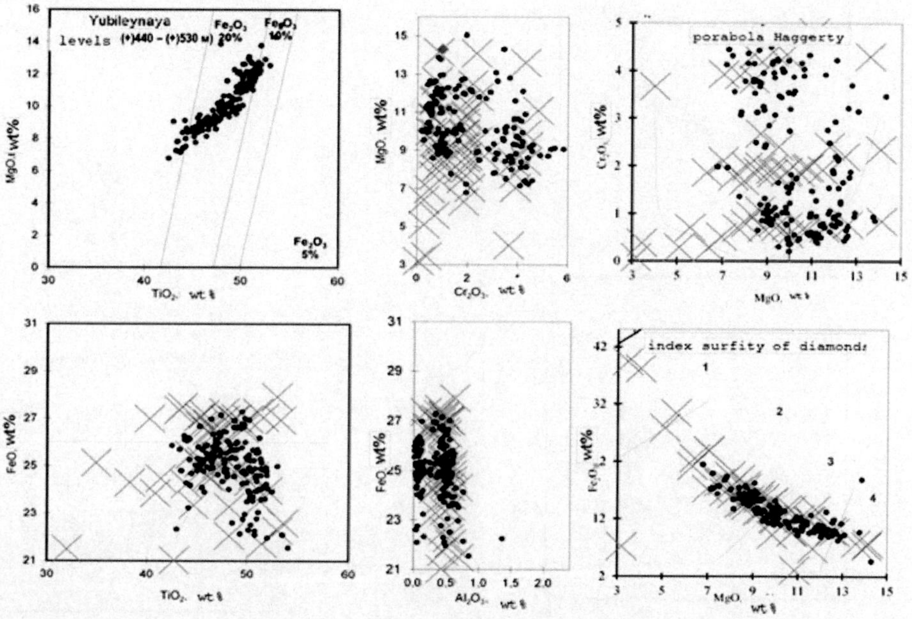

Fig. 10 A comprehensive diagram of the composition of the ilmenite from upper levels of the Yubileynaya pipe. Large crosses on this and subsequent diagrams are average cluster values of the indicator minerals compositions from kimberlites of the Yakutian province data bank

Chromespinelides are widely distributed in the Yubileynaya kimberlite pipe. The size of the grains is −2 + 0.5 mm, and predominate grain size is −1 + 0.5 mm. The chromespinelides of the pipe are characterized by a diverse morphology with a predominance of octahedral forms. The chemical composition of chromespinelides varies widely (Figs. 11 and 12), a more wider variety of spinelides compositions is observed in the kimberlites samples of upper levels, that is probably partly determined by a large number of analyses. In the studied samples of both levels are dominated chromespinelides the composition of which corresponds to chrompicotite. Less than 10% of the studied grains correspond to alumochromite. Chromespinelides of the diamond association accounted for 18% of the total number of grains studied in the upper levels and 33% in the kimberlites of the deep levels, which is clearly visible in the diagram Al_2O_3—Cr_2O_3 (Figs. 12 and 13) and indicates an increase of diamond content with the depth.

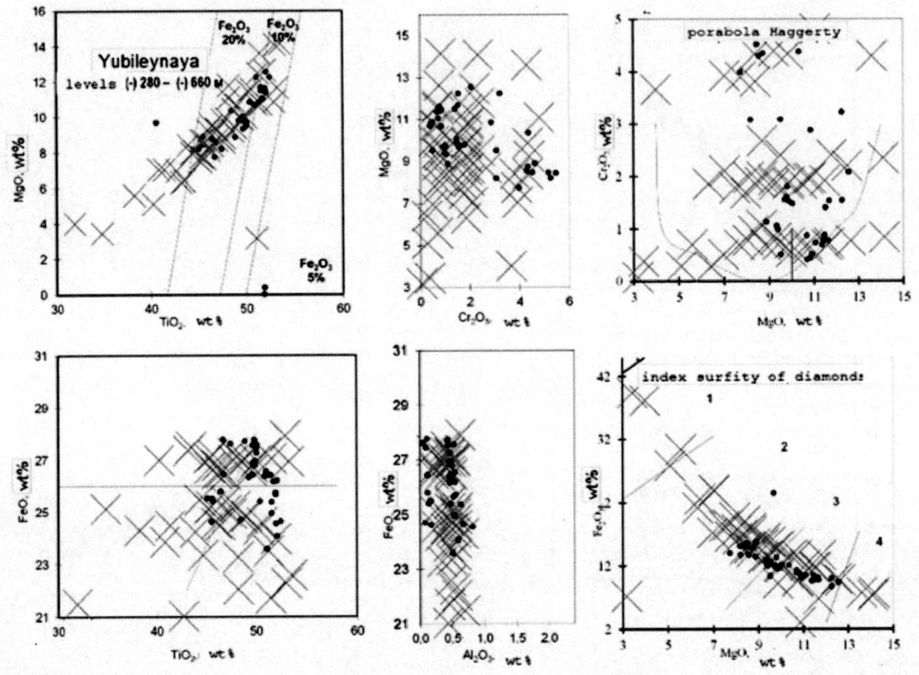

Fig. 11 A comprehensive diagram of the composition of ilmenite from deep levels of the Yubileynaya pipe

5 Conclusion

Thus, the Yubileynaya pipe kimberlites are characterized by the following characteristics of their material composition: relatively high indicator minerals content, predominantly garnet association of indicator minerals, predominance of crimson pyropes of the lherzolitic paragenesis, low frequency of occurrence of titanic garnets, presence of two composition types of ilmenites and chromspinelides. We can assume that these parameters are typical for the medium-grade kimberlite bodies. The data obtained (2018–2019 years) on the chemical composition of kimberlites of deep levels suggest that with the increasing of depth there is an increase in K_2O content in the bulk rocks. That, along with the proportion of garnets of eclogitic paragenesis and chromite of diamond associations, suggests a possible increase in diamond content with the depth of development and processing of the Yubileynaya pipe.

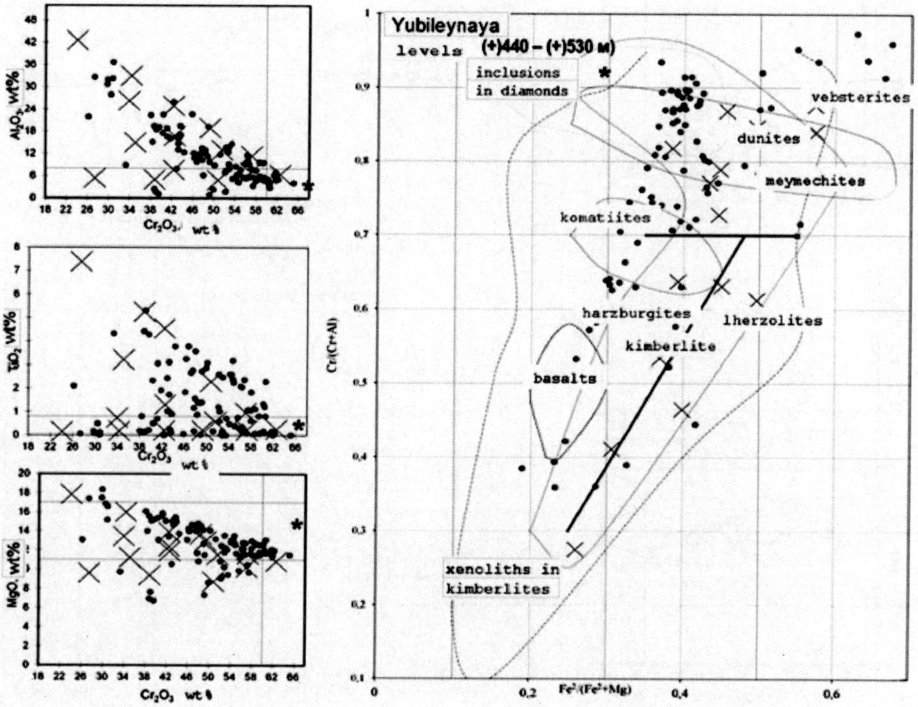

Fig. 12 A comprehensive diagram of the composition of chrome spinel from upper levels of the Yubileynaya pipe

The results of the study of garnet chemistry and data on the ratio of eclogitic and ultramafic garnets in the Yubileynaya pipe concentrate, taking into account the number of corresponding paragenetic associations of diamond paragenesis, indicate a high content of eclogite-type garnets. Respectively, we should note a significant contribution of eclogite paragenesis diamonds to the formation of the diamond population of this pipe. In the kimberlites of deep levels of the pipe, established that more than 6% pyropes belong to area of diamond eclogitic paragenesis. In our opinion, this explains the rather high diamond content of the Yubileynaya pipe, including the enhanced presence of large diamond crystals.

In conclusion, we consider it a pleasant duty to thank M. V. Shalkina for performing macropetrographic descriptions of drill core kimberlite samples and V. Ya. Batyr for the help in preparing tables and drawings.

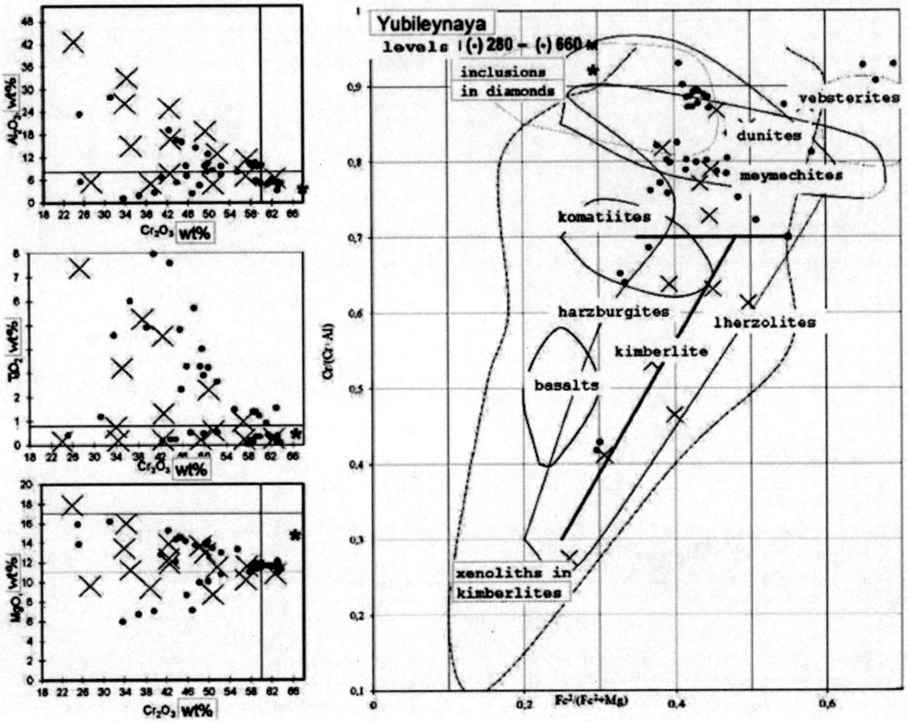

Fig. 13 A comprehensive diagram of the chrome spinel composition from deep levels of the Yubileynaya pipe

References

1. Dawson, J.B., Stephens, W.E.: Statistical analysis of garnets from kimberlites and associated xenoliths. J. Geol. **83**, 589–607 (1975)
2. de Bruin, D.: Report on the mineral chemistry of garnet and oxide (2004)
3. Kharkiv, A.D., Zinchuk, N.N., Kryuchkov, A.I.: Primary Diamond Deposits of the world. Nedra, Moscow (1998), 555p. (in Russian)
4. Kostrovitsky, S.I., Spetsius, Z.V., Yakovlev, D.A., Von der Flaass, G.S., Suvorova, L.F., Bogush, I.N.: Atlas of primary diamond deposits of the Yakut kimberlite province. Mirny, MTp printing house (2015), 480p. (in Russian)
5. Scott Smith, B.H., Nowicki, T.E., Russel, J.K., et al.: Kimberlite terminology and classification. In: Proceedings of 10th International Kimberlit Conference Vol. 2. J. Geol. Soc. India, pp. 1–17 (2013)
6. Sobolev, N.V.: Deep inclusions in kimberlites and the problem of the upper mantle composition. Novosibirsk, Nauka, 264 p. (1974) (in Russian)
7. Spetsius, Z.V., Aminov, A.V., Ivanov, A.S.: Peculiarities of kimberlites material composition from deep levels of the Yubileynaya pipe. In: Vladykin, N.V. (ed.). Proceeding of XV International Workshop “Deep-seated magmatism, its sources and plumes”. Irkutsk, pp. 183–203 (2019) (in Russian)

8. Vasilenko, V.B., Zinchuk, N.N., Kuznetsova, L.G.: Petrochemical models of diamond deposits in Yakutia. Nauka, Novosibirsk. SB RAS, 280p. (1997) (in Russian)
9. Zolnikov, G.V., Kovalsky, V.V., Zinchuk, N.N., and others: Features of rocks and conditions of formation of one of the pipes of the Alakit field. In: Mineralogy and Geochemistry of Ultrabasic and Basitic Rocks of Yakutia. Institute of Geology, Yakutsk, YAF SB RAS, p. 5–39 (1979) (in Russian)

Ustyansky Diamondiferous Region in the Central Part of the Russian Platform

S. M. Sablukov, L. I. Sablukova, A. V. Belov, N. S. Ermakov, and S. I. Leshukov

Abstract

Ustyansky perspective area is characterized by the combination of all favorable regional and local search prerequisites (morphological, tectonic, geological-stratigraphic, geophysical, morphometric, mineralogical) and search features, which may be due to the introduction of mantle diapir (plume) and may be accompanied by diamondiferous kimberlite magmatism. The content of the diamond indicator minerals with not distant moving: first of all, pyrope, less than picroilmenite, chrome spinel, olivine, and chrome-diopside in the alluvium of the region is 3–5 times higher than in the Zimny Bereg diamond-bearing region, and the share of G10 group diamond paragenesis pyrope is about 10%. Nine diamond crystals were found in panned samples of minimal volume from alluvial and quaternary sediments, of which six crystals (large octahedrons and dodecahedroids up to 3.8 mm in size and weighing up to 52 mg) were found in the predicted kimberlite field "Severnoye". In terms of morphological features and physical properties, all diamonds of the Ustyansky region occupy a completely isolated position and have virtually no analogues among the diamonds of the known deposits of Arkhangelsk, Finland, the Urals and Timan. This may indicate that the crystals are coming from a new, not yet known kimberlite source (or sources). This potential new root kimberlite source may have an increased diamondiferous capacity, high quality and size of diamonds. The discovery of this new kimberlite source will be the final confirmation of the predicted isolation of the Ustyansky Region as an area of plume activity. Signs of a possible similarity between the Arkhangelsk diamondiferous province and the Yakutian diamondiferous province in terms of kimberlite location patterns

S. M. Sablukov (✉) · L. I. Sablukova
RUSGEO Limited, Moscow, Russia
e-mail: sablukoff@rambler.ru

A. V. Belov · N. S. Ermakov · S. I. Leshukov
Arkhangelsk Diamonds LLC, Arkhangelsk, Russia

N. Vladykin (ed.), *Alkaline Rocks, Kimberlites and Carbonatites: Geochemistry and Genesis*, Springer Proceedings in Earth and Environmental Sciences,
https://doi.org/10.1007/978-3-030-69670-2_14

and morphological features of diamond crystals confirm the high prospects for diamondiferous activities in the central regions of European Russia as a whole, and, above all, in the Ustyansky promising region with the predicted kimberlite field "Severnoye".

Keywords

Russian Platform · Diamond · Pyrope · Picroilmenite · New Promising Region

1 Introduction

The study of plume formation products is of great scientific and practical interest, as it allows to estimate the nature of deep geological processes taking place at the level of upper and lower mantle. In addition, the products of plume magmatic activity is associated with the formation of large and unique deposits of different types of minerals.

However, the existence and position in the space of a particular "plume" is usually revealed already "post factum", after the discovery, subsequent comprehensive detailed study and establishment of the plume nature of magmatic objects.

At the same time, preliminary spatial localization of a particular area of possible and even known real development of plume processes of tectonic and magmatic activity on the basis of detection, comparison and evaluation of indirect geological prerequisites and features may allow to concentrate prospecting activities on the most favorable and promising areas for deposit detection. Such researches can be considered as "solution of inverse *geological* problem" (by the type of widely used term "solution of inverse *geophysical* problem"). Multiple interpretation of the results of such studies is eliminated by attracting independent data from all available geological, geophysical, mineralogical and other methods.

An example of such preliminary spatial localization of a particular area of plume activity are the results of the work on forecasting and assessment of diamond-bearing prospects of one of the areas in the middle part of the Russian Platform, given below.

2 Substantiation of Task

Diamond prospecting in the traditional areas of the North of European Russia over the last 24 years (after the opening of the Grib pipe in February 1996) has not yielded positive results, and the already known diamond deposits in Arkhangelsk either have limited open mining periods (Grib deposit), or due to extremely difficult engineering and geological conditions have indistinguishable prospects for cost recovery (Lomonosov deposit). *Therefore, there is a need to discover diamond deposits with higher quality and profitability of mining operations.* Such deposits may be discovered in new, unconventional areas, but generally in more promising exploration areas.

Currently, all known primary and alluvial deposits and mineral manifestations of diamonds are located in the ***peripheral*** northwestern part of the Russian Platform (Zimny Bereg, Tersky Bereg, Karelia and Finland), or in the east and northeast (Urals and Timan). However, the largest and richest primary diamond deposits are usually confined to the ***central*** parts of the ancient platforms. Therefore, of particular interest is the allocation of new promising areas in the central part of the Russian Platform.

Search work carried out by various organizations on a large part of the selected [5] predicted promising potential diamondiferous kimberlite fields, did not reveal any serious signs of confirmation of the presence of diamondiferous kimberlite magmatism in these areas (Nelidovo field, the area "Oredezh" Leningrad region, Dvinsko-Pinezhskaya morphological structure, the lower reaches of the Northern Dvina River (Orletsy village), Nizhne-Onezhsky region, Plesetsky region, different districts of Karelia, Kola Peninsula, etc.).

At present, in the central part of the Russian Platform there is one unique predictive-perspective area, where almost all possible regional and local preconditions and signs of probable development of mantle (plume) tectonic–magmatic activity with the manifestation of diamondiferous kimberlite magmatism are revealed. This area is the *Ustyansky perspective region* located in the south of the Arkhangelsk district in the interfluve of the Northern Dvina, Vaga and Sukhona rivers (Fig. 1). The study of this area is of undoubted scientific and practical interest. The main results of the work in Ustyansky promising region are given in this article.

3 Stages of Research Work

Independent tectonic and morphotectonic methods, developed on the basis of new geotectonic-metallogenic concepts, were added to the purely geophysical methods of isolating the supposed areas of plume activity in the 70s- early 80s: "NON-LINEAR METALLOGENY" (A.D. Shcheglov) and "TECTONICS OF FOCAL STRUCTURES" or "MORPHOTECTONICS OF CENTRAL TYPE STRUCTURES" [4, 13, 35]. These new methods have significantly increased the possibilities for forecasting promising areas. It was noted [30] that the areas of development of volcanic pipes of the Russian Platform's alkali-ultramafic composition (including melilitites of the Nenoxa and diamondiferous kimberlites of the Zimny Bereg) are confined to large centrally-type circular morphostructures that are distinctly distinguished in relief and in the drawing of the river network [13]. At the same time, assumptions were made and preconditions for the possible existence in the middle part of the Russian platform of manifestations of Mesozoic kimberlite

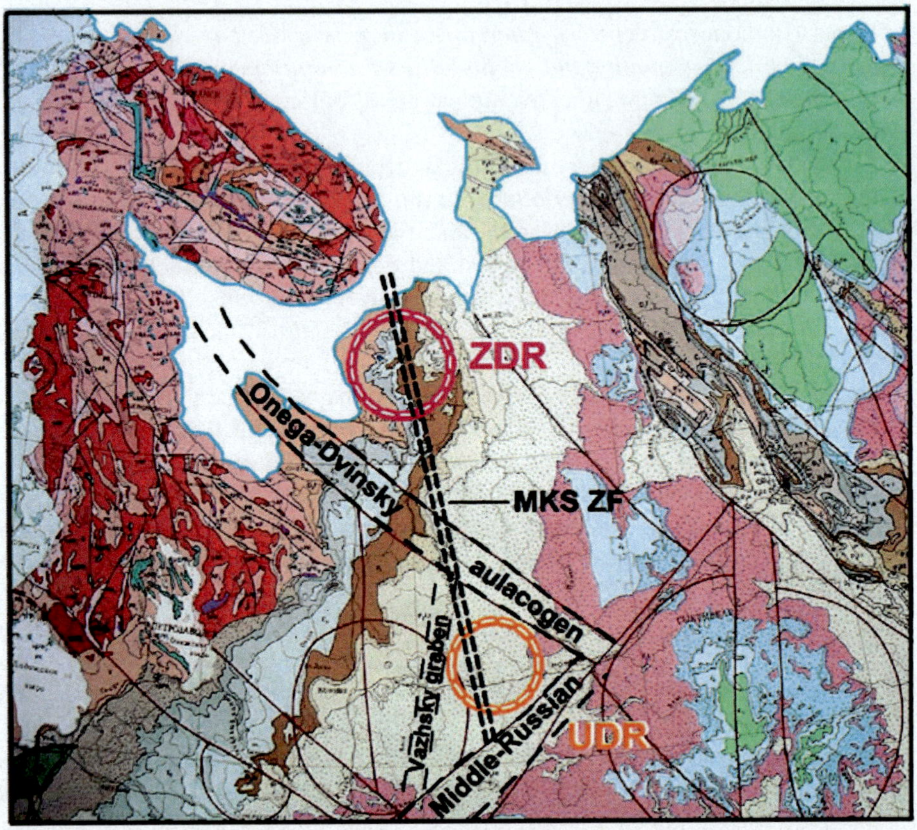

Fig. 1 General position of the Ustyansky diamondiferous region in the structures of the north of the Russian platform. Letters on the diagram indicate: ZDR—Zimneberezhny diamondiferous region; UDR—Ustyansky diamondiferous region; MKS ZF—Megra-Kepin-Shenkursk zone of deep faults

volcanism [30, 31], which is younger than Devonian kimberlites and melilitites of the Zimny Bereg, Onega Peninsula and Middle Timan [19–22, 24] were determined.

Taking into account these new data, we chose one of the morphostructures of the central type of the middle part of the Russian Platform—*Ustyanskaya morphostructure* (morphostructure "*Ustya*")—to carry out reconnaissance prospecting and evaluation of diamondiferous prospects.

The first panned sampling works carried out by S.M. Sablukov (Diamond Department of TsNIGRI MinGeo USSR) in September 1984 in the middle reaches of the Ustya river revealed an unusually high content of diamond indicator minerals in alluvium, including the presence of two grains of G10 pyrope group [3], typical for diamondiferous kimberlites. This allowed recommending the Ustyankaya structure for more detailed specialized diamond prospecting [19].

From 1990 to 1992, S. M. Sablukov and L. I. Sablukova carried out small-scale prognostic and search works on the Ustyanskaya morphological structure and its frame. The area of the works is 17000 km^2, scale of works is 1:1 000 000. Panned sampling, mineralogical studies of modern alluvial deposits and native bedrocks of Permian and Triassic (including petrographic and geochemical studies of native breccia rocks), detailed study of surface and chemical composition of diamond indicator minerals from all types of rocks were carried out. The analysis of regional magnetic and gravitational fields and tectonic, morphological and morphometric analysis of the territory were performed (by [15, 16]). These works confirmed the high prospects for the diamondiferous area [26].

From 2001 to 2017, general searches for diamond deposits at the Ustyanskaya structure at a scale of 1:50,000 were carried out by CJSC Arkhangelsk Diamonds (since 2011—LLC) in cooperation with RUSGEO Limited (mineralogical and petrographic study). There was carried out a panned sampling and small-volume sampling of modern alluvium, quaternary glacial deposits and indigenous rocks of Permian and Triassic (a total of about 20 thousand samples), area aeromagnetic and ground magnetic survey, ground detailing of geophysical anomalies, verification of geophysical anomalies by drilling holes.

The work carried out in 1984–2017 allowed us to characterize the Ustiansky morphological structure as the most promising area of the central part of the Russian Platform in terms of diamond bearing capacity—the "Ustiansky perspective region" [25, 27, 28]. This is indicated by the identified regional and local ***preconditions*** and ***signs of*** potential indigenous magmatic and alluvial diamondiferousness of this area.

4 Regional Search Prerequisites

The area is characterized by a full complex of favorable search prerequisites for identifying the area of plume tectonic–magmatic activity and, as a consequence, for selecting the brightest and most important element of such activity as kimberlite deposits of diamonds (Fig. 2).

4.1 Favorable Tectonic Prerequisites

- The central part of the Russian Ancient Platform *(usually the most significant primary diamond deposits are confined to the central parts of ancient platforms)*;
- The eastern part of the Shenkursky Achaean Craton *(according to the clarified rule of Clifford this is a necessary condition of potential diamondiferous nature of the area)*;

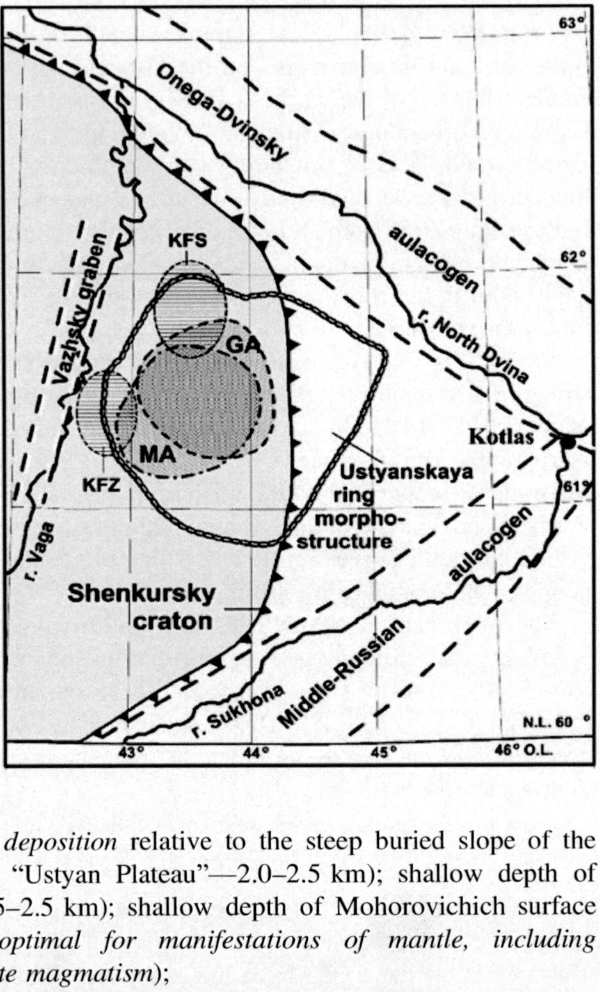

Fig. 2 Combination of favorable regional tectonic and geophysical prerequisites in the area of the Ustyansky prospective region. Letters on the diagram indicate: GA—gravitational anomaly, MA—magnetic anomaly; forecasted kimberlite fields: KPS—field "Severnoe", KPZ—field "Zapadnoe" [5]

- The zone of *step-by-step deposition* relative to the steep buried slope of the Baltic Shield (the buried "Ustyan Plateau"—2.0–2.5 km); shallow depth of crystalline foundation (1.5–2.5 km); shallow depth of Mohorovichich surface (33–35 km) (*which is optimal for manifestations of mantle, including diamond-bearing kimberlite magmatism*);
- The zone of influence of two avlacogens (Onegsky and Sredne-Russky) and also Vazhsky graben; block structure of the territory; presence of zones of deep faults of mantle formation (*that assumes the possibility of repeated deep tectonic and magmatic activation of the area*).
- Adjacent to the regional submeridional Meghra-Kepinskaya—Shenkurskaya deep faults zone of mantle structure (600 km *to the north, Zimneberezhny diamond-bearing region is located in the same deep faults zone*).
- Timing to 7-ball zone of the earthquake intensity at the western periphery of Sukhona epicenter of macroseismic field of Russian platform [2] (*completely similar to the same location of Zimneberezhny diamond-bearing region at the eastern periphery of White Sea 7–8-ball zone of macroseismic field*).

4.2 Favorable Morphological and Structural Prerequisites

- Large positive ring morphological structure of the central mantle type [4, 35] (*which is typical for the kimberlite magmatism regions of both Yakutia and Arkhangelsk*). The Ustyansky morphological structure is distinguished on the map of the modern stratoisogypsum relief in the form of a rounded dome elevation with an amplitude of 50–140 m and a diameter of about 120 km. The Ustyansky morphostructure has a complex block structure: by zones of deep faults in the NO and NW it is divided into two main parts—southwestern, relatively elevated, with a developed, dendriti form river network (region of the Archean craton) and northeastern, omitted, with a pattern of the river network close to linear (region of the Proterozoic belt). The dome-shaped uplift of the southwestern part of the morphological structure also appears on the geological map—outcrops of rocks from the Lower Tatar sub-stage surrounded by rocks from the Upper Tatar sub-stage of Upper Permian system.
- The presence of "subsidiaries", including local ring morphological structures of different order and size (*to which local magmatic manifestations can be timed*).
- The presence of zones of long-lived deep faults, emphasized by deep taps of straightened river sections of different order, and the deepest trench is the valley of the Ustya, which cuts the morphological structure in two radii. (Actually, the Ustianskaya morphological structure is named after the name of the river Ustya, which crosses the entire morphological structure with a broken line of several straightened sections and forms the main, radius, "frame" elements of the ring morphological structure).
- Spiral-vortex structure of the morphostructure (*which emphasizes the activity of its formation processes*).
- *The* area of neotectonic uplifts (*which indicates the possibility of the presence of Permian and post-Permian magmatism in the area*). On the map of neotectonic zoning and on the structural map of the surface of Pre-Pliocene sediments, the Ustianskaya morphostructure stands out as an isometric dome "Ustian Rise" with an amplitude of 50–120 m within the limits of the "Vazhsky Vault" [36]. (*Formation of such residual relief uplifts (against the background of general sinking of the territory) can occur due to reservation of the "body" of the morphostructure with a framework of magmatic manifestations (as a kind of piles*—Fig. 3*) (according to Offman P.E., Yermolenko Yu.P.).*

4.3 Favorable Geological and Stratigraphic Preconditions

- The presence of several stratigraphic and angular disagreements;
- A sharp change of tectonic regime in the late Permian era:

1. on the border of the Kazan and Tatar ages (rise of the territory, change of Kazan sea limestones to Tatar continental "basin" deposits);

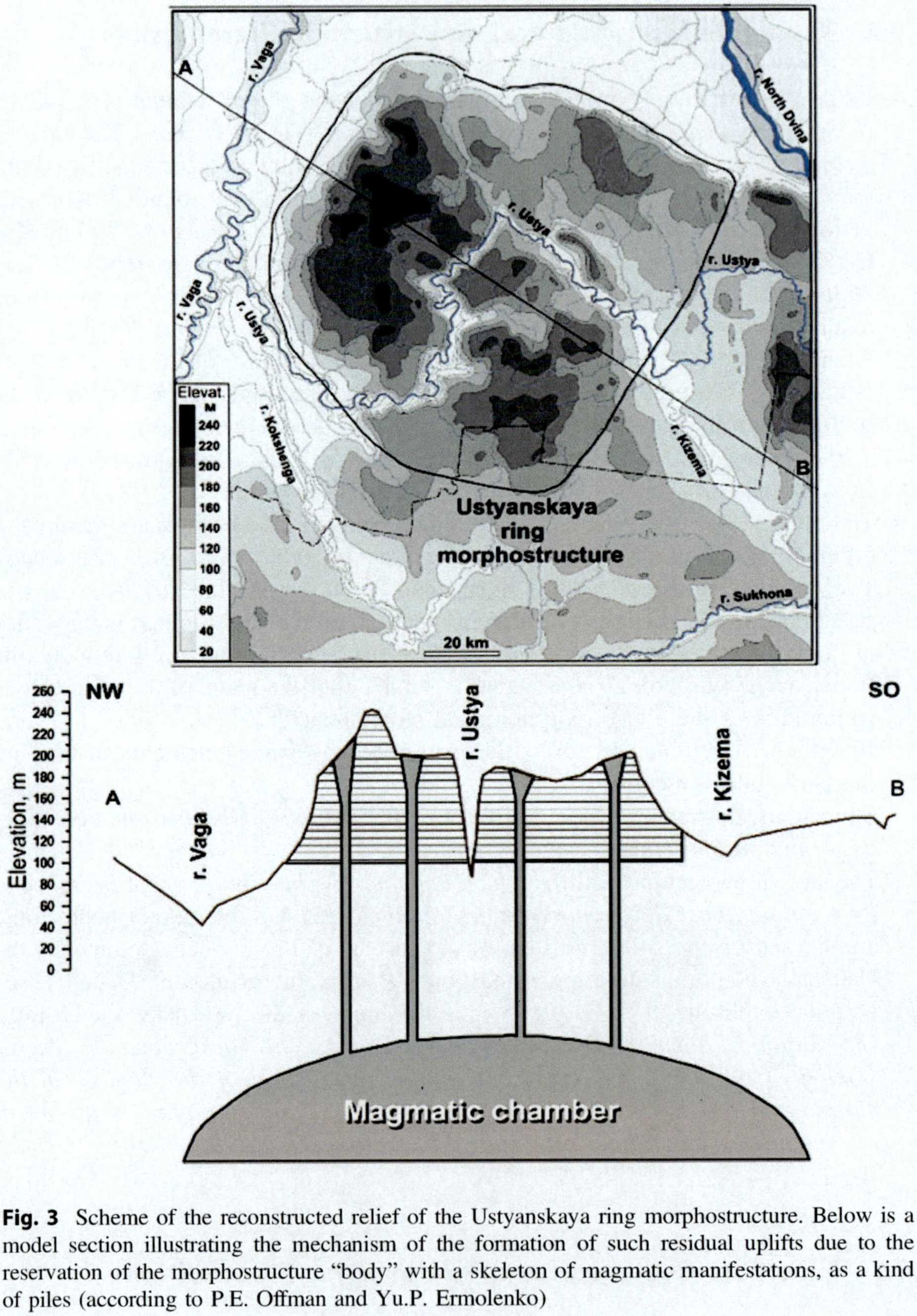

Fig. 3 Scheme of the reconstructed relief of the Ustyanskaya ring morphostructure. Below is a model section illustrating the mechanism of the formation of such residual uplifts due to the reservation of the morphostructure "body" with a skeleton of magmatic manifestations, as a kind of piles (according to P.E. Offman and Yu.P. Ermolenko)

2. on the border of early Tatar ages and late Tatar ages (change of the area of detritus from the northwest direction (from the Baltic shield) to the east direction (from the Ural Mountains).

– The area of long-term existence under the continental regime in late Permian and Mesozoic times (*"paleotectonic niche", typical for the areas of kimberlite activation*).

4.4 Favorable Geophysical Prerequisites

– *The* combination in space of a large gentle, brachiform positive *gravity* anomaly and a large low-intensity gentle, brachiform positive *magnetic anomaly* surrounded by sharp high-intensity alternating *magnetic anomalies of* linear and local type (*which may indicate the existence within the mantle diapir area*). The presence within this region of a combination of regional brachyform anomalies of local decompaction zones (which *may be directly related to the development of the mantle magmatic focus with the manifestations of kimberlite volcanism*).
– The presence of two potential kimberlite fields—"Akichkinsky" ("Severnoye") and "Vazhsky" ("Zapadnoye") (by [5]).

An important role in confirming the connection of favorable geological, geophysical and tectonic settings (prerequisites) with manifestations of mantle, including diamond-bearing kimberlite magmatism is played by mineralogical preconditions—the presence of anomalous diamond indicator minerals content of kimberlite origin in the prospective area.

4.5 Favorable Mineralogical Prerequisites

– Alluvial deposits of the area contain a large number of diamond indicator minerals of kimberlite origin—hundreds of thousands of pyrope and chrome spinel grains, less chrome-diopside, olivine and picroilmenite (the *total content of pyrope in alluvium is* 10–30 *times higher than in the diamondiferous area of the Zimny Bereg and any other area of European Russia*).
– Initial, not worn out kimberlite indicator minerals: pyrope, chrome spinel, chrome-diopside, olivine, and picroilmenite are present in the alluvium of the area 3–5 times higher than in the diamondiferous area of the Zimny Bereg, including in the class of +1 mm (pyrope and especially picroilmenite);
– High (about 10%) share of near demolition pyrope grains of diamond association of G10 group (*which is typical for areas with high-diamondiferous kimberlite volcanic rocks*).

The combination of all favorable regional conditions in space (Fig. 2) may be due to the introduction of mantle diapir and may be accompanied by kimberlite, including diamondiferous kimberlite magmatism within the Ustyanskaya perspective area, primarily in its western ("Zapadnoye" field) and north-western ("Severnoye" field) parts.

According to the aggregate and combination of favourable geological, geophysical and mineralogical prerequisites and features, the Ustyansky Diamond-bearing Region (UDR) is almost a complete analogue of the Zimny Bereg Industrial-Diamondferrous Region (ZDR), differences only in the age of kimberlite magmatism: ZDR is the set age of D_3 (late Devonian [22, 24]) and UDR is the assumed age of PZ_3 (Late Permian)—MZ.

5 Indirect Search Features

Due to the specificity of the mineral composition of diamondiferous kimberlites and very low content of a useful component—diamond (0.01–0.2 g/t, rarely up to 1 g/t), the practice of diamond prospecting uses not only ***direct search features, i.e. the*** presence of diamond crystals themselves in the samples, but also ***indirect search features, i.e. the*** presence of diamond indicator minerals (hereinafter—DIM) in the samples, first of all—the *diamond paragenesis* DIMs with not distant moving.

Search features (both direct and indirect) have a mineralogical basis, so of great interest are the mineral composition of ancient sedimentary (native) bedrocks, which could supply the material in the Quaternary sediments (including as possible intermediate collectors of diamond). All these sediments have continental (for the Upper Permian—"basin") origin and their mineral composition reflects conditions and character of their formation.

Main, mineralogically studied stratigraphic divisions of the district:

Upper Permian:

Kazan Stage *(sea limestone, no heavy fraction).*
Tatar Stage *(continental, "basin" carbonate-terrigenous rocks):*
Lower Tatar Substage, Urzhumsky Horizon:
Nizhneustyinskaya Formation P_2nu, **Sukhonskaya Formation** P_2sh.
Uppertatarskiy Substage:
Severodvinsky Horizon P_2sd, **Vyatsky Horizon** P_2vt.

Lower Trias:

Vetluzhsky *supragorizon* T_1vtl *(continental terrigenous rocks).*

The total mineral composition of the root rocks of the district reflects the location of the sources of demolition and the direction of movement of the detritus material. Figure 3 clearly records sharp tectonic reorganization of the territory that took place between Sukhonsky and Severodvinsky times and changed the north-western direction of demolition (from the Baltic shield) to the eastern direction of demolition (from the Ural Mountains). Terrigenous rocks of the Urzhumsky horizon Lower Tatar substage of the Upper Permian's with erosion lie on the limestone of the Kazan Stage, and the "mature" zircon-ilmenite-garnet mineral association of the heavy fraction (hereinafter—so called—h.f.) of the sandstones of the Nizhneustyinskaya Formation is replaced by almost purely garnet mineral association of the Sukhonskaya Formation. But, since the Severodvinsky time, the mineral composition of the heavy fraction of terrigenous rocks becomes quite different: the sediments of the Severodvinsky and Vyatsky horizon is characterized by "immature" magnetite-amphibole and magnetite-amphibole-epidote association. A similar, chrome spinel-epidote mineral association is also noted for the above rocks of the Vetluzhian supra-horizon of the Lower Triassic, coarse-clastic differences which are also characterized by an abundance of pelleted fragments of Urals jasper.

Terrigenous rocks of all the above subdivisions contain potential diamond indicator minerals (DIM): pyrope + chrome spinel (P_2nu, P_2sh) or chrome spinel only (P_2sd, P_2vt, T_1vtl), but only two of them have DIM content quite high.

Judging by the high, even hurricane content of diamond indicator minerals (up to 40% of the heavy fraction!), sediments of the ***Vetluzhsky horizon of the Lower Triassic*** may well be a potential intermediate collector of diamonds, but they found only chrome spinels, which have a far-reaching character (small size (−0.25 mm) high degree of rounded) and a non-kimberlite root source (this is indicated by the peculiarities of their morphology, chemical composition and intergrows with uvarovite grains). The root source of these chrome spinels are most likely hyperbasite massifs of the Urals (eastern direction of demolition).

The most interesting potential intermediate collector of diamonds in the area are the ***sandstones of the Nizhneustyinskaya Formation of the Lower Tatarsky Substage of the Upper Permian,*** where the high content of kimberlite indicator minerals has been noted: pyrope (15–236 grains/10 L, up to 15,000 grains/1 kg of the h.f.) and chrome spinel (6–125 grains/10 L, up to 24,000 grains/1 kg of the h.f.). Pyrope, and chrome spinel are characterized by the same morphological features (Fig. 4a, b), which indicate at least two stages of their exogenous morphogenesis (in different proportions and sequences):

(1) intensive mechanical wear under conditions of coastal-lake sedimentation (small size, high degree of rounded, polished and/or rough surface); (2) hypergenic dissolution (for pyropes—dissolution of "cuboid" type) under conditions of weathering crust. Chrome spinels are characterized by a wide range of chemical composition (20–50%, up to 60% Cr_2O_3) with predominance of moderate chromium, low titanium grains. The color of pyrope is mainly violet, crimson pale tones; reddish-colored and orange-colored pyrope are extremely rare. The overwhelming part of pyrope of Nizhneustyinskaya Formation belongs to low chromium (1–4% Cr_2O_3) and low titanium variety of lerzolite paragenesis of G9 group

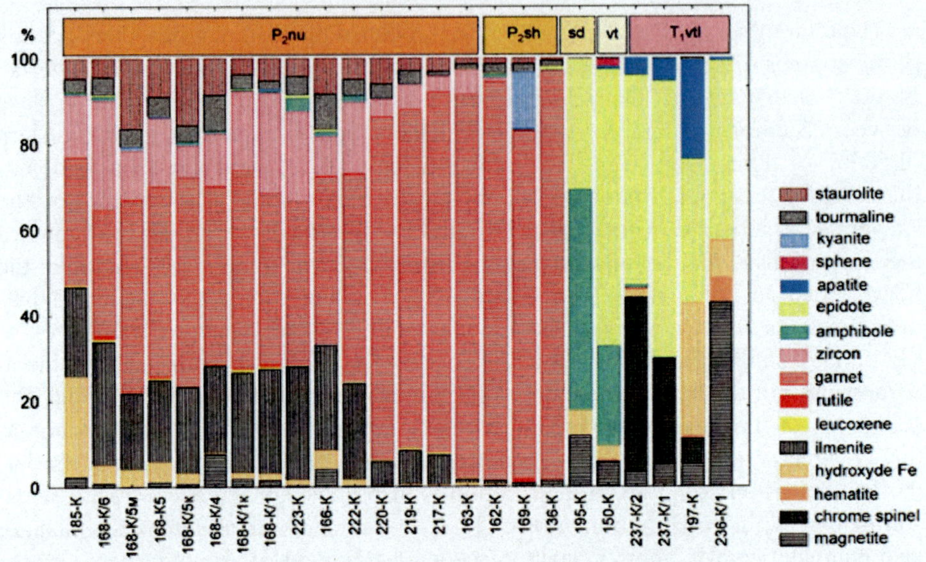

Fig. 4 The general mineral composition of the bedrock of the Ustyansky region according to the study of crushed samples of the Permian and Triassic rocks

(according to [3], more chromium pyrope (up to 7% Cr_2O_3) are single (Fig. 4c). At the same time, a noticeable number of low chromium low-calcium pyropes of dunite-harzburgite paragenesis of G10 group (of diamond association according to [6]) is noted. Their presence allows to consider rocks of the Nizhneustyinskaya Formation as possible intermediate collector of diamonds, the root source of which could be ***low-diamondiferous kimberlites of post-Kazan—pre-Tatar age of Late Permian***.

Modern alluvial deposits and Quaternary sediments of glacial origin are much more diverse with respect to diamond indicator minerals. Most of the prospecting mineralogical information in the region was obtained by studying slug samples from modern alluvium. In this case, panning sampling of modern riverbed alluvium in the region is, in fact, testing the material of the ***Quaternary glacier cover,*** enriched in favorable hydrodynamic conditions of modern watercourses ("glacial background") with the addition of root rock material, which is diluted in the immediate vicinity of the point of testing ("abnormal component"). This allows interpretation of the results of the panned sampling not as a linear (strip) panned sampling of watercourses, but as an ***area testing (panned sampling) of glacial deposits of the territory through an uneven network***.

Alluvial deposits of the region contain hundreds of thousands of grains of diamond indicator minerals (pyrope, chrome spinel, chrome-diopside, olivine, picroilmenite), the total content of pyrope is 10–30 times higher than in the diamond-bearing region of the Zimny Bereg and any other region of European Russia. Moreover, the distribution of pyropes from alluvium in terms of grain size, grain surface type, and chemical composition has a distinct bimodal character, which indicates that they come from at least two types of native sources (the third type of native source—Lower Triassic deposits—contain only far-bearing chrome spinels of non-kimberlite origin from Ural hyperabasites and are unlikely to be of search interest).

Type 1. The most part of low chromium pyropes and chrome spinels is well rounded and has a particle size class of −0.5 + 0.25 mm and −0.25 mm (pyropes and chrome spinels of the group P3 and X3, respectively, according to classification [26], and by all characteristics is identical to these minerals from the intermediate collector—sandstones of the Nizhneustyinskaya Formation of the Upper Permian (P_2nu), characterized above (Fig. 4). The distinct maximum concentrations of these pyropes and chrome spinels in the Upper Permian sandstones (and subsequently in alluvium) are observed in the lower reaches of the section of the Nizhneustyinskaya Formation in the west of the Ustyansky Region.

Type 2. The uncoated diamond indicator minerals of near-moving (or DIM of the 1st group, further DIM-1 according to the classification [26]) form a completely separate group and have a special search value. They ***directly*** indicate the possibility of presence kimberlites in the area: they are, first of all, pyrope and picroilmenite (Fig. 5), as well as chrome spinel, chrome-diopside and olivine.

The content of uncoated ***pyrope*** grains in alluvium of the district is 3–5 times higher than in the diamond-bearing district of the Zimny Bereg. The pyrope of close demolition sharply differs from the pyrope of the Upper Permian Nizhneustyinskaya Formation collector by its surface character ("primary-magmatic" surface of the 1st and 2nd types by [1], absence of signs of mechanical wear, larger sizes (particle size class −1 + 0.5 and −2 + 1 mm), high chromium content (2–12%, up to 14% Cr_2O_3) and high percentage of group G10 diamond association pyrope grains (about 10%), which is typical for diamondiferous kimberlites (Fig. 5c), including the diamondiferous kimberlites of the Zimny Bereg.

Picroilmenite is contained in alluvium in smaller quantities than pyrope, but large grain sizes (class −4 + 2 mm and −2 + 0.5 mm), primary magmatic surface (corrosive, microbumpy, "studded"), The presence of leukoxene edges and the formation with altered carbonate rock, as well as features of the chemical composition (4–16% MgO, up to 3.5% Cr_2O_3) indicate a direct connection with the local indigenous diamondiferous kimberlite source. The composition of picroilmenite (Fig. 6), on the one hand, is close to the composition of ilmenite from the high-diamondiferous Grib pipe of the Zimny Bereg area.

On the other hand, a significant proportion of low-magnesian (4–9% MgO) picroilmenite with variable chromium (0.1–2.5% Cr_2O_3) differs sharply from picroilmenite from all known kimberlites of the Zimny Bereg area, but has similarities with picroilmenite from the high- diamondiferous pipe Mir in the Malo-Botuobinsky area (Yakutia).

Chromium spinelide of distinct kimberlite origin is rarely found in alluvium, but among its grains there are grains of diamond paragenesis with Cr_2O_3 content up to 69%.

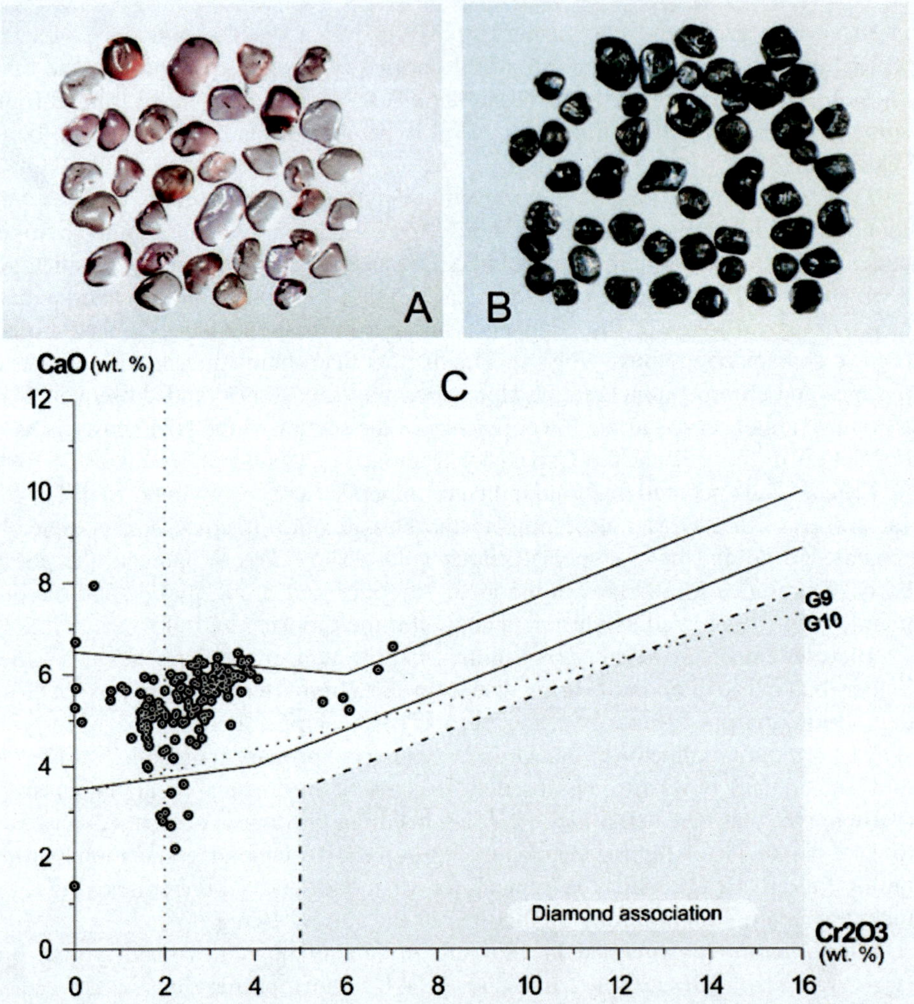

Fig. 5 Diamond indicator minerals from the rocks of the Nizhneustyinskaya Formation of the upper Permian of the Ustyansky region. Above is a general view of grains of pyrope (**A**) and chrome-spinels (**B**). Bottom (**C**)—diagram of pyrope composition in Cr_2O_3–CaO coordinates (according to [34])

The distribution of small, rounded low chromium pyrope of group P3 in alluvial sediments of Ustyansky region (in signs for 1 kg of heavy fraction) as a whole very well corresponds to the area of reaching the prequaternary surface of sandstones of Nizhneustyinskaya Formation sandstone collector of Upper Permian (P_2nu), moreover, the location of maximum concentrations of these pyrope in Upper Permian rocks exactly repeats (corresponds) the location of maximum concentrations of these pyrope in alluvium. The similar correspondence is noted between the area of development of rocks of the Lower Triassic (T_1vtl) and the maximum

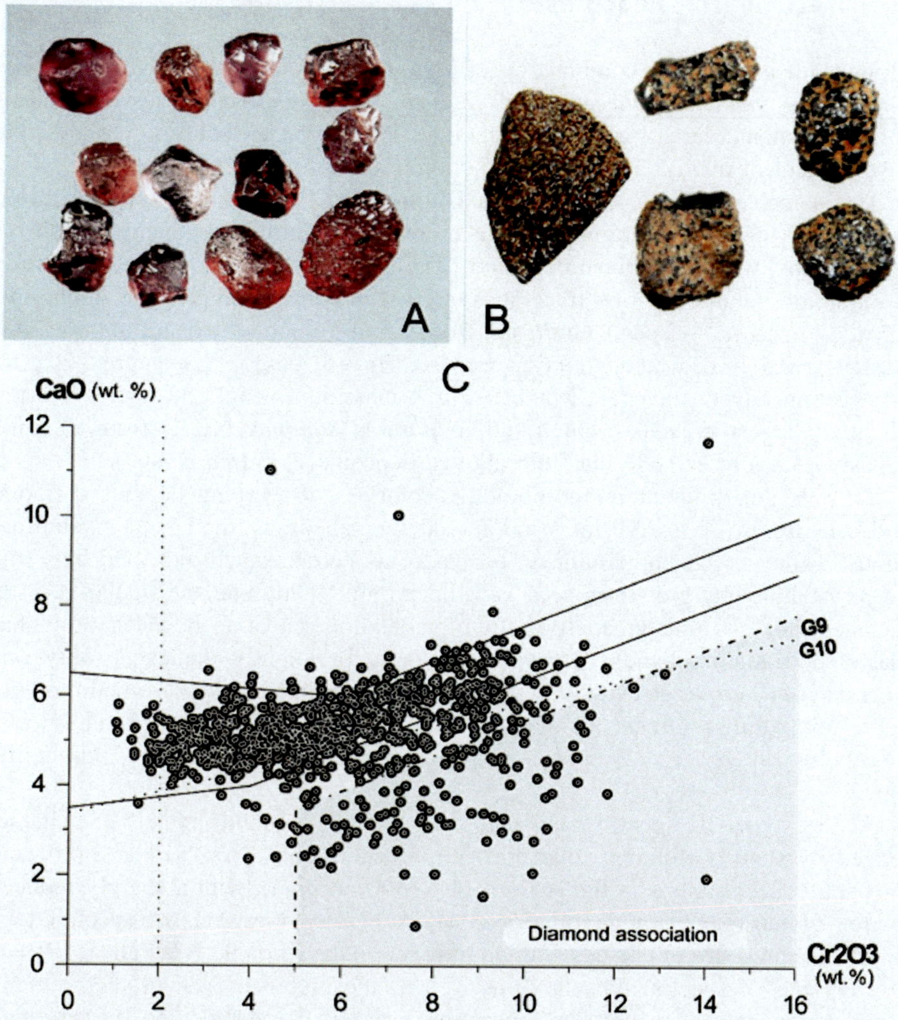

Fig. 6 Diamond indicator minerals from Quaternary deposits of the Ustyansky region. Above is a general view of grains of pyrope (**A**) and picroilmenite (**B**). Bottom (**C**)—diagram of pyrope composition in Cr_2O_3–CaO coordinates (according to [34])

concentration in alluvium of chrome spinels characteristic for rocks of the Triassic. The fact that typomorphic kimberlite indicator minerals in alluvium in essence mark the areas of development of known indigenous sources (rocks P_2nu and T_1vtl) allows us to expect that both pyrope and picroilmenite, and other diamond indicator minerals with not distant moving (DIM-1) in alluvial sediments of the Ustiansky region can also largely reflect the composition of underlying indigenous rocks, including the expected kimberlites.

6 Direct Search Features

Along with a favorable combination of *search prerequisites* and *indirect search features, the immediate,* ***direct search features***—the presence of diamond crystals in the promising area—are of paramount importance to confirm the possible development of mantle, including diamondiferous kimberlite magmatism.

The **search works of Arkhangelsk Diamonds LLC and RUSGEO Limited established the fact of diamondiferous modern alluvium and glacial deposits of the Ustyansky diamondiferous region** [27]. **Nine diamond crystals** were found in minimum volume samples (6 crystals in 20–40 l panned samples and samhles of 0.5 m^3; 1 m^3; 2 m^3—3 pcs.). Grain size from 0.5 to 3.8 mm, the distribution by size classes: −0.5 + 0.25 mm—1 pcs., −1 + 0.5 m—5 pcs., −2 + 1 mm—2 pcs., −4 + 2 mm—1 pcs.; average weight 8.3 mg, maximum weight 52 mg (1/4 carat). The two largest crystals were found in ***panned samples*** (40 l!) from moraine deposits (3.8 mm crystal) and from alluvial deposits (2.65 mm crystal).

Crystals are different in morphology: colorless, to varying degrees distorted dodecahedroids (up to 2.65 mm) and flat-sided octahedrons (the largest crystals are almost 4 mm in size) predominate; octahedroids, tetrahexahedroids, and only one polycrystalline intergrowth (Fig. 7) have been found. Only one, the smallest crystal (octahedroid) has rare green spots of pigmentation, and one dodecahedroid has plastic deformation bands. Crystals are whole or slightly damaged (only one fragment of a very large crystal), with no signs of mechanical wear, sometimes with ochre admixtures. Almost all diamonds are of gem quality. In general, such a morphological variety of crystals is typical for high-diamondiferous kimberlite pipes.

According to IR spectroscopy data, the studied diamond crystals are characterized by sharply different total nitrogen concentrations (46–1514 at. ppm) and nitrogen concentrations in the A-form (46–1300 at. ppm), with a sharply reduced degree of nitrogen aggregation (%NB = 8.4–17.8 − 4 crystals of spectral type IaAB), the presence of crystals with no nitrogen in the form of "B" at all (%NB = 0 − 4 crystals of spectral type IaA), as well as the presence even in such a small sample of a large (2.4 mm) "nitrogen-free" crystal of spectral type IIa (Fig. 8a). A common feature of almost all crystals, along with a sharply reduced degree of nitrogen aggregation, are also very low concentrations of platelets (P-center = 0.0–3.8 cm^{-1}) and hydrogen (0.00–1.07 cm^{-1}) (Fig. 8b, c). Despite the sharp morphological differences, these crystals form a single, quite homogeneous set in the distribution of the impurity centers. In terms of the content of separate structural admixtures of nitrogen and hydrogen, the crystals studied have similarities with rare varieties of diamonds from the Middle Urals, the Grib deposit (Arkhangelsk region), and Finland. However, by the aggregate distribution of the main syngenetic structural impurities and structural defects, the studied diamond crystals occupy a completely separate position.

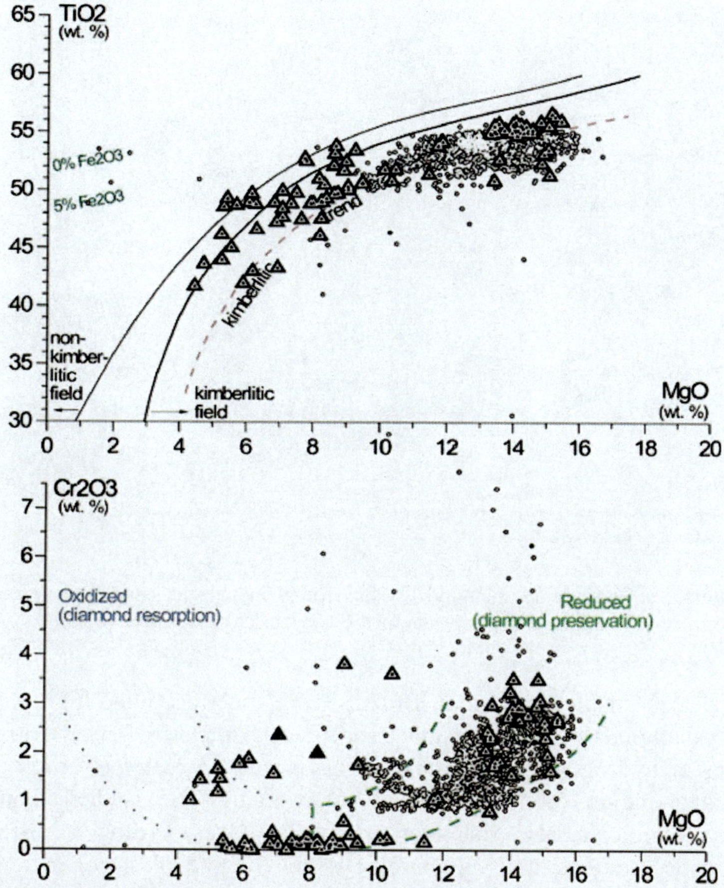

Fig. 7 Composition of picroilmenites from alluvial deposits of the Ustyansky region: MgO–TiO_2 (diagram [40] above) and MgO–Cr_2O_3 (diagram [7, 8] below)

Relatively low concentrations in nitrogen, hydrogen, and platelets crystals, as well as a very low degree of nitrogen aggregation, indicate rather slow crystallization under conditions close to equilibrium (early and main stages of natural diamond formation). The crystals are relatively low-temperature (at the conventional age of 3 billion years the temperature in [39] does not exceed 1100 °C). Judging by the nitrogen distribution in the crystals, we can assume an ultramafic association (paragenesis) of diamonds.

It is interesting to note that by the nature of distribution of structural impurities in the World there are not many original and isolated, bright, "reference" groups of diamonds [11, 12, 32]. For example, most of the kimberlites of both Yakutia and Arkhangelsk differ slightly in these indicators, and are sharply isolated mainly in the kimberlites and placers of South America (Brazil and Venezuela). The most extreme position in terms of these indicators is occupied by kimberlites and

Fig. 8 External appearance of diamond crystals found in panned samples and small-volume samples from alluvial and moraine deposits of the Ustyansky prospective region

diamond placers of the Juina field, Brazil. These most deep and high-temperature (1250 °C) diamonds of lower mantle origin are characterized by the lowest concentrations of hydrogen, platelets and nitrogen with an extremely high degree of nitrogen aggregation (about 90% N_B on average). The studied diamonds of Ustyansky Region are also characterized by the lowest hydrogen, platelets, low nitrogen concentrations, but an extremely low degree of nitrogen aggregation (about 5% N_B on average) and the lowest formation temperatures (less than 1100 ° C)! That is, by the nature of structural impurities distribution, the studied diamonds of Ustyansky Region occupy the same "extreme" position, but directly opposite to Juina diamonds. The Ustyansky diamonds form a peculiar and isolated, bright, "benchmark" group. (The only thing left to do is to find and study an unusual root kimberlite source of the Ustyansky diamonds to give a name to this new "reference" group of diamonds).

The only crystal not included in this sufficiently homogeneous and original set is characterized by a high total nitrogen content (1413 at. ppm) and an increased degree of aggregation of nitrogen (%NB = 30.8). In these terms, it is close to the crystals of the diamond from the Lomonosov deposit (Arkhangelsk diamond province). However, a very high concentration of platelets (28.1 cm^{-1}—3 times higher!) and a completely unusual, extremely high hydrogen content (35.9 cm^{-1}—an order of magnitude higher!) indicate that this crystal also comes from an unknown (not yet discovered) root primary source!

Thus, in terms of distribution features, complex morphological features, and the ratio of the main structural impurities of nitrogen, hydrogen, and platelets, the studied diamonds occupy a completely separate position and practically have no analogues among diamonds of the known deposits and ore manifestations of the Russian Platform and the Urals. This may indicate that the crystals studied are coming from a new, not yet known kimberlite source. This potential new root kimberlite source may have a higher diamondiferous capacity, higher particle size and high diamond quality. For most of these features, the closest analogues of the studied diamonds are diamonds from the pipes of the Malo-Botuobinsky district of Yakutia province (first of all, the Mir pipe diamonds).

In no other region of European Russia (including areas of deposits and ore manifestations of diamonds of the Urals, Timan, Arkhangelsk and Finland) such diamond finds in the plaster samples were not *(usually single diamonds are found in large-volume samples of* 30–50 m^3*)*. Such large diamond crystals in panned samples of minimal volume are found only in industrial-high-diamondiferous regions of Yakutia (Daldyno-Alakitsky, Malo-Botuobinsky, Sredne-Markhinsky and Anabarsky regions).

7 Local Search Prerequisites and Features. Allocation of the Most Promising Sites.

In terms of combination and overlapping of favourable regional geological, geophysical and mineralogical prerequisites and search features, the Ustyansky Diamondiferous Region (UDR) is almost a complete analogue of the Zimny Bereg High-Diamondiferous Region (ZDR), where about 80 kimberlite and related magmatic bodies have been discovered, including the name Lomonosov and name Grib diamond deposits. The differences are manifested only in the age of the known and predicted kimberlite magmatism: for the Zimny Bereg area is actually a late Devonian (D_3), for the Ustyansky area is presumably a late Permian, the Tatar age —Mezozoic (P_2t-MZ). The total area of the Ustyansky perspective region is about 15,000 km^2, but some of the available search prerequisites and features allow us to localize the most likely areas of kimberlite magmatism and identify the most promising areas for detailed work.

The most important for localization in the prospective area of the proposed kimberlite field are ***geophysical*** prerequisites. As noted above, within the Ustyansky area within the region of combination of regional brachiform magnetic and gravitational anomalies there are separate local zones of decompaction, which may be directly related to the development of deep magmatic foci and manifestations of kimberlite volcanism. Based on the analysis of geophysical information within the area, two local areas of potential kimberlite fields—"Akichkinskoe" ("Severnoye") field and "Vazhskoe" ("Zapadnoye") field [5] (Fig. 2) were identified.

Fig. 9 Physical properties of diamonds found in the Ustyansky promising region. Distribution of the total impurity of nitrogen (**A**), platelets (**B**), and hydrogen (**C**) depending on the degree of aggregation of nitrogen in diamonds. Diamonds of the Ustyansky region are stars, a large star is the average value for 8 crystals, the shaded ellipse is the standard deviation area. Reference fields of the composition of diamonds of the World according to the data: [11, 12, 32]. Isotherms (Fig. A) according to [39]

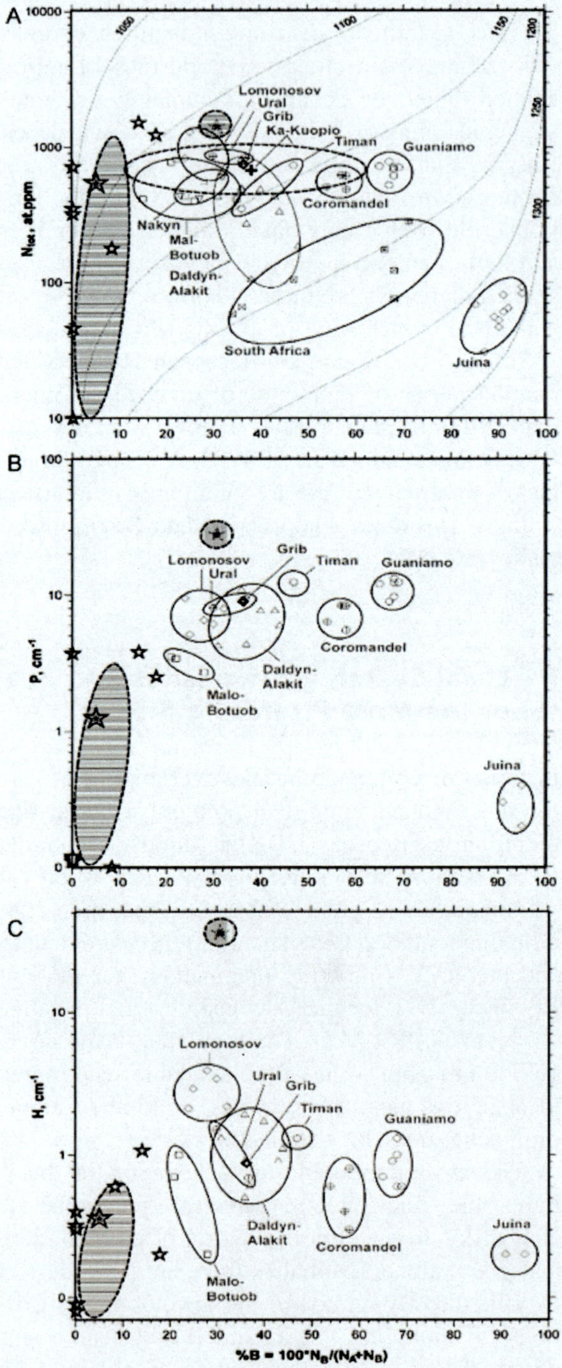

Other independent indicators for localization of the most promising areas of magmatism development are preconditions of ***morphometric*** characteristics of the territory relief. The most interesting areas are singled out as areas of overlap of the maximum area of anomalous zones of density, isotropic and specificity of the river network [15, 16]. Such areas are distinguished primarily in the north-western, western and central parts of Ustyansky Region (Fig. 9). The presence of "daughter" ring morphological structures of different order and size (whole circle, big circle, small circle, smallest circle) in the prospective area is also a favorable indicator; the most interesting in this respect is the ***northwestern part of Ustyansky Region***.

The ***mineralogical*** characteristics of the areas—indirect and direct local search features, i.e. the presence of anomalous kimberlite indicator ***mineral*** concentrations of near demolition and diamond crystals themselves, respectively—are very important in terms of confirming the reliability of potential kimberlite fields within the Ustyansky Region. As noted above, despite the complex nature of the formation, the mineral composition of modern alluvium inherits the mineral composition of quaternary glacial sediments and largely reflects the mineralogical characteristics of underlying rocks. The possibility of relative localization in the area of possible indigenous kimberlite sources, i.e. the real formation of panned-mineralogical "near moving halos", is indicated by some peculiarities of properties and character of distribution in the rocks of diamond indicator minerals and diamonds themselves.

Thus, the signs of near moving of the material are complete or partial preservation of primary magmatic surfaces and absence of mechanical wear of Kimberlite Indicator Minerals (KIMs), presence of polymineral (complex) halo of KIMs and absence of particle-size sorting of their grains (classes of grain size from −0.5 mm to +2 mm), and also ratio of different kinds of KIMs in halo similar to such ratio of minerals in supposed initial root source.

These very peculiarities of kimberlite indicator minerals distribution are observed in alluvial sediments, first of all ***in the northwestern part of Ustyansky Region,*** where anomalous pyrope and picroilmenite content of near moving are recorded, forming low-contrast halos of scattered type in a limited area. Here, the full set of kimberlite indicator minerals and the presence of diamond grains are noted in individual samples, as well as the increased grain size of these grains (up to −4 + 2 mm) and the predominance of picroilmenite over pyrope (which is typical for most kimberlite pipes of the World). In addition, the detected diamond crystals have sharply different sizes: from 0.5 mm to 3.8 mm (from the particle size class −0.5 + 0.25 mm to −4 + 2 mm), which is also typical of with not distant moving halos and is not typical of far-bearing sorted ("calibrated") material (including material from intermediate collectors).

Mineralogical prerequisites, indirect search features and, finally, direct search features determine as the most promising exactly ***north-western part of Ustyansky Rgion.*** Most of the diamonds (6 out of 9 crystals) were found in alluvial and moraine sediments in the northwestern part of the Ustyansky Region, at an area of about 400 km^2. within the Zimny section, of which 5 diamond crystals (including 3 of the largest, including 2 octahedrons) were found as a single chain of north-western strike at an area of less than 100 km^2. At the same time, a very rare phenomenon for the glacial region was noted here: the repeatability of the diamond finds and the reproducibility of the results, which indicates elements of the natural character of diamond distribution and allows at this stage of prospecting to estimate the predicted resources of the primary and alluvial diamond deposits in the Ustyansky perspective area in general and the predicted kimberlite field in particular.

The location of local preconditions and signs of diamondiferousness within Ustyansky Region is quite definite and logical (Fig. 10). The most interesting area of combination of haloes of near moving kimberlite indicator minerals in alluvial and glacial sediments, maximum concentrations of picroilmenite and red–orange

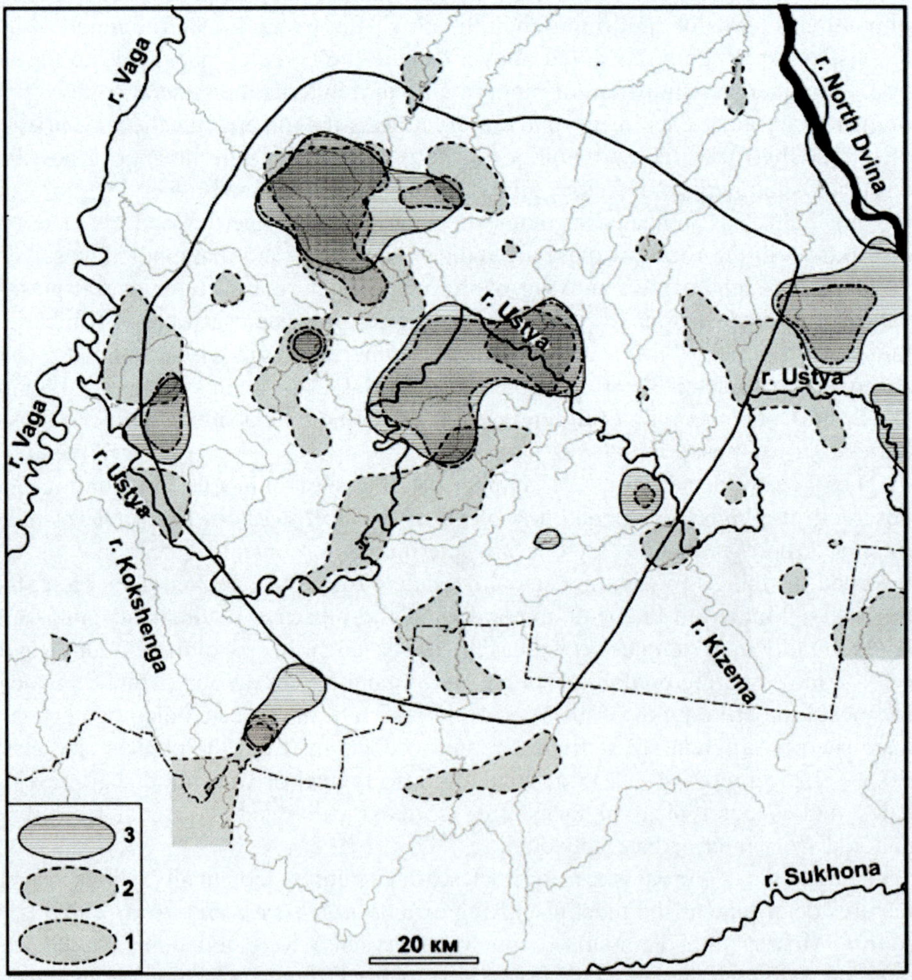

Fig. 10 Results of a morphometric study of the territory of the Ustyansky ring morphostructure. Scheme of combining anomalous zones of density (1), isotropy (2), and specificity (3) of the river

titanium pyrope, finds of the majority of diamonds, including all major diamonds, the maximum area of favorable anomalous zones of density, isotropy, and specificity of the river network coincide with the potential kimberlite field "**Severnye** " [5]. It is this section of the Zimnyaya and Kodima River basins and the upper reaches of the Volyuga River that is the most promising for the detection of full-fledged primary diamond deposits of ***late Tatar-Mesozoic age***.

The possible localization of a fundamentally different indigenous kimberlite source is indicated by the presence of low chromium pyrope and chrome spinels and the natural character of their distribution in an intermediate collector—sandstones of the Nizhneustyinsky Formation of the Upper Permian described above. This ***indirectly*** confirms the possible presence within the Ustiansky area of the second new kimberlite field, whose kimberlites should be of late Permian (post-Kazan—pre-Tatar) age. Despite the high degree of rounded of KIM grains, the localization of their root source is quite probable and possible, which is indicated by the continental nature of Permian deposits, the joint presence of pyrope and chrome spine in the rocks, as well as sharp but logical differentiation of the content of these minerals in the samples in both the section and the area of work.

The distinct maximum concentration of these pyrope and chrome spinels in the sandstones of the Upper Permian (and after them—in alluvium) is noted at the bottom of the section of the Nizhneustyinsky Formation in the west of the Ustyansky region, in the zone of junction of the three rivers: Vaga, Ustya and Kokshenga and precisely coincides with the location of the predicted by geological and geophysical features kimberlite field "**Zapadnoye**" [5], as well as with the area of local anomalies of morphometric characteristics (density, isotropic and specificity of the hydropower network (Figs. 9 and 10). Judging by the low chromium composition of pyrope and chrome spinels, and even despite the formal presence in the sandstones of the Nizhneustyinsky Formation of pyrope group G10 of the Diamond Association (according to [6]), the prospects for industrial diamondiferous of the predicted ***Late Permian (post-Kazan—pre-Tatar)*** kimberlite field "**Zapadnoye**" are completely unobvious and, most likely, unlikely. The possibility of kimberlite magmatism development in the area of this field is also indicated by the elements of magmatic (or geological?) preconditions revealed here.

Within Ustyansky Region, there are indigenous breccia rocks, which may have various (including volcanic) origins [26]. Most of the breccias studied form ***sub-consistent*** geological ***bodies***, which by petrographic, mineralogical and geochemical characteristics ***are syngenetic to the host sedimentary rocks of the*** Permian. There is only one exit of a ***secanting subvertical body*** with sharp contacts, 2–3 cm in thic and 40 cm in length. This body is formed by yellowish-green, fully clayed fine psephite breccia, which consists of 60% of clay debris, rarely sandstone, cemented with silty clay. There are no distinct traces of magmatic material. In terms of structure and conditions of occurrence, this body may be eluvial-deluvial deposits that have filled the subvertical crack of the side rest ("neptunic dike"?). However, another type of formation of this rocks cannot be completely excluded. According to the results of the instrumental neutron activation analysis for this breccia a very high content of chromium—**520 ppm** (!!!) is observed, which is comparable to the content of chromium in ultramafic rocks and requires more detailed study of this breccia in the field and laboratory conditions. This sectional body of high chromium breccia (sample G-168-K/S) is located in the area of maximum concentrations of pyrope and chrome spinels in the sandstones of the Nizhneustyinsky Formation of Permian—at the mouth of the Kokshenga River, within the predicted kimberlite field **"Zapadnoye"**.

8 Discussion of Results

The research confirms that Ustyansky promising region has been identified as an area of plume tectonic and magmatic activity. As shown above results of forecast and prospecting works, Ustyansky perspective region is characterized by combination in the space of ALL favorable regional search conditions (morphological, tectonic, geological-stratigraphic, geophysical, morphometric, mineralogical) and search features (presence of diamond crystals) that can be caused by mantle diapir and can be accompanied by diamond-bearing kimberlite magmatism. The location and combination of local search prerequisites and features allow us to identify two predicted kimberlite fields within the western (the "Zapadnoye" field) and northwestern part of the Ustyansky morphological structure (the "Severnoye" field). The expected kimberlites of these two fields should differ sharply in composition (petrologic type), diamondiformity and age.

Judging by the set and composition of the kimberlite indicator minerals (low chromium and low titanium pyrope + chromium spinel), ***kimberlites of late Permian, post-Kazan—pre-Tatar age,*** which supplied the material to the rocks of the Nizhneustyinsky Formation, may be close in composition to some of the ilmenite-free kimberlites of the alumina series (Al-series [23]) of the Zimny Bereg area (similar Lomonosov deposit) and kimberlites of group 2 of South Africa [33]. These kimberlites may at best be weakly diamondiferous, so the **prospects for** industrial-diamondiferous predicted kimberlite **field "Zapadnoye" is completely unobvious**.

On the contrary, relatively younger ***kimberlites of late Permian (Late Tatar)—Mesozoic age were*** expected to supply material only to the Quaternary rocks, judging by the set and composition of kimberlite indicator minerals (high chromium pyrope + high magnesia and high chromium picroilmenite), as well as the presence in the samples of a minimum volume of large diamond crystals (including a significant proportion of octahedrons) may be close to the composition of ilmenite-bearing kimberlites of iron–titanium series (Fe–Ti-series [23]) of the Zimny Bereg area (including pipe named V.P. Grib) and kimberlites of group 1 in South Africa [33] and Yakutia. The **prospects for the** industrial-diamondiferous **potential of the** projected late Permian—Mesozoic kimberlite **field "Severnoye" is very high**. Wide distribution, increased particle size and high gem quality of the discovered diamonds allow us to expect high diamond content and value in the proposed kimberlite pipes and placers of this field.

In terms of combination and combination of favourable geological, geophysical and mineralogical prerequisites and search features, as well as confinement to the same Megra-Kepinsko-Shenkursky deep fault zone, Ustyansky promising region is almost a full analogue of the Zimny Bereg high diamond-bearing area, and the projected kimberlite field "Severnoye" is almost a full analogue of the fields of diamond deposits named after Grib and (or) named after Lomonosov diamond fields.

On similarity Ustyansky perspective region and Zimny Bereg high-diamondiferous region specifies also a present in them variety petrological type kimberlites (kimberlite groups 1 and groups 2)—prospective (Ustyansky region) and already known (Zimneberezhny region).

It is interesting to note full analogy of Zimneberezhny high-diamondiferous region and Ustyansky perspective region in macroseismic field of Russian Platform [2] (Fig. 11). Zimneberezhny region is confined to the eastern periphery of VII-VIII ballt epicenter of the White Sea maximum macroseismic field connected with Onego-Dvinsky avlacogen (activated in late Devonian), and Ustyansky region is confined to the western periphery of VII-ball epicenter of Sukhona maximum macroseismic field connected with Sredne-Russky avlacogen (activated in Permian-Mesozoic time). It means that the existence of the Devonian White Sea mantle diapir (plume) and the Permian-Mesozoic Sukhona plume is still manifested, including the intensive tectonic life of the regions.

Therefore, there is probably only one important and obvious difference between Zimneberezhny and Ustyansky regions—the younger age of the alleged kimberlite magmatism in Ustyansky region. On the Russian Platform, the age of the known kimberlite and related volcanism, as well as basalts, is defined as Late-Middle Devonian for the Zimny Bereg diamond-bearing region, the Nenoksky region of the Onega Peninsula and the Umbinsky region of the Middle Timan [19–22, 24], the Tersky Bereg of the Kola Peninsula [10] and the Zhlobinsky field of Belorussia [18]. An older, Early Paleozoic, Late Proterozoic age of such magmatism for the Baltic Shield and its framing (kimberlites of Finland, Kimozero, etc.) has also been established. The manifestations of younger, Late Permian-Mesozoic basic ultramafic magmatism are known so far only from the eastern framing of the Russian Platform in Timan and the Urals. The presence of hyabissal intrusion of melaleucetites found in the root outcrop of the Sukhona River in the Vologda region [14], our further work has not been confirmed [29]. In spite of this, the obvious tectonic rearrangement (activation) of the Russian Platform in the Late Paleozoic–Mesozoic era should have been accompanied by synchronous magmatic activation. Therefore, the Mesozoic magmatism could well have manifested itself on the Russian Platform in general and in Ustyansky Region in particular. Kimberlite (including diamond-bearing) and related volcanism of Mesozoic age actively manifested itself in Yakutia, Africa, North and South America, Australia, etc. At the same time, kimberlite pipes of different ages (from Proterozoic to Mesozoic) can be formed not only in the neighboring territories but even within the same area (South Africa, Australia, Canada, India, Finland, northern Yakutia, etc.).

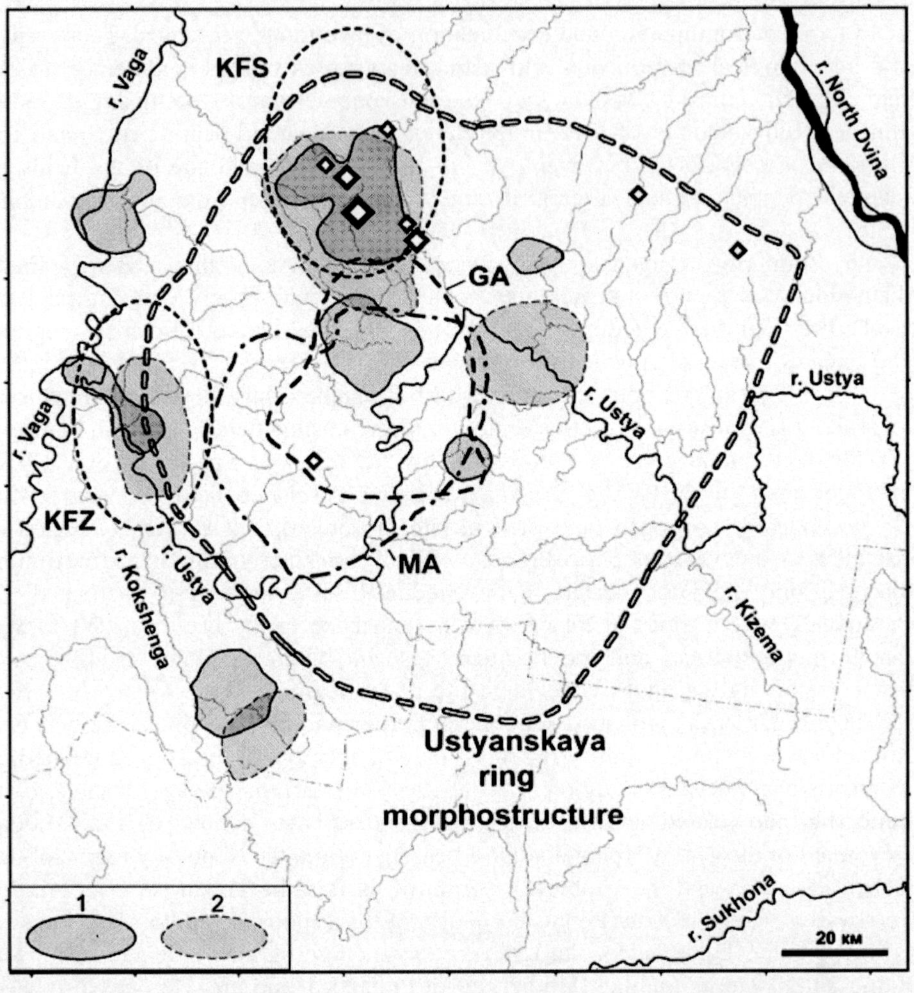

Fig. 11 Scheme of combining in space all favorable regional and local prospecting prerequisites (morphostructural, tectonic, geophysical, morphometric, mineralogical) and prospecting signs. Letters on the diagram indicate: GA—regional gravity anomaly, MA—regional magnetic anomaly; forecasted kimberlite fields: KPS—Severnoye field, KPZ—Zapadnoye field. Legend: 1—placer halos (maximum concentrations) of diamond indicator diamond indicator minerals of near-demolition (DIM-1) in Quaternary deposits; 2—areas of the most favorable combination of anomalous zones of density, isotropy and specificity of the river network. Rhombuses—points of finding of diamond crystals in Quaternary deposits (the size of the symbols is proportional to the size of the diamond crystals)

Development within the Ustiansky region manifestations of Mesozoic diamond-bearing magmatism (confirmed by the combination of the local area of all possible favorable conditions and signs) is much more likely and logical than the hypothetical movement of glaciers diamondiferous kimberlite material from the

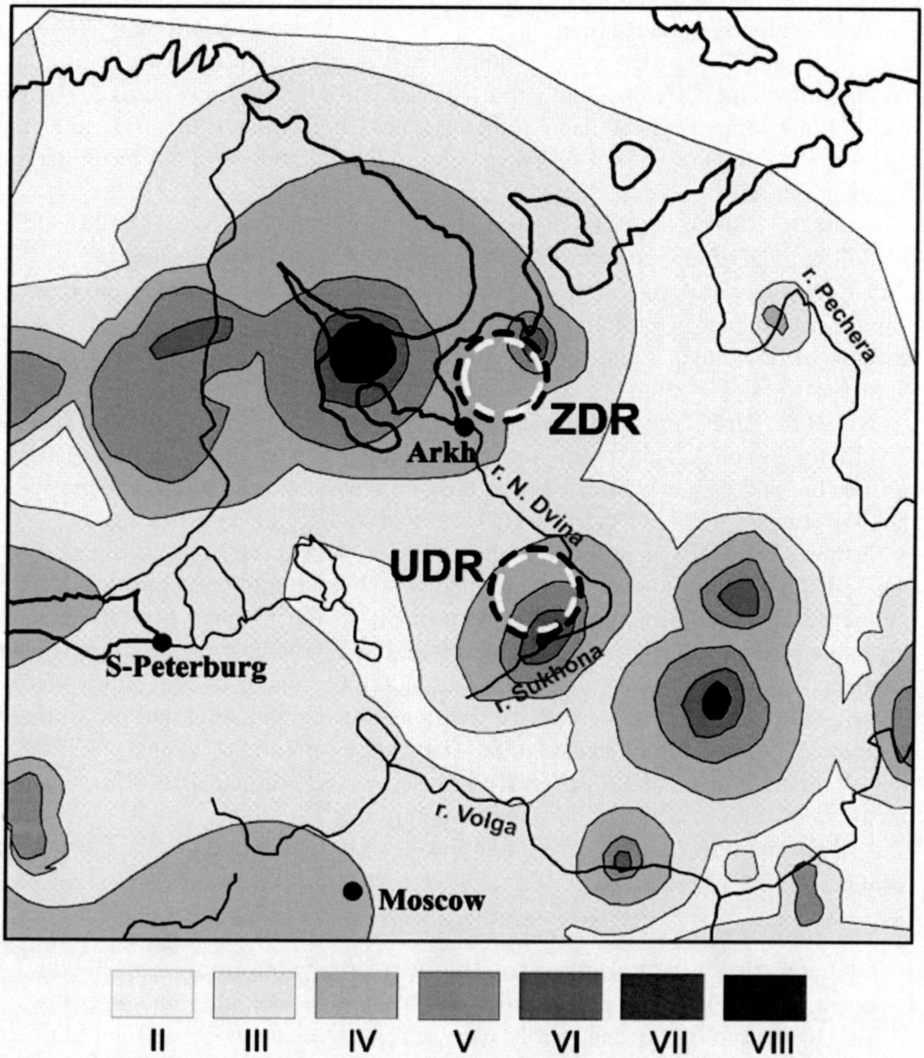

Fig. 12 Macroseismic field of the north of the Russian platform [2]. The contours of the generalizing isoseismals of earthquakes are shown. Legend (bottom): earthquake intensity in points (balls)

hypothetical unknown kimberlite bodies of the Baltic Shield and "random" unprecedented concentration of this far-bearing material precisely within the predicted for all independent signs of kimberlite field. In glacial environments at a distance from the primary root source there is an obvious rapid dilution of mineralogical halos rather than their enrichment. This is fully confirmed by search operations in Finland and Canada, and just by common sense. Nothing like panned halos of kimberlite indicator minerals and diamonds of the Ustyansky region has not been found on the Baltic Shield neither in Karelia, nor on the Wind Belt, nor in Finland, let alone on the Russian Plate Square, so we have to admit that("Vetrenny Poyas") in the Ustyansky region should exist so far unknown on the Russian platform of diamondiferous kimberlites of late Permian—Mesozoic age. Panned halos of the kimberlite indicator minerals and the diamonds themselves in the Ustyanskaya structure are not distant, random ("allochthonous"), but local, natural ("autochthonous").

As for the alluvial deposits, the prospecting works have already established the fact of alluvial diamondiferous modern alluvium and glacial deposits of the Ustyansky diamondiferous region. At the same time, as noted above, the modern alluvial complex of precipitation of the region was almost entirely formed by the washout of fluvioglacial and moraine deposits. Glacial deposits and modern alluvium form a single complex of sediments in the ancient valleys, which can be united by the term "productive sands" in terms of the prospects of alluvial diamondiferous nature. This promising placer model can be used to plan detailed prospecting and estimate forecast resources of alluvial diamondiferous deposits in the "Severnoye" field and the Ustyansky prospected region as a whole.

Such bright, clearly manifested and combined in the space of all possible search prerequisites (morphological, tectonic, geological-stratigraphic, geophysical, morphometric, mineralogical) and even more direct search features (the presence of large diamond crystals in the samples of minimal volume) are nowhere on the Russian platform, including the high-diamondferous region of the Zimny Bereg. This makes it possible to characterize the mineralogical situation and the situation of diamond prospecting in general at the Ustyansky structure (Ustyansky region) as unique for the whole of European Russia (the whole Russian Platform), and the prospects of search works as very encouraging.

At the same time, it should be noted that such difficult conditions for searching for diamond deposits do not exist anywhere else in the World! This:

1. The short (only post-glacial) time of formation of the modern river network (although there is a clear inheritance of the river network from preglacial time—rivers, especially the mouths, deeply cut through the root rocks and probably use the channels of preglacial, partly early glacial river network) did not allow to form a distinct linear mineralogical halo in alluvium and even more so classical contrast and compact alluvial diamond deposits.
2. The absence in the area overlapping the supposed primary sources of the intermediate terrigenic collector of continental origin, which could form halos of KIMs near moving, much wider in area than kimberlite bodies, but having a direct connection with them, which is actually to establish accessible, limited volumes of drilling and mineralogical works. How it helps in one way or another to search for pipes on the Zimny Bereg or in the Malo-Botuobinsky and Nakyn fields of Yakutia.
3. Almost complete cloak-shaped overlapping of the host rocks of the powerful, complex thickness of the Quaternary glacial deposits of different genesis (moraine, fluvioglacial, lake and glacial, etc.) and different glaciation centers

(Scandinavian and Bolshezemelsky), and thus different direction of movement of the detritus material with complex patterns of location in the area. This leads to a sharp weakening of the intensity of the mineralogical "signal" from the root rocks on the surface—the level of area search and non-obviousness of the direction of movement of the material from the primary root source, as well as the presence of numerous "false" geophysical (magnetic) anomalies, etc.
4. Essentially clayey section of ore-bearing rocks of late Permian, which greatly complicates the application of geophysical (primarily electrical exploration) methods.
5. Possible weak contrast manifestations of the alleged kimberlites in geophysical fields due to their weak magnetism (diamondiferous kimberlites are sometimes completely non-magnetic), clogging xenogenic material, the possible presence of powerful sedimentary crater parts on the kimberlite pipes, which by definition should be relatively young and, therefore, probably poorly eroded.
6. It is also impossible to completely exclude the possible dyke-core nature of some manifestations of kimberlite magmatism in this area.

Taking into account such complexity of conditions for prospecting works, we have to admit that the way to discovering diamond deposits on the Ustyansky structure cannot be easy—it will take forces, time, means and, first of all, application of new, most modern (maybe even non-traditional) geophysical and mineralogical methods of prospecting and interpretation of results. But the prospects for diamondiferous Ustyansky promising region with the expected (predicted) kimberlite field "Severnoye" is very promising.

9 Conclusions

The performed forecast and prospecting works are an example of "inverse *geological* problem solution" for preliminary spatial localization of a particular area of plume activity, forecasting and assessment of diamond-bearing prospects of one of such areas in the central part of the Russian Platform.

Ustyansky perspective region is characterized by the combination of all favorable regional and local search **prerequisites** (morphological, tectonic, geological-stratigraphic, geophysical, morphometric, mineralogical) and **search features**, which may be due to the introduction of mantle diapir and may be accompanied by diamondiferous kimberlite magmatism.

The content of near-moving diamond indicator minerals (primarily pyrope (1206 grains), less than picroilmenite, chrome spinel, olivine, chrome-diopside) in the alluvium of the region is 3–5 times higher than in the diamond-bearing region of the Zimny Bereg, and the share of G10 group diamond paragenesis pyrope is about 10%.

Nine diamond crystals were found in alluvial and Quaternary sediments, of which 6 crystals (large octahedrons and dodecahedroids up to 3.8 mm in size and up to 52 mg in weight) were found in the predicted kimberlite field “Severnoe”.

According to the complex of morphological features and physical properties, all diamonds of Ustyansky region occupy a completely separate position and practically have no analogues among diamonds of known deposits and manifestations of Arkhangelsk, Finland, the Urals and Timan. This may indicate that the studied crystals come from a new, not yet known kimberlite source (or sources). This potential new primary root kimberlite source may have an increased diamondiferous capacity, high quality and large size of diamonds. The discovery of this new kimberlite source will be the final confirmation of the predicted isolation of the Ustyansky region as an area of plume activity.

Signs of a possible similarity between the Arkhangelsk diamondiferous province and Yakutsky diamondiferous province in terms of patterns of kimberlites location and morphological features of diamond crystals confirm the high prospects for diamondiferous activities in the central regions of European Russia as a whole, and, above all, in the Ustyansk promising region with the “Severnoye” forecast field.

At the same time, the probable overlapping of diamondiferous tubes with terrigenous crater sediments and powerful strata of far-bearing glacial sediments makes it very difficult to identify real traces of kimberlites in modern surface sediments available for study and testing. This is why the path to discovering diamond deposits cannot be an easy one. But the prospects for the diamondiferous Ustyansky promising region are very high.

Taking into account the told above, systematic realization of a full complex of geology-search works told above in Ustyansky perspective region (and in promising kimberlite field “Severnoye” in particular) not only quite is proved, but also it is completely necessary for revealing new primary deposits of diamond in the central part of the European Russia.

References

1. Afanasyev, V.P., Zinchuk, N.N., Pokhilenko, N.P.: Morphology and Morphogenesis of Indicative Kimberlite Minerals. Novosibirsk: Branch “Geo” of the Siberian Branch of the Russian Academy of Sciences, Manuscript Publishing House, 276 p. (2001) (in Russian)
2. Ananin, I.V.: Connection of Russian platform seismicity with modern tectonic movements // Modern Earth crust movements. M. № 3. (1968) (in Russian)
3. Dawson, J.B., Stephens, W.E.: Statistical analysis of garnets from kimberlites and associated xenoliths. J. Geol. **83**, 589–607 (1975)
4. Ezhov, B.V., Andreev, V.L.: Instrumentation in Central Type Morphological Structures of Mantle Deposit. Moscow: Nauka, 126 p. (1989) (in Russian)
5. Feldman, A.A et al.: Report: Develop geological and geophysical criteria for forecasting areas and fields of kimberlite magmatism on the Russian platform and justify the direction of geological prospecting. Moscow: TsNIGRI funds. (1988) (in Russian)
6. Gurney, J.J.: A correlation between garnets and diamonds in kimberlites. In: Glover, J.E., Harris, P.G. (eds.) Kimberlite occurrence and origin. University of Western Australia, Geol. Dept. Publ. No 8. pp. 143–166 (1984)

7. Gurney, J.J., Moore, R.O.: Kimberlite garnet, chromite and ilmenite compositions, applications to exploration. In: Sheahan, P., Chater, A. (eds.) Diamonds: exploration, sampling and evaluation, pp. 109–146. Prospectors and Developers Association of Canada, Toronto (1991)
8. Haggerty, S.E.: The chemistry and genesis of opaque minerals in kimberlites; p. 295–307, in Physics and Chemistry of the Earth, v.9, L.H. Ahrens, J.B. Dawson, A.R. Dunkan and A. J. Erlank editors, Pergamon Press, Oxford, 940 pp. (1975)
9. Ignatiev, V.I.: Tatarsky Stage of Central and Eastern Regions of the Russian Platform. Part 1. Stratigraphy. Kazan: Kazan University Publishing House, 334 p. (1962) (in Russian)
10. Kalinkin, M.M., Arzamastsev, A.A., Polyakov, I.V.: Kimberlites and related rocks of Kola region. Petrology. T.1, N2, pp. 205–214. (1993) (in Russian)
11. Kaminsky, F.V., Khachatryan, G.K.: Characteristics of nitrogen and other impurities in diamond, as revealed by infrared absorbtion data. Can. Mineral. **39**, 1733–1745 (2001)
12. Khachatryan, G.K., Kaminsky, F.V.: Eqilibrium" and "non-eqilibrium" diamond crystals from deposits in the East European platform, as revealed by infrared absorbtion data. Can. Mineral. **41**, 171–184 (2003)
13. Map of morphostructures of the central type of the USSR territory m-ba 1:10 000 000. Under the editorship of Solovyov V.V. Leningrad. VSEGEI (1981) (in Russian)
14. Masaitis, V.L., Trufanov, A.I.: Melaleicitites from the banks of the Sukhona River (Northeast of Moscow Syneclise). Regional Geology and Metallogeny. **36**, 49–60 (2008). (in Russian)
15. Milashev, V.A.: Structures of kimberlite fields. Leningrad: Nedra. 184 p. (1979) (in Russian)
16. Milashev, V.A.: Cracks, block structure of platform cover and diatrem localization. SPb.: VNIIOceangeology. 145 p. (1997) (in Russian)
17. Pakhtusova, N.A.: To stratigraphy of upper perm deposits of the north of the Russian platform. In: Materials on geology and minerals of the North-West of RSFSR. Episode. 5. Leningrad: Nedra. pp. 30–86 (1966) (in Russian)
18. Nikitin, E.A., Levyi, M.G., Drozdov, V.A., Stefan, L.V., Khomich, P.Z.: Pipes of explosion of kimberlite-lamproite rocks of Belarus and the problem of their diamond-bearing capacity. Lithosphere **1**(168–175) (1994) (in Russian)
19. Prokopchuk, B.I., Feldman, A.A., Franzesson, E.V., et al.: Report: "East-European Platform Diamond Prospects Assessment to Identify Areas for Specialized and Accompanying Prospecting of Diamond Deposits". Moscow: TsNIGRI funds (1985) (in Russian)
20. Sablukov, S.M., Dudar, L.P., Dovzhikova, E.G., Fishman, M.V.: Formation phases and age of the kimberlite pipes of the Middle Timan. Proc. Of TSNIGRI. Ep. **182**, 64–74 (1983). (in Russian)
21. Sablukov, S.M.: To the question about the phases of formation and the age of explosion pipes of the Onega peninsula. Dokl. of the USSR Academy of Sciences. T.277. № 1. pp. 168–170 (1984) (in Russian)
22. Sablukov, S.M.: About the age of the ultramafic explosion pipes. Proceedings of TsNIGRI. **218**, 24–27 (1987). (in Russian)
23. Sablukov, S.M.: On Petrochemical Series of Kimberlite Rocks. Reports of the USSR Academy of Sciences, T.313, N 4, pp. 935–939 (1990) (in Russian)
24. Sablukov, S.M.: Zimny Bereg Volcanism and Petrological Criteria of Diamondiformity of Kimberlites: Author's thesis P.h.d.: MOSCOW: TSNIGRI, 24 p. (1995) (in Russian)
25. Sablukov, S.M.: Pantry of diamonds of the Russian North. Sci. Russia **1**, 82–88 (2001)
26. Sablukov, S.M., Sablukova, L.I.: Report: "Advanced prognostication and reconnaissance studies within the Ustyansky ring morphological structure of the central type for the purpose of creation of the schlich-mineralogical basis for large-scale specialized panned searches and preliminary preparation of new search areas" in 2 books. Funds of "RUSGEO" Limited, Moscow (2001) (in Russian)
27. Sablukov, S.M.; Belov, A.V.; Leshukov, S.I.; Sablukova, L.I.; Khachatryan, G.K.: Unique diamonds of the Ustyansky potential diamond-bearing region (south of the Arkhangelsk district). In: Proceedings of the Jubilee Congress of the Russian Mineralogical Society "200 years of the RMO", St. Petersburg. T.2, pp. 149–151 (2017) (in Russian)

28. Sablukov S.M., Ermakov N.S., Belov A.V., Leshukov S.I., Sablukova L.I.: Unique prospects of diamond bearing capacity of Ustiansky region in the light of new geological data of the Company "Arkhangelsk Diamonds". In: Proceedings of the VIII International Scientific and Practical Conference "Scientific and Methodological Basis of Forecasting, Prospecting and Evaluation of Diamond Deposits, Noble and Non-Ferrous Metals", M, CNIGRI (2018) (in Russian)
29. Sablukov S.M.: Report: "Results of the revision study of rocks of problematic origin, originally diagnosed as "early Mesozoic alkaline ultrabasic magmatic rocks—melaleicitites" (Sukhona River, Purtovino village, Vologda district). Moscow: funds of OOO INPC RUSGEO (2009) (in Russian)
30. Shchukin, V.S., Efremov, I.V., Grib, V.P.: New data on the geological structure of the Northern Dvina and Pinega interfluve. Proc. Of TsNIGRI. Ep. **188**, 33–39 (1984). (in Russian)
31. Sinitsyn, A.V., Ermolaeva, L.A., Stankovsky, A.F., et al.: On Mezozoic Tectonic and Magmatic Activation of Northern Part of Russian Plate (in Russian). DAN USSR. T. 286. № 6. (1986) (in Russian)
32. Smart, K.A., Cartigny, P., Tappe, S., O'Brien, H., Klemme, S.: Lithospheric diamond formation as a consequence of methane-rich volatile flooding: an example from diamondiferous eclogite xenoliths of the Karelian craton (Finland). Geochim. Cosmochim. Acta **206**, 312–342 (2017)
33. Smith, C.B., Gurney, J.J., Skinner, E.M.W., et al.: Geochemical character of Southern African kimberlites. A new approach based on isotopic contains. Trans. Geol. Sos. S. Afr. **88**, 267–280 (1985)
34. Sobolev, N.V.: Deep inclusions in kimberlites and the problem of the upper mantle composition. Novosibirsk: Science. 264 p. (1974) (in Russian)
35. Solovyov, V.V.: Structures of the central type of the USSR territory according to geological-geomorphological analysis data. Leningrad.: VSEGEI. 111 p. (1978) (in Russian)
36. Structure of the platform cover of the European part of the USSR. Edited by V.A. Dedeev: Science, Leningrad Branch. 200 p. (1982) (in Russian)
37. Taylor, Milledge H.J.: Nitrogen aggregation character, thermal history, and stable isotope composition of some xsenolith-derived diamonds from Roberts Victor and Finch. In: 6-th IKC. Russia. Novosibirsk. Exstented Abstracts. pp. 620–622 (1995)
38. Trufanov, A.I., Masaitis, V.L.: The first discovery of early Mesozoic alkaline ultrabasic magmatic rocks in the north of the Russian plate. Regional Geology and Metallogeny **30–31**, 57–61 (2007). (in Russian)
39. Upper Permian and Lower Tertiary sediments of the Moscow Syneclise. Edited by S. V. Tikhomirov: Nedra. 140 p. (1984) (in Russian)
40. Wyatt, B.A., Baumgartner, M., Ancar, E., Grutter, H.: Compositional classification of "kimberlitic" and "non-kiberlitic" ilmenite. In: 8th International Kimberlite Conference Selected Papers, vol. 2, Lithos 77, pp. 819–840 (2004)